Springer Series in Physical Environment 1

Managing Editor
D. Barsch, Heidelberg

Editors
I. Douglas, Manchester · F. Joly, Paris
M. Marcus, Tempe · B. Messerli, Bern

Advisory Board
F. Ahnert, Aachen · R. Coque, Paris · Y. Dewolf, Paris
W. Eriksen, Hannover · O. Fränzle, Kiel
A. Godard, Meudon · A. Guilcher, Brest
H. Leser, Basel · J. Nicod, Aix-En-Provence
H. Rohdenburg, Braunschweig · A. Semmel, Frankfurt
G. Stäblein, Berlin · H. Svensson, København
H. Bremer, Köln · J. R. Mather, Newark
A. R. Orme, Los Angeles

Richard J. Huggett

Earth Surface Systems

With 162 Figures

Springer-Verlag
Berlin Heidelberg New York Tokyo

Dr. Richard J. Huggett
University of Manchester
School of Geography
Manchester, M13 9PL
England

ISBN 3-540-15421-3 Springer-Verlag Berlin Heidelberg New York Tokyo
ISBN 0-387-15421-3 Springer-Verlag New York Heidelberg Berlin Tokyo

This work is subject to copyright. All rights are reserved, whether the whole or part of the material is concerned, specifically those of translation, reprinting, re-use of illustrations, broadcasting, reproduction by photocopying machine or similar means, and storage in data banks. Under § 54 of the German Copyright Law where copies are made for other than private use a fee is payable to 'Verwertungsgesellschaft Wort', Munich.

© by Springer-Verlag Berlin Heidelberg 1985
Printed in Germany.

The use of registered names, trademarks etc. in this publication does not imply, even in the absence of a specific statement, that such names are exempt from the relevant protective laws and regulations and therefore free for general use.

Typesetting: K + V Fotosatz GmbH, Beerfelden.
Offsetprinting and bookbinding: Konrad Triltsch, Graphischer Betrieb, Würzburg
2132/3130-543210

For my parents

Preface

Discussions of "systems" and the "systems approach" tend to fall into one of two categories: the panegyrical and the disparaging. Scholars who praise the systems approach do so in the belief that it is a powerful and precise method of study. Scholars who try to shoot it down fail to see any advantage in it; indeed, many deem it pernicious. Van Dyne (1980, p. 889) records a facetious comment he once heard, the gist of which ran: "In instances where there are from one to two variables in a study you have a science, where there are from four to seven variables you have an art, and where there are more than seven variables you have a system". This tilt at the systems approach is mild indeed compared with the comments of an anonymous reviewer of a paper by myself concerned with the systems approach as applied to the soil. The reviewer stated bluntly that he or she had no time for an approach which falsifies and belittles work that has been done and is of no use for future work. My summary of the paper opened with the seemingly innocuous sentence "The notion of the soil as a system is placed on a formal footing by couching it in terms of dynamical systems theory". However, the reviewer thought that the term notion summed it up and proceeded to list synonyms from the Oxford English Dictionary – general, fancy, invention, small, cheap, speculative, abstract. Regardless of the merits or demerits of the paper in question, the dismissal of the systems approach in such an atrabilious and uncompromising a fashion is unjustified.

Not all members of the anti-systems lobby are so unreasonable and unconstructive. Nonetheless, they do see in the systems approach a number of contradictions and difficulties. An aim of this book is to show that these contradictions and difficulties are more apparent than real and arise from misinterpretations and misunderstandings of systems concepts and terms. Two main sources of misunderstanding can be identified. The first is the erroneous belief that the systems approach is new-fangled: taken in wide perspective, the systems approach has a long pedigree dating back at least to the sixteenth-century scientific revolution. A second source of misunderstanding is the erroneous belief that systems are invariably complex: systems can be viewed and defined as either simple or complex structures. Bearing in mind these sources of misunderstanding, and in an attempt to show that the systems approach is inextricably linked with the traditional scientific approach, the systems and systems methods dealt with in this book are, when first met with, placed in an historical setting.

The scope of a book on Earth surface systems is broad indeed. Only a small sample from the rich literature in the field can be offered here. In taking the sample, I have tried to pick case studies from the full range of systems research carried out by scientists with an interest in the surface of the Earth. Inevitably, however, I shall have left myself open to the charge of partiality in giving the lion's share of attention to topics with which I am most familiar and in which my main interests lie. I trust that any bias in the selection of topics does not detract from the development of the general theme of the book.

I should like to thank a number of people without whose assistance the book would never have appeared: for asking me to write the book and for help and encouragement at all stages in writing it, Ian Douglas; for producing the typescript, the secretaries in the Geography Department at Manchester University; for drawing most of the diagrams, Nick Scarle and Graham Bowden; and lastly, for never succeeding in stopping me from writing for more than a few hours at a time, my family.

Macclesfield, June 1985   RICHARD HUGGETT

Contents

Part I Introduction

Chapter 1 Systems and Models 3

1.1 Defining Systems 3
    1.1.1 Systems as Form and Process Structures 4
    1.1.2 Systems as Simple and Complex Structures ... 4
    1.1.3 Systems and Their Surroundings 5
    1.1.4 A Problem of Scale 7
1.2 Models of Systems 7
    1.2.1 Conceptual Models 9
    1.2.2 Scale Models 10
    1.2.3 Mathematical Models 11

Part II Conceptual Models

Chapter 2 Simple and Complex Systems 17

2.1 Simple Systems 17
2.2 Systems of Complex Disorder 20
    2.2.1 Irreversible Processes 20
    2.2.2 Accounting Models: the Laws of
           Thermodynamics 22
2.3 Systems of Complex Order 25
    2.3.1 Nonequilibrium Systems 25
    2.3.2 Systems Far from Equilibrium 26
    2.3.3 Open Systems at the Earth's Surface 27

Chapter 3 Form and Process Systems 30

3.1 Models of System Form 30
    3.1.1 Models of System Constitution 30
    3.1.2 Models of System Geometry 30
3.2 Models of System Process 32
    3.2.1 Land-Surface Cascades 33
    3.2.2 Solid-Phase and Liquid-Phase Cascades 35

3.3	Models of System Form and Process		37
	3.3.1 Concepts of Landscape Development		39
	3.3.2 Concepts of Soil Development		41
	3.3.3 Concepts of Soil-Landscape Development		43

Part III   Mathematical Models

Chapter 4   Deductive Stochastic Models 47

4.1	Introduction to Probability		47
	4.1.1 The Classical View of Probability		48
	4.1.2 The Relative Frequency View of Probability		48
	4.1.3 Axioms of Probability Theory		49
4.2	Independent Events in Time		51
	4.2.1 Binomial Processes		51
	4.2.2 Poisson Processes		53
4.3	Independent Events in Space		55
	4.3.1 Point Patterns		55
	4.3.2 Line Patterns		57
	4.3.3 Area Patterns		60
4.4	Random-Walk Models		60
	4.4.1 Stream Networks		61
	4.4.2 Alluvial Fans		65
4.5	Markov Chains		70
	4.5.1 Transition Probabilities		70
	4.5.2 Sedimentary Sequences		72
	4.5.3 Volcanic Activity		73
4.6	Entropy Models		74
	4.6.1 Entropy Maximization		74
	4.6.2 Entropy Minimization		75
	4.6.3 Developments of the Thermodynamic Approach		76

Chapter 5   Inductive Stochastic Models 78

5.1	Box and Jenkins's Models: an Introduction		78
	5.1.1 System Definition		81
	5.1.2 Stages in Systems Analysis		82
5.2	Autoregressive Moving-Average Models of Time Series		83
	5.2.1 Model Formulation		83
	5.2.2 Modelling Procedures		85
	5.2.3 The Lagan Rainfall Series		87
5.3	Autoregressive Moving-Average Models of Distance Series		88
	5.3.1 Model Formulation		88
	5.3.2 River Meanders		89

	5.3.3	Landforms	92
5.4	Transfer Function Models		95
	5.4.1	Model Formulation	95
	5.4.2	Rainfall and Runoff in the Lagan Drainage Basin	97
	5.4.3	Channel Form in the Afon Elan, Wales	97
5.5	Problems of Inductive Stochastic Modelling		100

Chapter 6 Statistical Models ... 102

6.1	Simple Regression and Correlation		103
	6.1.1	The Regression Line	103
	6.1.2	The Correlation Coefficient	104
	6.1.3	Problems of Correlation	106
	6.1.4	Linear Relations	107
	6.1.5	Linear Versus Nonlinear Relations	108
6.2	Multiple Regression		112
	6.2.1	"Simple" Multiple Regression	113
	6.2.2	Trend Surface Analysis	114
	6.2.3	Stepwise Regression	117
	6.2.4	Problems of Multiple Regression	121
6.3	Correlation Systems		122
	6.3.1	Principal Component Analysis	123
	6.3.2	Principal Coordinate Analysis	125
	6.3.3	Factor Analysis	125
	6.3.4	Canonical Correlation	129
	6.3.5	Problems with Correlation Systems	134

Chapter 7 Deterministic Models of Water and Solutes ... 135

7.1	Ice		137
	7.1.1	Glaciers	137
	7.1.2	Ice Sheets	138
7.2	Water		142
	7.2.1	Overland Flow	142
	7.2.2	Open Channel Flow	144
	7.2.3	Flow in Porous Media	145
	7.2.4	Unsaturated Flow	148
7.3	Solutes		151
	7.3.1	Seas and Lakes	151
	7.3.2	Solutes in Groundwater	154
	7.3.3	Solutes in Soils	157

Chapter 8 Deterministic Models of Slopes and Sediments ... 161

8.1	Discrete Component Models		161
8.2	Analytical Models		162
	8.2.1	Heuristic Models	162

	8.2.2	Models Based on the Continuity Equation	173
8.3	Simulation Models		180
	8.3.1	Landscape Simulation	181
	8.3.2	Drainage Basin Simulation	185
	8.3.3	Nearshore Bar Formation	192
	8.3.4	Sand Dune Formation	193

Chapter 9 Dynamical Systems Models 198

9.1	Model Building		198
	9.1.1	State and State Change	198
	9.1.2	Transfer Equations	201
9.2	System Stability		202
	9.2.1	State Space	202
	9.2.2	Sensitivity Analysis	204
9.3	Biogeochemical Cycles		206
	9.3.1	The Global Cycle of Phosphorus	206
	9.3.2	The Global Cycle of Carbon Dioxide and Oxygen	206
	9.3.3	Strontium and Manganese in a Tropical Rain Forest	214
	9.3.4	Water in Soils	216
	9.3.5	Nutrients in Lake Erie	218
9.4	Dissipative Structures		222
	9.4.1	Bifurcations and Catastrophes	222
	9.4.2	Thresholds	226
	9.4.3	Dominance Domains	231

Chapter 10 Conclusion and Prospect 233

10.1 Models as a Complement to Field Studies 234
10.2 Models as a Testing Ground for Long-Term Change . 238
10.3 Models as Good Predictors of Complex Situations ... 239

References 243

Subject Index 263

Part I Introduction

CHAPTER 1
Systems and Models

1.1 Defining Systems

Hillslopes extend from the crests of interfluves, along smooth or cliffed valley sides, to sloping valley floors. They occupy much of the land surface forming distinctive and ubiquitous landforms. But are they systems?

The word "system" is derived from the Greek $\sigma\nu\sigma\tau\eta\mu\alpha$ meaning a whole compounded of many parts or an arrangement. Some modern definitions of systems retain this classical clarity and simplicity. To Schumm (1977, p. 4), a system is simply "a meaningful arrangement of things". By this definition, a hillslope is a system. It consists of things (rock waste, organic matter, and so forth) arranged in a particular way. The arrangement is meaningful because it may be explained in terms of physical processes. Other definitions of systems refer to a set of objects, and attributes of the objects, standing together to form a regular and connected whole. Chorley and Kennedy write:

"A system is a structured set of objects and/or attributes. These objects and attributes consist of components or variables (i.e. phenomena which are free to assume variable magnitudes) that exhibit discernible relationships with one another and operate together as a complex whole, according to some observed pattern". (Chorley and Kennedy 1971, pp. 1–2)

The objects and attributes associated with a hillslope may be described by variables such as particle size, soil moisture content, vegetation cover, and slope angle. These variables, and many others, interact to form a regular and connected whole: hillslopes, and the mantle of debris on them, record a tendency towards mutual adjustment among a complex set of variables. These variables include rock type, which influences rates of solution, the geotechnical properties of the soil and mantle, and rates of infiltration; climate, which influences slope hydrology and so the routing of water over and through the hillslope mantle; tectonic activity, which may alter base level; and the geometry of the hillslope which, acting mainly through slope angle and distance from the divide, influences the rates of processes such as landsliding, creep, solifluxion, and wash. Changes in any of the variables will tend to cause a readjustment of hillslope form and process.

Chapman (1977, p. 79) remarks that the word "system" has become loose with use and has come to mean all things to all men. In the natural and Earth sciences, however, laxity in the use of the word "system" is uncommon. True, natural systems are defined in a variety of ways, but this is because different kinds of system are recognized and not because definitions are lax. As far as

Earth surface systems are concerned, two chief typologies of system are used, each with its own definitions and assumptions. They are a typology based on form and process facets of systems; and a typology based on the level of system complexity, as reflected in the degree of regularity and connectedness of the whole.

1.1.1 Systems as Form and Process Structures

This typology of systems was proposed by Chorley and Kennedy (1971) and adopted and adapted by Terjung (1976) and Strahler (1980). According to these authors, systems may be defined at a number of distinct levels, of which the following are relevant to the study of the natural environment: morphological systems, cascading systems, and process-response systems.

Morphological, or form, systems are conceived as sets of morphological variables which are thought to interrelate in a meaningful way in terms of system origin or system function. An example is a hillslope represented by variables pertaining to hillslope geometry, such as slope angle, slope curvature, and slope length, and to hillslope composition, such as sand content, moisture content, and vegetation cover, all of which are assumed to form an interrelated set.

Cascading systems or, if the terminology of Strahler (1980) is preferred, flow systems are conceived as "interconnected pathways of transport of energy or matter or both, together with such storages of energy and matter as may be required" (Strahler 1980, p. 10). An example is a hillslope represented as a store of materials: weathering of bedrock and wind deposition add materials to the store, slope processes transfer materials through the store, and erosion by wind and fluvial erosion at the slope base remove materials from the store. Other examples of cascading systems include the water cycle, the biogeochemical cycle, and the sedimentary cycle, all of which may be identified at scales ranging from minor cascades in small segments of a landscape, through medium-scale cascades in drainage basins and seas, to mighty circulations involving the entire globe.

Process-response systems or, if the terminology of Strahler (1980) is preferred, process-form systems are conceived as an energy flow system linked to a morphological system in such a way that system processes may alter the system form and, in turn, the changed system form alters the system processes. A hillslope may be viewed in this way with slope form variables and slope process variables interacting. Thus Small and Clark (1982, p. 6) see a hillslope as a natural system within which there are numerous complex linkages between "controlling" factors, processes, and form. The atmosphere can be regarded as a process-form system in which the morphological variables (air temperature, pressure, wind, humidity, and turbidity) interact with flows of energy, mass, and momentum (Terjung 1976).

1.1.2 Systems as Simple and Complex Structures

A second typology of systems, attributable to Weaver (1958), has been introduced into environmental science by Wilson (1981) and into geomorphology by

Thornes and Ferguson (1981). It recognizes three main kinds of system: simple systems, complex but disorganized systems, and complex and organized systems. The first two conceptions of systems have a long and illustrious history of study. Since at least the sixteenth-century revolution in science, astronomers have referred to a set of heavenly bodies connected together and acting on each other in accordance with certain laws as a system. The solar system is the Sun and its planets. The Saturnian system is the planet Saturn and its moons. These structures may be thought of as simple systems. In Earth science, a few boulders resting on a hillslope can be regarded as a simple system. The conditions required to dislodge the boulders and their fate once they have been dislodged can be predicted from mechanical laws involving forces, resistances, and equations of motion, in much the same way that the motion of planets around the Sun can be predicted from Newtonian laws.

In the complex but disorganized conception of systems, a vast number of objects are seen to interact in a weak and haphazard manner. An example is gas in a jar. This system could consist of upwards of 10^{23} molecules colliding with each other. In the same way, the countless individual particles in a hillslope mantle could be regarded as a complex but rather disorganized system. In both the gas and the hillslope mantle, the interactions are rather haphazard and far too numerous to study individually, so aggregate measures must be employed.

In a third and somewhat more recent conception of systems, objects are seen to interact strongly with one another to form systems of a complex and organized nature. Most biological systems and ecosystems are of this kind. Many structures at the Earth's surface, as well as biological and ecological structures, display high degrees of regularity and rich connexions and may be thought of as complexly organized systems. A hillslope represented as a process-response system could be placed in this category. Other examples include soils, rivers, and beaches.

1.1.3 Systems and Their Surroundings

Systems of all kinds may be regarded as open, closed, or isolated according to how they interact, or do not interact, with their surroundings. An isolated system is, traditionally, taken to mean a system that is completely cut off from its surroundings and which cannot therefore import or export matter or energy. This is in distinction to a closed system which is, traditionally, taken to mean a system which exchanges energy but not matter with its surroundings, and an open system which exchanges both matter and energy with its surroundings. Isolated and closed systems are difficult to find in nature. Even the planet Earth, a strong candidate for closed system status, imports material in the form of meteorites and possibly cometary debris (Velikovsky 1950; Silver and Schultz 1982) and viruses (Hoyle and Wickramasinghe 1979), and exports hydrogen ions and material in space launches. Nonetheless, isolated and closed systems are useful conceptual frameworks within which to consider a number of problems in physics, chemistry, and Earth science. It should be noted that a few authors, including Wilson (1981) and Chorley and Beckinsale (1980), use the term closed system in the same sense as isolated system, that is, cut off from the environ-

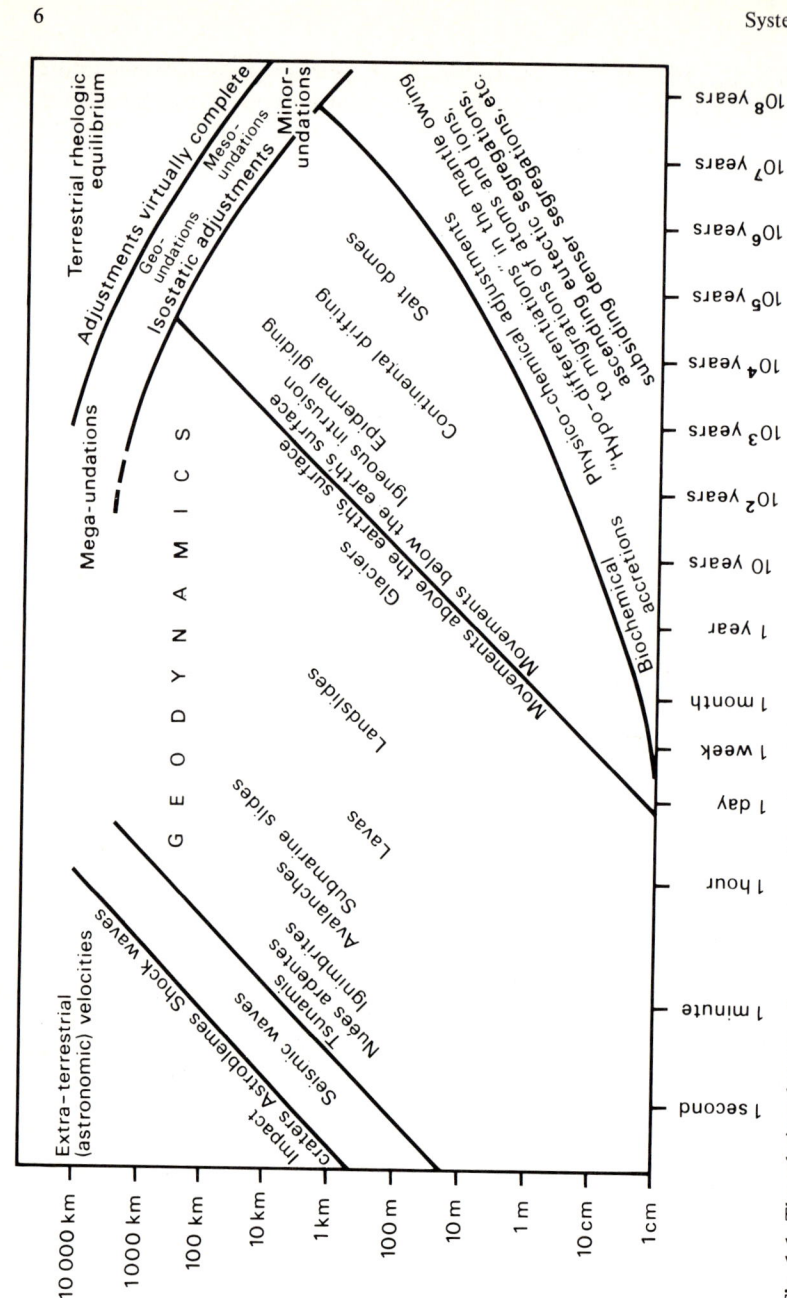

Fig. 1.1. The relations between the parameters of length and time in Earth Science (Van Bemmelen 1967).

ment. Although this may appear a minor point, it can be confusing to those unversed in the development of thermodynamic concepts.

1.1.4 A Problem of Scale

Hillslopes may be very short in gully systems and very long in large drainage basins and down mountainsides. The size of some other Earth surface systems is even more varied, ranging from a square metre or less, through areas of continental proportions, to the entire globe. Processes which produce the multifarious Earth systems operate at an equally varied range of rates. It is usual to recognize an hierarchy of Earth surface systems in which smaller systems are nested within larger ones. Harbaugh and Bonham-Carter (1970, p. 3) offer the example of a large system, the total coastline (consisting of sea, beaches, deltas, and cliffs), within which smaller systems, such as deltas, may be recognized in their own right. On a smaller scale still, a single distributary within a delta can be thought of as a system. The "scale problem" is whether the explanation of system form and function at one level of resolution is applicable to the explanation of system form and function at a higher or lower level. In the example, the problem is whether an understanding of the forms and processes in a delta enables much to be said about distributaries or an entire coastline. This is really the question that Schumm and Lichty (1965) addressed when they argued that variables which are dependent at higher order scales may become independent if the order of scale is reduced. For instance, as a variable in a river channel system, relief is a dependent variable if long periods of time (say a millenium or more) are considered, but is an independent variable during shorter spans of time.

There is not much that can be done about the scale problem; the important thing is to be aware of its existence. Schemes have been proposed, with varying degrees of success, which seek to express the scale of systems in a "simple" framework of distance (or area) and time (for example, Sugden and Hamilton 1971). The main function of these schemes is to draw attention to the vast range of time and space scales with which Earth scientists are concerned. In this regard, the scheme devised by Van Bemmelen (1967) is particularly revealing (Fig. 1.1). Length and time, when combined, describe distances travelled by moving objects over a given period; in other words, they specify the velocity of Earth processes. A selection of process rates, for both surface and subsurface processes, are shown in the diagram (Fig. 1.1).

1.2 Models of Systems

The terms "system" and "model" are not synonymous. Strahler (1980) takes pains to clarify the distinction between them. He explains that a system is assumed to exist in the real world and to possess unique attributes whereas a model is an attempt to describe, analyse, simplify, or display a system. Moreover, no model can ever be fully correct and achieve identity with the system it represents. The corollary of this is that the nature of a system can never be truly or thoroughly known, but will always remain a matter of conjecture. Thus, in a sense, a system is itself a conceptual model, the existence of which in reality "often rests upon little more than shared intuition" (Strahler 1980, p. 1).

The starting point of a model is some conception of the system itself. On this point, it is salutary to underscore Strahler's assertion that systems are out there in the real world to be modelled but that their existence can only be surmised. Charitably and justly, Strahler talks of general agreement over the existence of systems based upon shared intuition. Less charitably, but no less justly, it can be argued that a system, whether or not its existence is agreed upon, is a conception, a conjecture, a picture of what may exist. This point is quite fundamental but almost invariably neglected. The danger is that, because of general agreement, the existence of a system becomes established fact rather than hypothesis. For example, textbooks explain that the Earth consists of several spheres (technically shells). Büdel (1982) recognizes the lithosphere, decomposition sphere, hydrosphere, cryosphere, pedosphere, atmosphere, and biosphere. There is without doubt logic behind these distinctions. A fact commonly overlooked, however, is that the view of the Earth as a collection of shells is just a view, albeit a generally held view. It is just one of many possible conceptual models. Like all models, it is an hypothesis about how some portion of reality is put together. The validity of the model can be tested and should not be taken for granted. Complacency in science is a bad thing. Cherished beliefs are occasionally knocked down by outrageous hypotheses. For example, as Kennedy (1983) explains, the dependence of all ecosystems on solar energy was taken as a fact until a weird ecosystem was discovered in the Galapagos rift system in the eastern Pacific ocean. The newly discovered ecosystem consists of giant clams and tubeworms dependent on autotrophic bacteria which derive their energy from the outpourings of sulphur-laden gases issuing at 380°C from vents known as smokers (Speiss et al. 1980). It is powered by energy stored in the Earth's interior since the planet first accreted from a cloud of dust and gas and not by energy from the Sun (cf. Dole 1970). Another widely held belief which has almost become dogma is that the shells of the Earth interact through several, virtually endless recirculatory processes – the rock, water, and biogeochemical cycles. Again, it cannot be stressed too strongly that the existence of these cycles is not an established fact but an hypothesis which is at present in favour. As Lerman (1979, p. 1) stresses, geochemical cycles of chemical species are conceptual models. Recently too, the long-accepted view of eustacy has been radically revised (Mörner 1983). Changes of sea level were long believed to give rise to vertically displaced, but parallel, sea levels and shorelines. Geodetic measurements of the surface of the oceans from space have revealed an uneven topography which closely approximates the form of the geoid. Thus past sea levels and their associated shorelines cannot have been parallel because the geoid is always deforming and adjusting to changes occurring. A consequence of this new view of eustacy is that eustatic curves are valid at a local, but not at a global, level.

It cannot be stressed too strongly that a system is a concept. Like beauty, a system lies in the eyes of the beholder. It is an idea, an hypothesis about how some portion of reality is assembled and how it works. In this light, it is less important which typology is used to define and describe a system than that the system should be defined in a form that is susceptible of disproof. For instance, an alluvial fan may be conceived, among other things, as a cascading system. It is of the utmost important that this concept can be tested. For example, viewing an alluvial fan as a cascading system, Hack (1965a) and Denny (1965, 1967) proposed that growing fans will in time reach a state where the amounts of sediment delivered to and eroded from a fan balance so that no overall change will take place. This notion can, in theory at least, be put to the test. Rival hypotheses exist. Lustig (1965) and Bull (1975) are of the opinion that an alluvial fan is a depositional landform that will always increase in volume.

1.2.1 Conceptual Models

A conceptual model is a mental image of a natural phenomenon (Krumbein and Graybill 1965, p. 19) in which the supposedly essential details of the phenomenon are retained and the supposedly extraneous details are omitted (Chorley 1964). Conceptual models are always, to varying degrees, abstractions or simplifications of Earth surface phenomena. They represent attempts to capture and explain aspects of reality, the "true" nature of which are always elusive. They can be expressed in words, as in the Davisian model of landscape development; as some form of diagram, commonly, through not always, as a box-and-arrow diagram; or in a mathematical form, as Bagnold (1960) and Pickup and Rieger (1979) have shown.

The term "box-and-arrow diagram" covers a multitude of schemes for depicting systems. The schemes range from simple diagrams consisting of boxes with arrows indicating links to sophisticated creations in which an attempt is made to indicate the assumed role of various parts of the system. The preassembled parts that fill a standard role are represented by canonical structures, that is, basic functional units whose operation need not be resolved at a lower level. The variety of languages using canonical structures is great and there seems little hope of ever agreeing upon a standard form. A symbolic language which has caught on in the Earth sciences was devised by Odum (1971, 1983). The basis of the language, called energese, is that certain modules represent a particular structure and function within a system and lines represent pathways of energy. Another form of symbolism, with its origins in the hydrological literature, was developed by Chorley and Kennedy (1971) to depict physical process-response systems and has been used in subsequent publications (for example Terjung 1976; Bennett and Chorley 1978). Strahler (1980), unimpressed by Chorley and Kennedy's (1971) scheme, evinces a preference for a modified form of Odum's energese which was developed by Strahler and Strahler (1973, 1974) to present ecosystem energetics, and further modified by Strahler (1980) in an attempt to meet the needs of the wide range of systems of physical geography. However, the range of symbolic languages is wide and attempts to standardize them stand little chance of success. In any case, it is more important what is portrayed than how it is portrayed.

Many conceptual models can be quantified to a degree by going into the field and measuring the storage, flows, forces, and any other descriptors which are thought relevant to the system under scrutiny. Studies of this kind have led to detailed descriptions of many, though not all, Earth systems. Particularly well documented are those systems for which a "decade of study" has been instigated. A conceptual model that has been quantified by field measurements can generate hypotheses, especially if measurements have been assembled on a regional basis. These hypotheses may be tested by building an appropriate mathematical model or scale model, or they may lead to semi-quantitative models of a rather descriptive nature which remain conceptual in form. A number of important conceptual models have been inspired by careful observation of phenomena at the Earth's surface. Jenny's classic formulation of the relationship between soil nitrogen and climate grew out of a substantial collection of data on soils across the Great Plains acquired during the 1920's (Jenny 1980, p. X). The inspiration for conceptual models does not always come directly from fieldwork. Chorley (1967a, p. 85) argues that on close examination, the different conceptual approaches to geomorphology espoused by Gilbert, Davis, Penck, and climatic geomorphologists are all based upon gross intuitive assumptions regarding the significant behavioural patterns of landform assemblages and they are not based, though they give the impression that they are, on a detailed knowledge of geomorphic processes.

Conceptual models are normally built as a first step in the study of some phenomenon. The next step is to elaborate the model by carrying out a quantitative investigation of an analytical kind. In the Earth sciences, three broad analytical approaches are generally followed. One approach is to make a scale model of a system in the laboratory and carry out experiments on it. Another approach is to make a detailed field survey of a sample of system variables and subject the data collected to some form of statistical analysis. A third approach is to consider the problem from a purely mathematical point of view and build a theoretical model of a system. These three approaches are quite general and can be applied to the study of most Earth surface systems, through with varying degrees of success.

1.2.2 Scale Models

Scale models of Earth surface systems have been useful in studying a limited variety of problems. Chorley (1967a, p. 63) remarks that it is a source of surprise that scale models have not proved more useful than they have in geomorphological research. He adds that, with a few notable exceptions such as Friedkin's (1945) laboratory model of river meanders, it is probably true to say that the most valuable scale models are those which are basically parts of unscaled reality, closely circumscribed and examined in much detail. In geomorphology, two examples of this are Bagnold's (1954) wind tunnel experiments on sand dunes and Schumm's (1956) study of erosional forms and changes on the badlands of Perth Amboy, New Jersey. In pedology, an example is the experiments on soil columns rigged up in a laboratory (for instance, Yaalon 1965). Scale models have occasionally provided new insights into the development of Earth surface systems including alluvial fans (Schumm 1977, pp. 255–264) and deltas

(Schumm 1977, pp. 309–314). Further examples are discussed by Mosley and Zimpfer (1978).

Pitty (1979) argues that the artificiality of scale models limits their applicability to natural situations. Certainly, not all the dimensions of natural Earth surface systems can be scaled down with ease. Particle size in particular causes problems of scaling. Silt-sized particles, for instance, behave differently to, and cannot easily be used as miniature versions of, sand-sized particles. Attempts to reproduce the development of whole landscapes in miniature, such as Wurm's (1935) study of change in three parallel ridges moulded in a cement mixture on a 70 × 70 cm tray, are seriously hampered by difficulties of scaling.

A way of circumventing the scaling problem is to alter some of the properties of an Earth surface system as well as the size. When this is done an analogue model is produced. An example of an analogue model is Lewis and Miller's (1955) reconstruction of a valley glacier in which kaolin is used as a substitute for ice. Under carefully controlled conditions, many features of valley glaciers, such as crevasses and step faults, develop in the clay. Difficulties arise in this kind of analogue model, however, not the least of which is finding a material which has mechanical properties comparable to the material in the natural system. The reason why there are no analogue models of the development of rocky coasts is that no suitable substitute for rock has been found (Sunamura 1975), although Shepherd and Schumm (1974) used a kaolin-sand mixture to simulate bedrock in their experimental study of the incision of a river into bedrock.

Electrical analogues are used to study some Earth surface systems. Ground water systems lend themselves to electrical analogue modelling, a topic discussed at length by Rushton and Redshaw (1979). An excellent example is Getzen's (1977) electrical analogue of the regional, three-dimensional groundwater reservoir of New York. In glaciology, MacKay (1965) used an electrical field analogue to simulate the pattern of flow of large ice sheets during the Wisconsin glaciation. In ecology, Odum (1960, 1971, 1983) has developed electrical analogue circuits of ecosystems.

1.2.3 Mathematical Models

In a mathematical model, the features of a system are represented by abstract symbols and subjected to the rigour of mathematical argument. A mathematical model is usually an abstraction of a conceptual or scale model in which components and relations are replaced by an expression containing mathematical variables, parameters, and constants (Krumbein and Graybill 1965, p. 15). Mathematical models of Earth surface systems form the bulk of this book. The justification for keeping discussion of conceptual models brief and for merely mentioning scale models is a personal belief that future advances in Earth science will arise from mathematical modelling. Scale models are useful in some special, usually practical, circumstances but, because many systems of interest to Earth scientists operate over time scales which cannot be reproduced in the laboratory and because of the difficulties of scaling the properties of Earth materials, they have rather limited applications. Mathematics, on the other hand, offers a powerful tool of investigation limited only by the creativity of the human mind. Of all modes of argument, mathematics is the most rigorous. It provides a means of describing a system in a symbolism which is universally understood. Mathematical models seem capable of providing the deepest insight into how systems work, of giving the best basis for predicting change in systems, and of affording the most trustworthy indications of how best to manage or control systems. This is not to say that mathematics can replace the intuition and inspired guesses of the Earth scientist. Rather, it provides a way of formalizing thoughts and ideas. Of course, the act of quantification, of translating ideas and observations into symbols and numbers, is in itself nothing. As Craig and Labovitz (1981, p. 1) remark, the art and science of geomathematics is to uncover the kernal of the Earth system and to express it in symbols with generative power. Explanation and prediction validate the act of quantification. It is the predictive power of mathematical models that, in a sense, sets them apart from their conceptual model counterparts. There is no formal way in which an unquantified conceptual model can be tested — it remains a body of ideas. A mathematical model, on the other hand, can be tested by matching predictions against the yardstick of observations. By a continual process of mathematical model building, model testing, and model redesign, better and better explanations of the forms and processes recognized as Earth surface systems should emerge (cf. Thomas and Huggett 1980, p. 6).

Three chief classes of mathematical model are of interest to Earth scientists. They are stochastic models, statistical models, and deterministic models, the last including classical and modern varieties.

Stochastic models have a random process built into them which describes a system, or some facet of it, on the basis of probability and not on a purely deterministic basis. Stochastic models may be subdivided into deductive and inductive types. Deductive stochastic models, which will be discussed in Chapter 4, include independent events models, random-walk models, models based on Markov chains, and models based on the principles of statistical mechanics. Inductive stochastic models, which will be presented in Chapter 5, include the Box and Jenkins (1976) models that are popular with Earth scientists at present and have been applied to a variety of time and distance series data.

Statistical models, like stochastic models, have random components. In statistical models, the random components represent the unpredictable fluctuations in laboratory or field data which may arise from measurement error, equation error, or the inherent variability of the objects being measured. In statistical models, a body of inferential statistical theory determines the manner in which the data should be collected and how relationships in the data should be managed. The models discussed in Chapter 6 include simple tests to compare mean values of measured data sets, simple correlation and regression models, and sophisticated multivariate techniques designed to explore complex interrelations within large sets of measured data. Statistical models are, in a sense, a second best to deductive models; they can only be applied under strictly controlled conditions, suffer from a number of deficiencies, and are perhaps most useful when the "laws" determining system form and process are poorly understood.

Deterministic models are conceptual models which have been translated into the language of mathematics but which contain no random components. They can be derived from physical and chemical principles without recourse to direct experiment. It is sound practice, therefore, to test the validity of a deterministic model by experiment. Thus Stoke's Law, a deterministic model which relates the settling velocity of a small sphere in a viscous fluid to the radius of the sphere, was derived in 1851 directly from hydrodynamic principles. It has subsequently been verified by experiment and been found to apply exceptionally well to spheres ranging in size from 0.0005 mm to 0.08 mm settling in water. Smaller spheres are influenced by Brownian effects and larger spheres are affected by the turbulence they create. Deterministic models of the classical kind, such as Stoke's model, are dealt with in Chapters 7 and 8. Dynamical systems models, a modern form of deterministic model which have developed alongside the theory of open systems, are discussed in Chapter 9.

Part II   Conceptual Models

CHAPTER 2
Simple and Complex Systems

The appropriate mathematical model to use in describing and analysing a particular system depends on how the system is conceived — whether it is viewed as a simple system, a cascading system, or whatever. The theory and assumptions of the available mathematical models differ so fundamentally that before looking in detail at mathematical models of Earth surface systems it is salutary to consider the different ways in which Earth surface systems are conceived. To this end, the present chapter and the next will look at conceptual models of Earth surface systems, discussion being divided into models based on the simple-complex typology of systems and models based on the form-process typology of systems.

2.1 Simple Systems

The first systems to be studied in science consisted of a few objects which could be described by correspondingly few variables. The study of such simple systems dominated science until late in the nineteenth century. In a simple mechanical system, as studied by Isaac Newton, the motion of each system component, say a few billiard balls moving on a billiard table or planets revolving around the Sun, can be described by deterministic equations of motion. These equations define the exact movement of the balls or planets with respect to the forces acting upon them. In the case of the billiard balls, just four variables are needed for each ball — position and velocity in a plane. Taken together, the equations for individual balls form a set of simultaneous differential equations which enable the prediction, for a given initial distribution of balls and known forces applied from outside by a billiard cue, of system changes, of the future location of all balls. The dynamics of virtually all simple systems, including balls rolling down inclined planes, levers, cog-wheels, bodies colliding with each other, and billiard balls bouncing off cushions, can be studied and predicted using the classical methods of Newtonian mechanics (Waddington 1977, p. 64). The dynamics of many simple systems have been remarkably well verified, witness the success of space travel, which requires accurate computation of the dynamical trajectories of spacecraft, and the achievements in artillery and ballistics where targets can be hit with frightening accuracy from great distances. The method is, however, limited to objects of moderate mass, the velocities of which are much less than the speed of light — ordinary stars, planets, boulders, sand grains, and other sedimentary particles. Very small objects (atoms and elementary particles) are studied by quantum mechanics and hyperdense objects (neutron stars and black holes) are studied by relativistic dynamics.

In the Earth sciences, the Newtonian conception of dynamics has proved a useful and powerful method of analysing simple systems. Its application to the study of the movement of material at the surface of the Earth has been particularly rewarding. Erosion and weathering are often regarded, in true Newtonian spirit, as the result of external forces or stresses applied at the Earth's surface and internal forces resisting them. Strahler (1952a) classified Earth materials according to their response to an applied stress caused by either gravity or by molecular processes. The literature dealing with the mechanics and dynamics of Earth surface systems is extensive. Statham (1977) and Davidson (1978) provide recent general accounts. A more theoretical exposition is offered by Scheidegger (1970), whose detailed reference to early work on the subject is illuminating. More specific statements and discussions include Nye's (1951) application of plasticity theory to the flow of ice sheets and glaciers; Bagnold's (1954) classic study of the physics of blown sand and desert dunes, his (Bagnold 1956) classic paper on the bedload stresses set up during the transport of cohesionless grains in fluids, and his (Bagnold 1966) approach to the problem of sediment transport from the viewpoint of general physics. Other examples are Yalin's (1977) treatment of the mechanics of sediment transport, W. H. Graf's (1971) account of the hydraulics of sediment transport, the papers by W. E. H. Culling (1983a, b) on soil creep mechanisms, and the book by Richards (1982), which discusses a wide variety of work on the dynamics and mechanics of rivers from a geomorphological stance. Applications of mechanical and dynamical principles to specific field cases are legion. There are studies of earthflows (Carson 1981), landslides (Skempton 1964; Statham 1975), talus slopes (Kirkby and Statham 1975), the movement of sediment by splash (de Ploey and Savat 1968), and many others. The theoretical and empirical relationships established for simple Earth surface systems are commonly fed into more complex models. For instance, various relationships between variables contributing to soil erosion are used in overall soil erosion models (for example, Beasley and Huggins 1981).

Relativistic dynamics is, presumably, irrelevant to the study of Earth surface systems; quantum mechanics is not. Formulated in the mid-1920's, quantum mechanics provides a way of describing the behaviour of atoms and molecules. It is not a field of study with which Earth scientists normally bother themselves. However, Weisskopf (1975) believes that quantum mechanics, and its statement of the electrical attraction between electrons and nuclei, in conjunction with the gravitational attraction of massive objects, enables an explanation to be given of all material phenomena which occur under terrestrial conditions. Many readers will no doubt think that this view is reductionism carried to a ridiculous limit. But Weisskopf's ideas, risky approximations and generalizations though he deems them to be, are original, by no means as outrageous as they may seem, and certainly deserve studying.

Weisskopf (1975) argues that all the relevant magnitudes which characterize the properties of matter can be expressed by six magnitudes: M, the mass of the proton; m, the mass of the electron; e, the electrical charge of the electron; c, the velocity of light; G, Newton's gravitational constant; and h, the quantum of action defined as $h = h^*/2\pi$ where h^* is the conventional value of Planck's constant. In addition, Z, the atomic number, and A, the atomic weight, are used because Ze determines the charge and AM the mass of the nucleus. In conjunction with certain rules governing the relations between nuclei, atoms, and molecules, these magnitudes enable answers to be given to questions such as: Why is a piece of rock as heavy as it is? And why are mountains as high as they are? In answer to the first question, Weisskopf argues that the density of a compact matter such as rock is the weight of the atom or molecule divided by the volume it fills. In compact matter the atoms are contiguous and the density, ρ, is given, approximately, by

$$\rho = \frac{AM}{f^3 \frac{4\pi}{3} a_0^3},$$

where A and M are as above, a_0 is the Bohr radius, and f is a function which, as an approximation, may be defined as

$$f = 1.5 A^{1/5},$$

although fluctuations exist around this value which enter into the density formula in the third power. The numerical result is then

$$\rho \sim 0.80 A^{2/5} \, (\text{g cm}^3),$$

which indicates that the density of compact matter ranges between 1 and 10 g cm^{-3}.

Why are mountains as high as they are? To answer this question, Weisskopf represents a mountain as a block of silicon oxide resting on a plane surface also composed of silicon oxide. When the weight of the block is so great that the base matter starts to flow by plastic deformation, the mountain will sink. If H is the height at which the block starts sinking, then clearly H is also the maximum height the mountain can attain. On a planet where tectonic and vulcanic activity are very active, as on Earth, mountains will be of the same order of magnitude as H but somewhat lower. The mountain will begin to sink at that height where the energy gained by letting the mountain sink is equal to the energy necessary to bring about plastic flow. This energy is comparable to, and probably somewhat lower than, the energy required to melt the rock; for convenience, these energies are set equal. The amount of matter to be "liquefied" is roughly the same as the amount of mountain matter that sinks into the ground. So, roughly speaking, the amount of gravitational energy gained by lowering matter from a height H must equal the liquefaction energy of the same amount of matter. Weisskopf does the calculation for each molecule separately. The mass of the molecule is A proton masses, where A is about 50, the atomic number of silicon oxide. The relation then holds that

$$AMHg = E_1,$$

where g is the gravitational acceleration on Earth and E_1 is the liquefaction energy per molecule. The liquefaction energy per molecule is a small fraction of the binding energy, B, of the material:

$$E_1 = \xi B .$$

This is because in the process of liquefaction the directional stiffness of the binding is removed but the binding is not broken. For metals and minerals ξ is of the order of 0.05. Given that

$$B = \gamma Ry ,$$

where γ is about 0.2 for rocks and Ry is the Rydberg unit, defined as

$$Ry \equiv \frac{-me^4}{2h^2} ,$$

and determines the energy by which the electron is bound to the proton. The maximum height of the mountain may therefore be written as

$$H = \xi \gamma \frac{Ry}{AMg} \sim 26 \text{ km} .$$

The actual value is 10 km or thereabouts, but the result is reasonable given that the energy necessary to produce plastic flow should be somewhat less than the liquefaction energy.

2.2 Systems of Complex Disorder

2.2.1 Irreversible Processes

A problem with the classical conception of simple dynamical systems is that time, in the mechanical equations, is treated in a restricted way in that it can be reversed or run backwards. The fundamental laws of physics and chemistry are taken to be symmetrical in time: physical and chemical processes are seen as reversible. This means that, if the present state of a system is known, then all future and past states can be predicted. Classical mechanical and dynamical methods therefore do not apply to phenomena in which past and future do not play the same role, to phenomena in which time-orientated or irreversible processes occur. For instance, two liquids put in the same jar generally diffuse to form an homogeneous mixture which does not separate spontaneously; the diffusive process is irreversible. Likewise, if part of a macroscopic body is heated and then isolated thermally, then the temperature in the body will become uniform. The uniformity cannot be reversed unless further heat is applied.

In physics, the concept of irreversible processes was expressed in the second law of thermodynamics originally formulated by Rudolf Clausius in 1865. For an isolated system, Clausius's formulation implies the existence of a function, S, the entropy, which increases monotonically until it reaches its maximum value at the state of thermodynamic equilibrium when $dS/dt = 0$. In this classical concep-

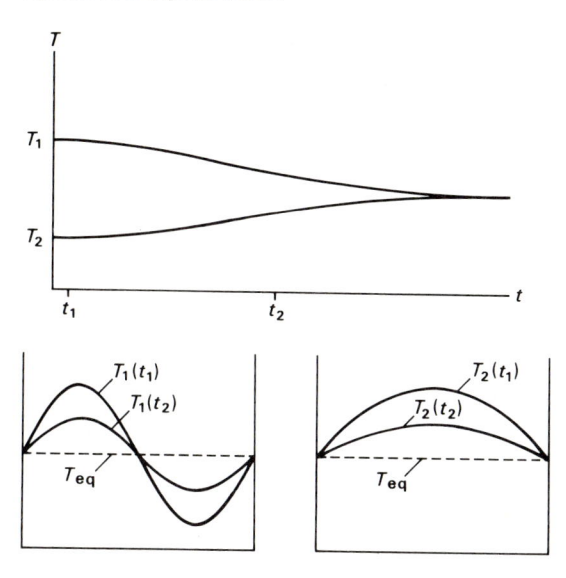

Fig. 2.1. A uniform temperature distribution as an attractor of all initial states. Different initial distributions such as T_1 and T_2 lead to the same temperature distribution. (Prigogine 1980)

tion, entropy is a measure of molecular disorder. Put more simply, the second law of thermodynamics states that an isolated system must attain a time-independent equilibrium. In modern parlance, the state of thermodynamic equilibrium is an attractor for all nonequilibrium states. Thus a uniform temperature distribution is an attractor for all initial, nonuniform temperature distributions (Fig. 2.1). The state of equilibrium is defined by maximum entropy and minimum free energy. The same is true of a closed chemical system in equilibrium. Chemical equilibria are true equilibria based on reversible reactions. When a reversible reaction stops, a state of chemical equilibrium has been reached in which entropy is maximized, free energy is minimized, and there is a constant ratio between phases.

The concept of entropy is crucial to an understanding of many aspects of Earth surface systems. Chorley and Beckinsale provide a lucid explanation of entropy which is paraphrased below with the term "closed system" in the original replaced here by the term "isolated system":

"Entropy is a measure of the extent to which energy within a system is no longer free, in the sense of being capable of performing work on the system. An isolated system with low entropy possesses within itself differentials in the distribution of energy such that the flow of energy from locations of high energy to locations of low energy is capable of performing work. As time passes, and the energy in the system becomes more equally distributed, the entropy increases, until, at the state of maximum entropy, all parts of the isolated system have the same energy levels, no flows of energy take place, and no work is being performed". (Chorley and Beckinsale 1980, p. 129)

Classical thermodynamics tackles systems which contain a large number of weakly or randomly interacting components, a huge number of billiard balls on a big billiard table for instance. Newtonian dynamics cannot tackle systems of this complexity because there are too many equations to handle. In the late nineteenth century, Boltzmann and Gibbs showed that systems that are complex but disorganized can be studied using a new kind of mathematics to find certain quantities of interest. The new mathematics was statistical averaging or, more technically, entropy maximizing (Wilson 1981, p. 39). By these methods, although the behaviour of any one particle of gas, or one billiard ball among many, cannot be predicted, the distribution of particles among energy states and aggregate measures such as the pressure and temperature of a gas, and the distribution of velocities of billiard balls and the average rate at which a ball strikes a cushion, can be predicted. The predictions made by this branch of classical thermodynamics apply to closed system at or very near to equilibrium.

Regarding Earth surface systems as systems of complex disorder opens up two possible avenues of enquiry. Firstly, systems can be handled in a probabilistic manner, the rationale for doing so being variously held to be the general analogy of the landscape with ideal isolated and closed systems (Chorley 1962; Scheidegger 1964a; Karcz 1980), the intrinsic randomness of Earth surface processes such as flow through porous media (Scheidegger 1954, 1961c) and soil creep (W. E. H. Culling 1963), and the aggregate randomness of complex landscape systems in which a large number of individually deterministic relationships interact (Leopold and Langbein 1962). Secondly, systems can be handled in a more dynamic way by applying principles of energy and materials accounting. The main idea here is that the gross amount of energy and matter moving into and through a system must be accounted for and each type of energy and each material must be subject to the same book-keeping rigour (Harbaugh and Bonham-Carter 1970, p. 51). Although the principles of accounting are simple, they can lead to sophisticated methods such as input-output analysis. They can also lead to statements of energy conservation and mass conservation which are used, in conjunction with principles of classical dynamics, to study a variety of Earth surface systems.

2.2.2 Accounting Models: the Laws of Thermodynamics

The laws of both Newtonian dynamics and thermodynamics are widely used in the study of Earth surface systems. These laws can be expressed as basic equations which are of four kinds: balance equations, physical-chemical state equations, phenomenological equations, and entropy balance equations (Isermann 1975).

Balance equations represent the laws of conservation. They indicate that what goes into a system must be stored, come out, or be transformed into something else: matter, energy, and momentum cannot suddenly appear or disappear in an unaccountable manner (Huggett 1980, p. 93). For energy transactions, the energy balance is defined by the first law of thermodynamics. Historically, this law is implicit in Newton's third law of motion, but is normally traced to the experiments of James Joule, who in 1843 firmly established the principle of equivalence between heat and work, a principle which is otherwise known as the first law of thermodynamics. A more general principle of energy transactions, that of the conservation of energy, was established by Herman von Helmholtz and Sir William Thompson (Lord Kelvin). This more general principle applies to all forms of energy including thermal, chemical, kinetic, and electrical. It may be expressed by saying that, in an isolated system, the sum of all forms of energy remains constant; or that energy may be transformed from one form to another but is neither created nor destroyed. The second definition is perhaps the more pertinent to Earth surface systems. An energy balance equation may be written

$$\frac{dE}{dt} = E_{in} - E_{out},$$

where E is energy, t is time, E_{in} is energy input, and E_{out} is energy output. If the equation is applied to a spatial system then spatial terms, for instance Cartesian co-ordinates x, y, and z, must be added. The principle of energy conservation has been fruitfully brought to bear on the study of ecosystems (Lindeman 1942; Odum 1971, 1983).

For mass transactions, a mass or materials balance is defined by the equation

$$\frac{dM}{dt} = M_{in} - M_{out},$$

where M is mass, t is time, M_{in} is mass input, and M_{out} is mass output. Historically, the idea of mass conservation goes back to Lavoisier, an eighteenth century scientist and victim of the guillotine, who stated that "Nothing can be created, and in every process there is as much substance (or quantity of matter) present before and after the process has taken place. There is only a change or modification of the matter". Conservation of mass is a general principle of nature and is applied in all models of Earth surface systems where an account is kept of all materials that enter and leave parts of the system. It applies to small-scale, short-lived systems such as deltas (Bonham-Carter and Sutherland 1968), to medium-scale systems such as sedimentary basins (Harbaugh and Bonham-Carter 1970, p. 373), and to global-scale systems such as the sedimentary cycle (Garrels and Mackenzie 1971) and world biogeochemical cycles (Lerman et al. 1975; Lerman 1979; Bolin 1981; Berner et al. 1983).

Applied to hydrodynamic systems, the law of mass conservation is expressed as the continuity equation as first developed by Laplace. In rectangular co-ordinates and in differential form, the continuity equation is written

$$\frac{\partial \varrho}{\partial t} = -\left\{ \frac{\partial \varrho u}{\partial x} + \frac{\partial \varrho v}{\partial y} + \frac{\partial \varrho w}{\partial z} \right\},$$

where ϱ is fluid density and u, v, and w are velocity components in the x, y, and z directions. It has wide-ranging implications in the study of Earth surface systems including the flow of groundwater (for example, Wang and Anderson 1982) and changes in atmospheric density (Hess 1959). In fact, in its present usage, the term "equation of continuity" is taken to include all cases of mass balance and not just fluids. Thus Kirkby (1971, p. 15), talking of hillslope models, describes the continuity equation as a statement to the effect that if more material is brought into a slope section than is taken out, then the difference must be represented by aggradation in the section; and conversely, if less material is carried into a slope section than is removed, then the difference must come from net erosion of the section. The same principle of continuity of sediment transport is applied to other Earth surface systems including rivers. The continuity condition, in the sense of a conservation of mass, also applies to solutes and colloids in the landscape and in soils (Boast 1973) and in ecosystems (Patten 1971).

For momentum transactions, the momentum balance is written

$$\frac{dI}{dt} = K_{in} - K_{out},$$

where I is momentum, K_{in} is an incoming force, and K_{out} is an outgoing force. The momentum, I, can be expressed as a product of mass and velocity. The basis of the momentum balance is Sir Isaac Newton's laws of motion, and particularly the second law of motion, which states that force is proportional to the rate of change of momentum in the body on which the force acts.

Physical-chemical-state equations express the dependence of one state variable upon another. A famous example is the gas law of Boyle and Gay-Lussac which states that

$$PV = RT,$$

where P is pressure, V is volume, R is a gas constant, and T is absolute temperature. This can be applied directly to Earth surface systems. For instance, the relation between pressure and stored mass in a gas reservoir is expressed by the equation

$$M = V\varrho = V\frac{1}{RT}P,$$

where M is stored mass, V is gas field volume, ϱ is gas density, P is gas pressure, R is the gas constant, and T is absolute temperature. Physical-chemical-state equations for heat and its relation to other state variables have been applied indirectly to landscapes where an analogy between temperature and height and change of heat and change of mass was recognized by Leopold and Langbein (1962) and affirmed by Scheidegger (1964a). The thermodynamic analogy for landscapes will be discussed in Chapter 4.

Phenomenological equations are used to describe irreversible processes. The word "phenomenological" is derived from a Greek word meaning "to appear", and refers to the fact that changes in observed macroscopic variables are described without trying to see what concomitant changes are occurring in atoms, molecules, or electrons. All phenomenological laws describe processes which tend to equalize the value of macroscopic variables throughout a system. They are tied up with the second law of thermodynamics, which is concerned with irreversible processes. Many statements of the second law have been given starting with the definition of Rudolph Clausius. The essence of Clausius' definition is that heat will not of itself flow from a colder to a hotter body. In other words, heat flow occurs spontaneously in a "downhill" direction and to make it flow "uphill" from a cold to a hot body work must be performed and the process driven. The phenomenological equation describing heat flow is Fourier's law of heat conduction which may be written (for one dimension)

$$q_x = -\lambda \partial T / \partial x,$$

where q_x is the heat flow rate, λ is a heat conduction coefficient, and $\partial T/\partial x$ is the temperature gradient. The flow of water in a porous medium may be described by Darcy's law, which for one dimension is written as

$$q_x = -K\partial\phi/\partial x,$$

where q_x is the water flow rate, K is hydraulic conductivity, and $\partial\phi/\partial x$ is the water potential gradient. The flow of a solute may be described by Fick's first law of diffusion which in one dimension may be written

$$J_x = -D\partial C/\partial x,$$

where J_x is the solute flow rate (mass flow density), D is a diffusion coefficient, and $\partial C/\partial x$ is the solute concentration gradient. The use of these equations, and other equations which describe processes in Earth surface systems, will be demonstrated and discussed in Chapters 7 and 8.

All irreversible processes change the entropy in a system and it may be necessary to keep account of the entropy balance. This is done using entropy-balance equations.

2.3 Systems of Complex Order

2.3.1 Nonequilibrium Systems

Systems that are regarded as complex and organized may be studied using the concept of open systems. Early ideas on open systems came from Lotka (1925). For many years, the leader in the field was Ludwig von Bertalanffy (1932, 1950), who considered general kinetic principles of open systems and their implications for biology. Later work by Denbigh (1951) looked at the kinetics of open reaction systems in industrial chemistry and compared the efficiency of batch and continuous reaction systems. At about the same time, Prigogine (1947) developed the thermodynamics of open systems.

From a thermodynamic viewpoint, open systems maintain themselves in a state of high statistical improbability, a state of order and organization. According to the second law of thermodynamics, the general trend of physical processes is towards increasing entropy, that is, towards increasing states of probability and decreasing order. Open systems maintain themselves in a state of high order and improbability and may even develop towards increasing differentiation and organization. The reason for this is given in the expanded entropy function proposed by Prigogine (1947). In a closed system, entropy always increases according to the Clausius equation. In an open system, in contrast, the total change of entropy can be written

$$dS = d_e S + d_i S$$

where $d_e S$ denotes the change of entropy by import and $d_i S$ denotes the production of entropy in the system due to irreversible processes such as chemical reactions, diffusion, and heat transport (Fig. 2.2). The term $d_i S$ is always positive, according to the second thermodynamic law; $d_e S$, entropy transport, may be

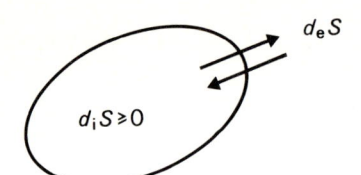

Fig. 2.2. An open system in which $d_i S$ represents entropy production and $d_e S$ represents entropy exchange between system and environment. (Prigogine 1980)

positive or negative. Entropy transport is negative when, for example, matter is imported as a potential carrier of free energy or "negative entropy". This is the basis of the negentropic trend in open systems and Schrödinger's (1944) statement that organisms feed on negative entropy.

Open systems, even simple ones, show remarkable characteristics. Under certain conditions, open systems approach a state which does not change with time. This is the steady state in which the system remains constant as a whole and in its macroscopic phases despite a continuous throughflow of component materials. A system in a steady state is not in equilibrium but is tending to attain it. Only in this condition can energy be won from the system. In a steady-state system, but not in a closed system, there is a steady flow of matter or energy which must be maintained and the energy content of which is transformed into work. Continuous working capacity is, therefore, not possible in a closed system, which tends to attain equilibrium as soon as possible, but it is in an open system. For the maintenance of a steady state, the rates of system processes must be precisely balanced; only then is it possible that certain components can be broken down, so liberating usable energy, while on the other hand, import prevents the system from attaining true equilibrium. In a steady state then, a system maintains a constant form in the face of continuous irreversible processes. A steady-state system is in a condition of minimum entropy production. As Prigogine (1980, p. 88) explains, the theory of minimum entropy production expresses a sort of inertial property of nonequilibrium systems; and when given boundary conditions prevent a system from attaining thermodynamic equilibrium (that is, zero entropy production), the system settles down in a state of "least dissipation".

2.3.2 Systems Far from Equilibrium

The theorem of minimum entropy production is valid only in the neighbourhood of equilibrium. Efforts to extend the theorem to systems far from equilibrium have recently yielded the surprising result that systems well away from equilibrium produce a quite different thermodynamic behaviour, a behaviour in fact exactly opposite to that predicted by the theorem of minimum entropy production.

Whereas a linear system will settle in a state of minimum entropy production when boundary conditions prevent it from reaching, but not by much, thermodynamic equilibrium (zero entropy production), a system far from equilibrium may act to increase entropy production. The system far away from equilibrium therefore becomes more ordered, the nonequilibrium being the source of order. Prigogine (1980, p. 88) offers the example of a convective cell. A horizontal layer of fluid sandwiched between two parallel planes is at rest. If the bottom plane is warmed and maintained at a higher temperature than the top plane, then, when the temperature gradient between the two planes becomes large enough, the state of rest becomes unstable and convection starts. Entropy production is then increased because the convection is a new mechanism for heat transport. What actually happens in this example is that the small convection currents which appear as fluctuations from the average state, even when the fluid is at rest, are damped and disappear beneath the critical gradient of temperature. Above the critical temperature gradient, some of the fluctuations are amplified and give rise to a macroscopic current. A new molecular order appears that basically corresponds to a giant fluctuation stabilized by exchange of energy with the outside world. Order of this kind is characteristic of what Prigogine (1980, p. 90) calls dissipative structures, that is, systems in which energy is dissipated in maintaining order.

The study of nonequilibrium systems has thrown light on the role played by irreversible processes in states far removed from equilibrium and has led to a deep insight into the evolution of chemical and physical systems (see Prigogine 1980; Haken 1982). Its potential application in the Earth sciences is starting to be realized (Beaumont 1982; Thornes 1980, 1982, 1983) and will be discussed in Chapter 9.

2.3.3 Open Systems at the Earth's Surface

The concept of the open system was introduced to geomorphology by Strahler (1950, 1952a), who took his inspiration directly from the writings of von Bertalanffy. A systems approach to the subject is much older, however. It can be traced to the writings of G. K. Gilbert who, in his monograph on the Henry Mountains, Utah, wrote

"The tendency to equality of action, or the establishment of a dynamic equilibrium, has already been pointed out ... but one of its most important results has not been noticed ... in each basin all lines of drainage unite in a main line, and a disturbance upon any line is communicated through it to the main line and thence to every tributary. And as any member of the system may influence all others, so each member is influenced by every other. There is interdependence throughout the system". (Gilbert 1877, pp. 123 – 124)

Chorley and Beckinsale (1980, p. 140) opine that Gilbert's legacy to contemporary dynamic geomorphology is in fact his anticipation of the systems approach to the discipline. Strahler's advocacy of systems theory merely ushered in a revival of Gilbertian thinking in geomorphology. Strahler's call to action was taken up by Hack (1960), Chorley (1962), Howard (1965), Schumm (1977), and many others.

The concept of the open system was introduced to soil science by Jenny (1958, 1961). Other statements championing a systems view of the soil followed. Nikiforoff (1959) emphasized the openness and dynamic nature of the soil. He saw that for a soil system to persist, incoming material must at least replace outgoing material. In the words of Buol et al. (1980, p. 11), "a soil is an evolving entity maintained in the midst of a stream of geologic, biologic, hydrologic, and meteorologic material". Simonson (1959, 1978) put forward a similar concept in his "outline of a generalized theory of soil genesis", while Yaalon (1960, 1971) discussed the soil system from the point of view of thermodynamic principles. More recently, the open system concept has been applied to soil-landscape units (Walker and Ruhe 1968, Ruhe and Walker 1968; Huggett 1975, 1982).

The ocean was originally conceived as a gigantic depository for materials washed from the land. This view has now been supplanted by the concept of the ocean as a steady-state system in which the amount of incoming materials continuously equals, more or less, the amount of outgoing materials. The first paper propounding this view was by Sillén (1961, 1963, 1967), who pointed out that the chemical composition of the oceans bears a striking resemblance to the composition of a theoretical solution brought into chemical equilibrium with a number of minerals found in the oceans, a solution which Sillén studied using the chemical thermodynamic principles of Josiah Gibbs. More recent studies of the ocean, for instance Mackenzie and Garrels (1966), take full account of the openness of the oceanic system and make allowance for dissolved species transported into the oceans by rivers and the precipitation of species from the ocean and cycling back to the continents through the atmosphere. Many models of lake systems also adopt open system concepts (for instance, Yuretich and Cerling 1983).

The term ecosystem was coined by Tansley (1935) but the ecosystem concept is implicit in much earlier writings. Charles Darwin, in the last paragraph of the *Origin of Species*, evoked "an entangled bank, clothed with many plants of many kinds, with birds singing on the bushes, with various insects flitting about, and with worms crawling through the damp earth ... [all] dependent on each other in so complex a manner", a statement which describes an ecosystem without giving it a name. The unsung hero of systems ecology in its modern form, however, is Lotka. Justly famous for his population equations, Lotka (1925) also considered variations in energy and particular chemical substances; he established a law of minimum energy flux which states that, if there is excess energy in a system, a species will evolve to use it and so increase the flux of energy through the system (Odum and Pinkerton 1955). This portion of Lotka's work lay follow until it was reworked by the systems ecologists (McIntosh 1980). The use of the second law of thermodynamics in ecology was urged by Haskell (1940) and soon came in Lindeman's (1942) classic, trophic-dynamic model of ecosystems, which emphasized energy flow and provided the conceptual seed for the subsequent flowering of systems ecology (Cook 1977).

Systems of complex order are aptly described as holistic. The holistic nature of complex systems is captured by aphorisms such as "everything is connected to everything else" (Commoner 1972) and "all systems are ultimately, if not intimately, related" (Holliman 1974). The holistic nature of Earth surface systems is recognized explicitly in the concept of the ecosystem. By definition, ecosystems have living and nonliving components. The ecosystem concept thus emphasizes the interaction between components of abiotic geological systems (land, sea, and air) and organisms. The far-reaching significance of recognizing biotic – abiotic interactions, of taking an ecological view of Earth surface systems, is only beginning to dawn. An example is in biogeochemical cycles where organisms appear to play an important role in the flow and storage of both major elements (for instance Lerman et al. 1975) and minor elements (Li 1981). Lovelock (1979) believes that the planetary environment is largely under the control of the biosphere. Views such as this are a departure from traditional discussions of Earth surface systems in which the abiotic world is generally thought to "control" the biotic world or in which the biotic world changes by a dynamic of its own (see Gould 1977). An ecological view stresses the mutual interaction of biotic and abiotic Earth systems, a theme considered at a conceptual level by Rich (1984).

CHAPTER 3
Form and Process Systems

3.1 Models of System Form

The word morphology was originated by Goethe in 1817, who proposed it to describe the study of the unity of type of organic form. Today, morphology deals with a wider study of structure and is either descriptive, using mathematical terms, or analytical, seeking to understand the forces which mould a particular form (March and Steadman 1974, p. 270). Terjung (1976) widens the definition of morphology even more to include the description of physical properties and occurrences as well as geometries. This broad conception of morphology has been adopted and amplified by Strahler (1980), who distinguishes morphological variables, which define the configuration and constitution of a system, from dynamic variables, which are mechanical properties representing the expenditure of energy and the doing of work. Strahler (1980) recognizes three groups of morphological variables: geometry variables, material-property variables, and mass-flow variables. Geometry variables have the dimension of length or are dimensionless ratios of length. They are used to describe a system in terms of size and shape. Material-property and mass-flow variables describe the constitution of a system.

3.1.1 Models of System Constitution

Where both geometrical and material-property variables are used to describe a system, it is usual to regard the system as "a web of correlations or feedbacks between the morphological components of part of a system or between systems" (Terjung 1976, p. 204). The relationships between the variables in such models are formalized by applying statistical models ranging from nonparametric tests of association, through simple regression and correlation, to sophisticated multivariate techniques. Figure 3.1 depicts the correlation structure of a valley-side slope system developed on sandy till in southern Manitoba, Canada as revealed by a statistical analysis of a sample of 25 slopes each with a stream moving away from its base. Models of this kind are dealt with in Chap. 6.

3.1.2 Models of System Geometry

Where the variables used to describe a system are purely geometrical, the system may be conceived either as a web of relationships between the geometrical

Models of System Form

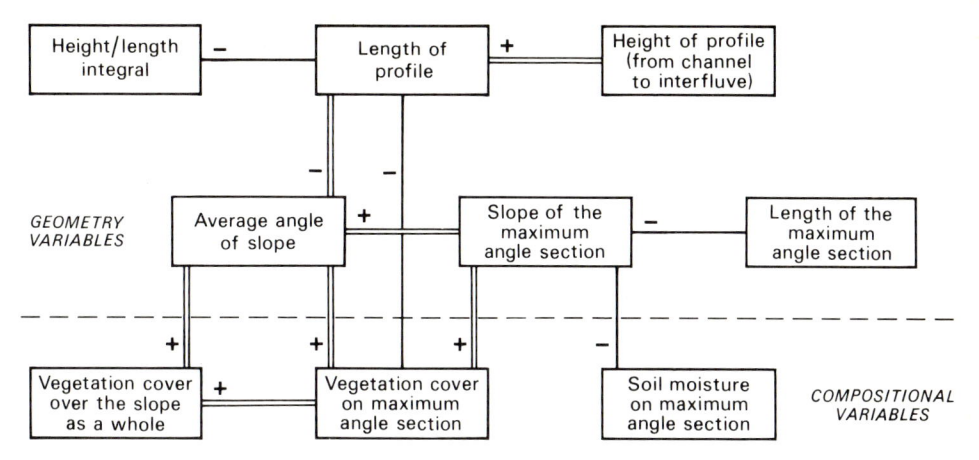

[Diagram: Morphological system of valley-side slopes showing relationships between GEOMETRY VARIABLES (Height/length integral, Length of profile, Height of profile (from channel to interfluve), Average angle of slope, Slope of the maximum angle section, Length of the maximum angle section) and COMPOSITIONAL VARIABLES (Vegetation cover over the slope as a whole, Vegetation cover on maximum angle section, Soil moisture on maximum angle section) connected by + and − signs.]

Fig. 3.1. Valley-side slopes in Manitoba represented as a morphological system. (After Chorley and Kennedy 1971)

variables or as an abstract pattern of points, lines, and areas. An elementary example of the "web of relationships" category of systems is an alluvial fan represented by just two geometrical variables: the area of the fan and the area of the source basin. These two variables are known to have a consistent relationship with one another over a wide range of sites in the western United States (Chap. 6.1.5) which, as Bull (1962), Hooke (1968), and Hooke and Rohrer (1979) have shown, may be interpreted in terms of fan development.

Many features of Earth surface systems may be represented by an abstract pattern of points, lines, or areas. The patterns can be applied across the full range of spatial scales encountered on the Earth. Point patterns have been used to represent plants, drumlins, karst depressions, volcanoes, and many other phenomena. Line patterns may be used to represent crack and joint patterns in rocks, rivers, valley trends, faults, coastlines, river networks, and any other feature that may be regarded as linear. Area patterns can be thought of as the "bits between the lines" in line patterns and include icewedge polygons, drainage basins, interbasin areas, and tectonic plates. The observed patterns of points, lines, and areas may be used in two ways. The first way is to test the observed patterns against patterns predicted by some probability process model (see Chap. 4). The second way is to use the observed patterns to set up conceptual schemes of morphometric units. This practice has proved popular with observers of hillslopes. Morphometric models of hillslope systems are based on the assumption that landforms all have slope as a lowest common denominator: every surface, whatever its origin, is made up of one or more slopes of variable inclination, orientation, length, and shape (Butzer 1976, p. 79). Most morphometric models of slope systems recognize a set of discrete slope units, each unit characterized by a distinct inclination, and are portrayed as block diagrams or geomorphological maps. One of the first models was proposed by Wood (1942), who recognized four units: the waxing slope, also called the convex slope or upper wash slope; the free face, also called the gravity or derivation slope;

the constant slope, also called the talus or debris slope if scree is present; and the waning slope, also called the pediment, valley-floor basement, or lower wash slope. Ruhe (1960) proposed a five unit model: summit, shoulder, backslope, footslope, and toeslope. Other schemes, each with its own terminology, have been put forward by Savigear (1965) and Dalrymple et al. (1968). Clearly, there is no shortage of attempts to construct morphometric models of slopes.

An alternative to conventional morphometric slope models, and the plethora of terms they entail, is the construction of digital terrain models. These are simply digital data files of elevations and are the digital equivalents of topographic maps. They represent landforms well enough to guide a cruise missile over a thousand miles of terrain. The missile senses the landform, converts it to digital form, searches its "remembered landform" for a match, and then computes position and makes course corrections (Craig 1982a). Digital terrain models can also be made to represent landforms in a form suitable for use by geomorphologists. By a method which will be discussed later (Chap. 10.1), Craig (1982b, c) has shown that digital terrain models can be designed in a way that enables a geomorphologist to infer the processes and the rates of processes responsible for the development of a landform in a fairly objective and precise manner.

3.2 Models of System Process

Many conceptual models of Earth surface systems focus on process rather than form. In process models, the correlation systems and geometrical patterns of form models are replaced by canonical structures representing cascades of energy, mass, and momentum which may be formalized by applying the principles of thermodynamics and mass and momentum conservation. Cascading or flow systems are conceived as "interconnected pathways of transport of energy or mass or both, together with such storages of energy and matter as may be required" (Strahler 1980, p. 10). The theory of open systems is an important conceptual tool in the building of process models because stress is laid upon the transfer of mass and energy and momentum from one part of a system to another.

The importance of flows in understanding phenomena is not a new idea. It dates from Heraclitus of Ephesus, a mystic who saw the process of change or flux as defining the underlying unity of nature. Heraclitus opined that human senses are deceptive and that objects which appear stable are in fact undergoing continual change. These views have been taken up by the physicist Bohm (1980), who envisages the universe made up, not of objects, but of an hierachy of flow patterns. In Bohm's conception, objects are self-maintaining and self-repeating features embedded in the flow which preserve a degree of invariance despite the fact the energy and matter are ever flowing through them; they are relatively constant abstractions amidst movement and transformation. An example is a cloud which, though moist air is continually flowing through it and continually evaporating and condensing, maintains a certain constancy of form. Similarly, emphasis in the conception of Earth surface systems as cascades is on the preservation of form in the face of inputs, throughputs, and outputs of energy and matter.

3.2.1 Land-Surface Cascades

A number of distinct cascades are recognized at or near the Earth's surface. Chorley and Kennedy (1971), for instance, distinguish between a debris cascade, involving a system of valley-side slope processes, and a stream channel cascade, involving a system of river processes. Although this separation is expedient, it is also a little misleading because weathered material and rock debris, in moving

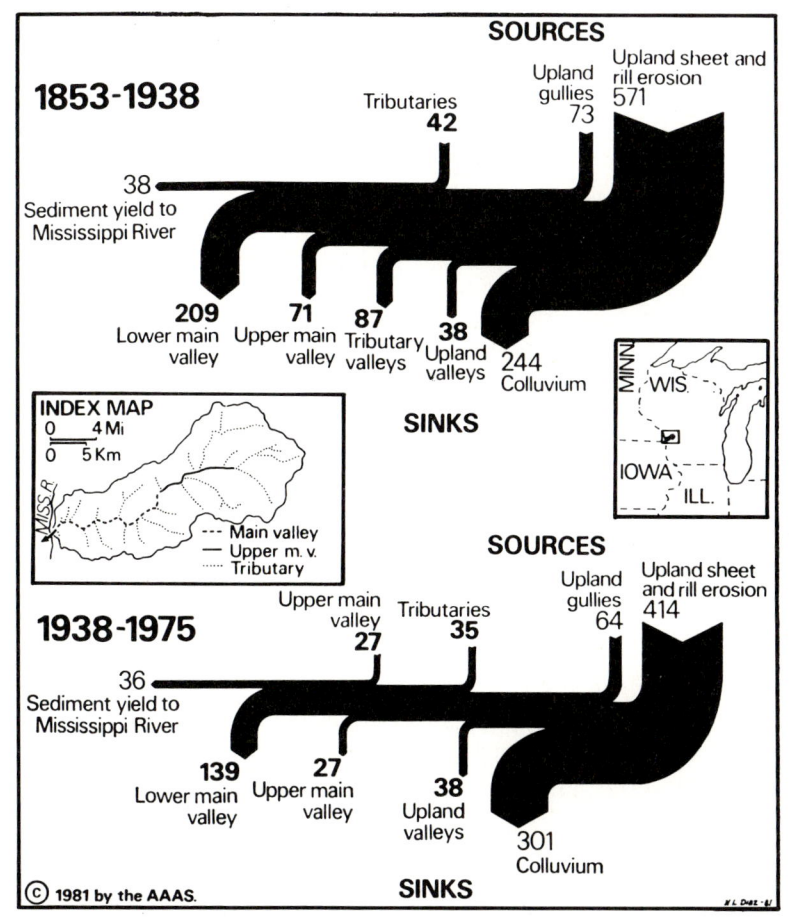

Fig. 3.2. The drainage basin as a unitary system: sediment budgets for Coon Creek, Wisconsin, 1853–1938 and 1938–1975. The basin is about 25 km southeast of La Crosse, Wisconsin, and has an area of 360 km². Numbers are annual averages for the periods in 10^3 Mg yr^{-1} and account for 3.6×10^7 Mg of sediment generated between 1853 and 1975. *Bold numbers* are measured; *other numbers* are estimated. Note that the main valley, which was a sink for sediment in the earlier period, has become a partial source of sediment in the later period. (Trimble 1981)

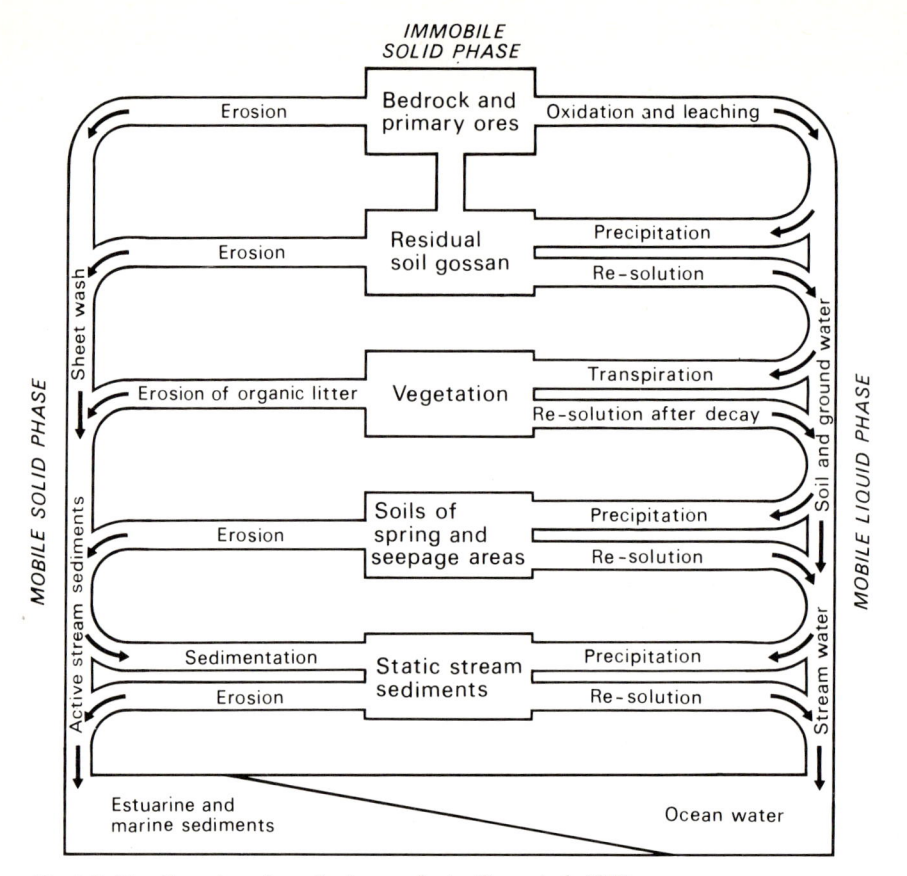

Fig. 3.3. The dispersion of weathering products. (Rose et al. 1979)

down a slope and along a stream, trace continuous lines of flow which form an hierarchy of drainage basins (Fig. 3.2). For this reason, land-surface cascades are best studied in the context of the drainage basin.

A case can be made for seeing the drainage basin as the fundamental unit of geomorphology (Chorley 1969a). Certainly, the drainage basin is regarded as a unitary system which is adjusted to transmit material down valley-side and stream slopes (Kirby 1971, p. 16). The unitary nature of the drainage basin was recognized long ago by Playfair (1802) and by Davis (1899), who wrote:

"Although the river and the hill-side waste sheet do not resemble each other at first sight, they are only the extreme members of a continuous series, and when this generalization is appreciated, one may fairly extend the 'river' all over its basin and up to its very divides. Ordinarily treated, the river is like the veins of a leaf; broadly viewed it is like the entire leaf". (Davis 1899, p. 495)

The changes of processes along lines of flow are related more to the relative proportions of water (or ice), the main transporting medium, to sediment, than to differences between slope positions. On valley-side slopes there is little, if any, water to a large body of sediment. In streams there is a large body of water in which some sediment is suspended and dissolved (cf. Davis 1899, p. 495). Between these extreme conditions, the relative proportions of water (or ice) to sediment form a more or less continuous series of flows ranging from fluvial transport, in both rivers and sheetwash, and glacial transport, through a variety of material flows, such as solifluxion, earth flow, mudflow, and debris avalanche, to the creep of rock, soil, and talus (Sharpe 1938).

The view that all processes in the landscape are part of a unitary system has been revived recently by a number of scholars incuding Ruhe and Walker (1968), Huggett (1975, 1982), Conacher and Dalrymple (1977), and Gerrard (1981). Huggett (1975) proposed the concept of the soil-landscape system as a unitary model for soil-landscape processes. He argued that materials in transit through landscapes form a more or less continuous series in terms of size, stability, and mobility. This idea has independently been put forward by Rose et al. (1979), who considered the dispersion of weathering products in the landscape (Fig. 3.3). Rose et al. regard bedrock as an immobile solid phase which weathers to form a mobile solid phase and a mobile liquid phase. Both mobile phases travel at varying rates from the site of weathering and in doing so interact with one another. Huggett's thesis in the model of soil-landscape systems was that the "dispersion" process over much of the land surface is organized within the framework of the erosional drainage basin.

3.2.2 Solid-Phase and Liquid-Phase Cascades

Cascades of sediment, solutes, water, and, in valley glaciers, ice can all logically be modelled in the framework of the drainage basin. For instance, sediment cascades in mountainous regions have been modelled in the context of drainage basins by Lehre (1982) (Fig. 3.4). Cleaves et al. (1970), Likens et al. (1977), Reid et al. (1981), and many others adopt the drainage basin as a unit in which to study solute cascades. The basin water cycle has long been a conceptual tool for understanding streamflow generation and forms the basis of mathematical models of land-surface hydrology (for instance, Dawdy and O'Donnell 1965). The study of valley glaciers too may be taken in the context of inputs, outputs, throughputs, and storages of water in a glacial catchment.

Of course, the cascades in drainage basins may be seen as but parts of larger-scale circulations of Earth materials. The larger-scale circulations may themselves be regarded as cascading systems and modelled as a set of flows and storages. As an example, the continental cascade of carbon is depicted as a quantified conceptual model in Fig. 3.5. At a global scale, it is usual to distinguish an exogenic cascade, involving the transfer and transformation of matter near the Earth's surface, from a slower and less well understood endogenic cascade involving processes in the lower crust and mantle (Lerman 1979, p. 1). The major reservoirs involved in the exogenic cascade, and the major fluxes between them, are shown in Fig. 3.6. This broad conception of global cascades forms the basis of the mathematical models of geochemical cycles that will be discussed in Chap. 9.

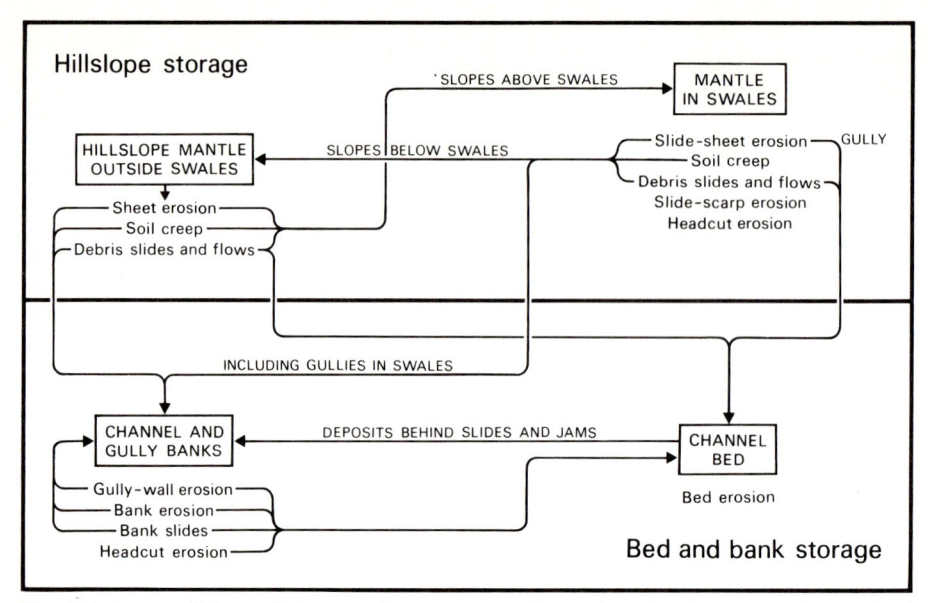

Fig. 3.4. General linkages between sediment storage sites and erosional processes in steepland drainage basins of the California Coast Range. (After Lehre 1982)

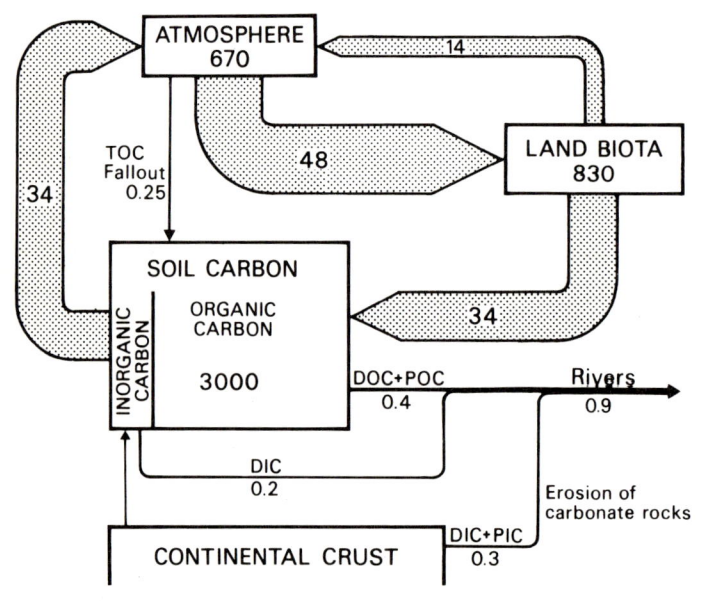

Fig. 3.5. Rivers and the natural continental carbon cycle. Annual fluxes and reservoirs are in 10^{15} g C (for the exoreic parts of the continents). D stands for dissolved, P for particulate, T for total, O for organic, I for inorganic, and C for Carbon. For sources of data see original article. (Meybeck 1982)

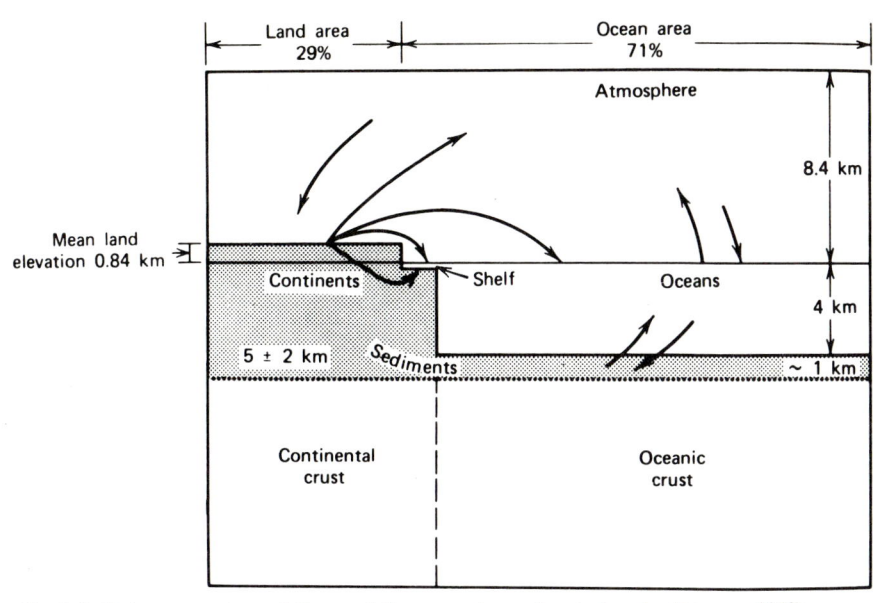

Fig. 3.6. Major reservoirs and fluxes of the exogenic geochemical cycle. (Lerman 1979)

Within drainage basins, some cascades may be picked out as objects of study in their own right. Brunsden (1973) recognizes a landslide cascade consisting of scree, landslide, and talus states all linked by slope processes. Individual hillslope profiles may be studied as cascades of sediment (Selby 1982, p. 2) and, in the case of soil catenas, as cascades of soil materials. Other examples include cliffs, screes, and stream channels, all of which can be viewed as cascading systems. However, the fact remains that all cascades will almost invariably interact with the landscape of which they are part. It is prudent at least to consider the possibility of interaction between landscape cascades for, as Chorley and Kennedy (1971, p. 84) hold, cascading systems differ considerably in size but all are ultimately interconnected.

3.3 Models of System Form and Process

The relation between form and process in systems may be conceived as a two-way interaction: process alters from and the changed form in turn modifies the process. In a process-form system the conjoint action of form and process is recognized. More specifically, a process-form system "links an energy flow system with a set of morphological variables in such a way that the latter are controlled by flow systems, while certain of the morphological variables can, in turn, regulate flows of energy and materials" (Strahler 1980, p. 11).

The foundations of process-form models in the Earth sciences were laid by Strahler (1952a) in the United States, who presented a system of geomorphology grounded in the basic principles of mechanics and fluid dynamics, and Cailleux

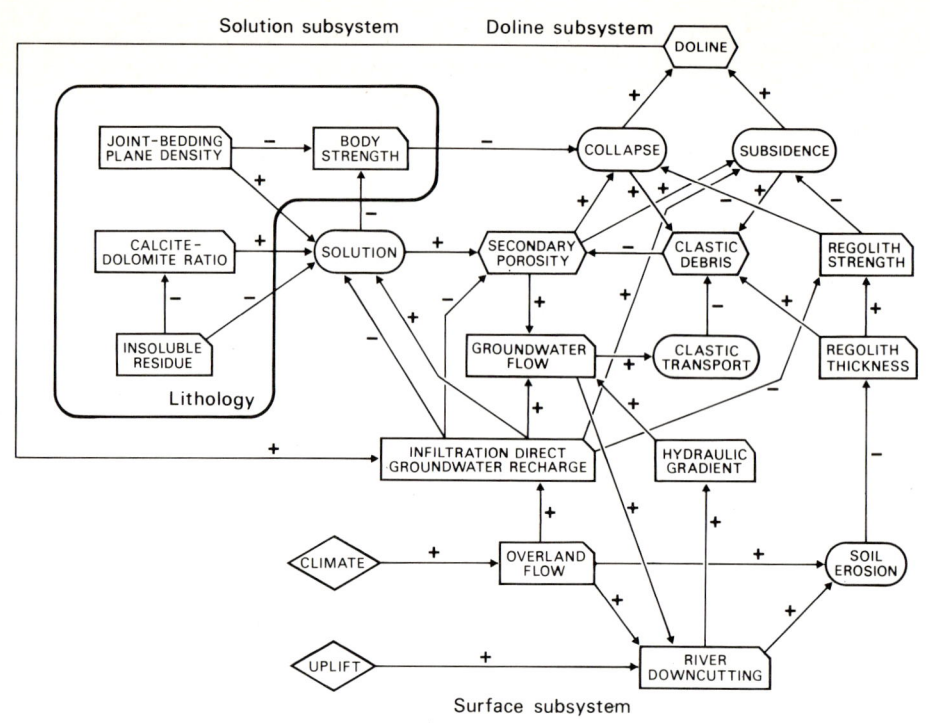

Fig. 3.7. Doline initiation represented as a process-form model. *Arrows* indicate direction of effect and *signs* indicate direct (+) and inverse (−) relationships. (Palmquist 1979)

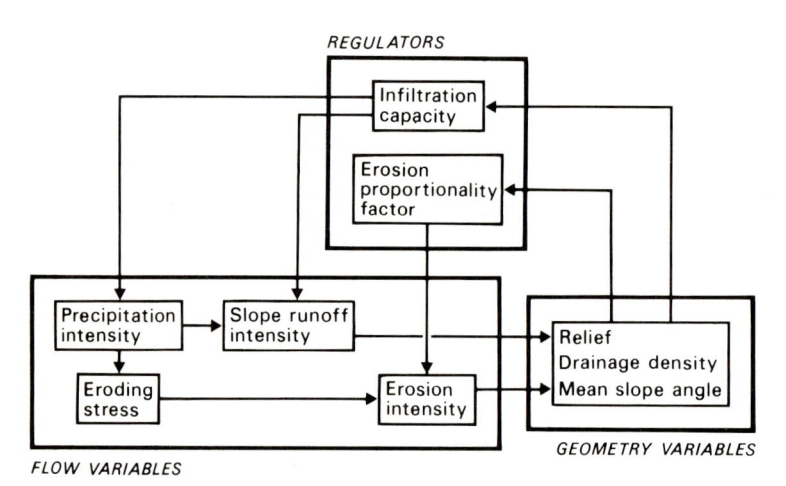

Fig. 3.8. An example of a process-form system. (After Strahler 1980)

and Tricart in France, who introduced the method of "dynamic geomorphology". From these beginnings, the modelling of process-form systems has moved in three directions, each of which takes a rather different conceptual stance. The first approach is essentially empirical and involves using statistical models to sort out the interrelationships within sets of morphological and dynamic variables. Examples include Harrison and Krumbein's (1964) process-response model of a beach which uses a multiple regression model; Melton's (1958b) analysis of correlation sets derived for a drainage basin system; and Palmquist's (1979) essentially conceptual model of doline initiation (Fig. 3.7). The second approach is to identify process-form systems by applying dimensional analysis and dimensionless numbers. For example, Strahler (1958) describes a process-form model of a slope system using two dimensionless numbers – the geometry number and the Horton number, and identifies feedback loops within the system (Fig. 3.8). The third approach leads to numerical-analytical, deterministic models in which the system is represented by a set of nonlinear, partial differential equations which enable the space-time variations of the morphological system components to be predicted. All these approaches will be discussed fully later in the book.

3.3.1 Concepts of Landscape Development

Most theories of landscape development, soil development, and ecosystem development are conceptual models. Some are purely verbal; others are expressed in a quantitative way. They all are concerned with the way in which form and process interact.

The first general theory of landscape development was proposed by Davis (1889, 1899). The Davisian model consists of two separate and distinct cyclical concepts, one for the progressive development of erosional stream valleys and another for the development of the whole landscape (Higgins 1975, p. 7). The "geographical cycle" was originally intended as a simplified conceptual frame applicable to landforms in humid temperate regions on rocks offering a uniform resistance to erosion and which had been rapidly uplifted and then subjected to prolonged wearing down by running water and mass wasting. As a concept, it has been extended to other landforms including arid landscapes (Davis 1903, 1905, 1930), glacial landscapes (Davis 1900, 1906), periglacial landscapes (Peltier 1950), to landforms formed by shore processes (Davis 1912; Johnson 1919), and to karst landscapes (Beede 1911; Cvijić cited in Sweeting 1981).

The major alternative theories to the Davisian system are the models propounded by Penck (1924), L. C. King (1953, 1963, 1967, 1983), and Crickmay (1959, 1960, 1975). The Penckian geomorphological system is rather difficult to understand but involves the development of a whole drainage basin as a result of the development of each individual slope within it (Scheidegger 1970, p. 27). Penck saw little reason to suppose with Davis that uplift and planation take place alternately through repeated cycles; instead, he regard them as phenomena which occur at the same time. To Penck, slope development is a differential process in which, at best, several relatively constant slope forms can be discerned. These discernible slope forms, which show a degree of persistence, are convex slope profiles, which result from waxing development (*aufsteigende Entwicklung*) produced by the rate of uplift exceeding the rate of denudation; straight slopes, which result from stationary development (*gleichförmige Entwicklung*) produced when the rate of uplift and the rate of denudation match one another; and concave slopes, which result from waning development (*absteigende Entwicklung*) produced when the rate of uplift is less than the rate of denudation. Perhaps the most notable facets of Penck's conception of landscape development are the recognition of slope recession as an ubiquitous phenomenon and the idea that each slope recedes individually according to the processes acting on it.

King's system of geomorphology, which is clearly presented in his mammoth book *The Morphology of the Earth* (1967), involves cycles of erosion variously styled "the landscape cycle", "the epigene cycle of erosion", and "the pediplanation cycle". Each cycle starts with a sudden burst of diastrophism, which is followed by diastrophic quiescence during which subaerial erosion produces a pediplain. King's pediplain is analogous to Davis's peneplain, but whereas a peneplain is formed by erosional downwearing, a pediplain forms by the coalescence of pediments that are enlarged by the headwards recession of scarps. Also, whereas peneplains cease developing and actually become rejuvenated when a new cycle is begun by crustal uplift, pediplains continue to grow headwards while their lower ends are eaten away by new receding scarps. In King's scheme therefore, a series of uplifts produces several pediplains at different elevations, each growing headwards and producing a landscape similar to the *Piedmont Treppen* envisaged by Penck. King's system of geomorphology in fact combines the best or most popular features of the Davisian system with some of the more accepted aspects of Penck's system (Higgins 1975, p. 10). In his latest book, King (1983) expands his idea of denudation cycles interrupted by tectonic episodes and reinterprets some of the evidence for plate tectonics to support the concept of "static" continents on an expanding Earth as favoured by Carey (1976) and Owen (1976, 1981).

Davis's, Penck's, and King's systems of landscape development all assume roughly equal activity of various slope-forming processes on individual slopes. Crickmay (1959, 1960, 1975) challenges the assumption of equal activity on the grounds that exogenic agents act unequally. In particular, Crickmay's thesis is that a slope will recede only if there is a stream (or the sea) cutting away at the slope base. To Crickmay therefore, the concept of a cycle of erosion has no meaning: as slope denudation is mostly achieved by lateral corrasion of rivers, some parts of a landscape may be virtually untouched by slope recession and retain "youthful" features until a very late stage.

Climatic geomorphology provides a rather different conceptual frame in which to view landscape development. It rests on the assumption that different climatic inputs in the denudation system will result in different landforms and that equilibrium landforms will differ between climatic regions (Stoddart 1969, p. 483). The concepts of climatic geomorphology were originated at the end of the last century by German and French geomorphologists and, indirectly, by Russian soil geographers. The German School, led in recent years by Büdel

(1982), sees morphoclimatic regions, each with a unique assemblage of characteristic landforms and sequence of development, determined by exogenic processes under the paramount control of climate. The French School, with Tricart and Cailleux (1972) among its leading and most persuasive advocates, also recognizes morphoclimatic regions but, in contrast to the German School, gives more weight to factors other than climate in the control of landscape form.

Peltier's (1975) "theory of quantitative geomorphology" is an attempt to bring together, to expand upon, and to express in a rational way the views of Davis and Penck concerning landform development. It is summarized as a "general landform equation":

landform = $f(m, dm/dl, de/dt, du/dt, t)$,

where m is geological material, dm/dl is the rate of change of geological material with distance (a structural factor), de/dt is the rate of erosion, du/dt is the rate of uplift, and t is the total time duration of the process. A fuller version of the equation is given in Peltier (1975), where its use is discussed in the context of climatic geomorphology. Peltier's general landform equation serves to draw attention to the full set of variables that are involved in the development of landscapes. The schemes of Davis, Penck, King, and Crickmay all make their own assumptions about rates of uplift and so forth and their relationship to time. As will be see later (Chap. 8), if the assumptions of Davis and the others are fed into mathematical models of slope development, then some of the broad patterns of landscape development envisaged by Davis and the others can be reproduced. As a final comment on the theme of landscape development, it is noted that the role allotted to tectonics in the development of landforms has recently been brought to the fore and has inspired discursive papers (for instance, Morisawa 1975), a book (Ollier 1981), and a conceptual model (Melhorn and Edgar 1975).

3.3.2 Concepts of Soil Development

Towards the end of the last century, Dokuchaev suggested the following equation to represent soil formation (Jenny 1980, p. 203)

$s = f(cl, o, p) t_r$

where s is soil, cl is the regional climate, o is vegetation and animals, p is the geological substratum, and t_r is the relative age. Hilgard, an American contemporary of Dokuchaev, also recognized the importance of soil-forming factors in the development of soils but he did not develop an equation of soil formation. Independently of Dokuchaev, Shaw (1930) saw the soil as a function of climate, parent material, vegetation, and time, as well as erosion and deposition. The most famous and now classic statement on the functional conception of the soil as a system is undoubtedly the one made by Jenny (1941). Jenny's original concept of the soil states that the soil is a function of five soil-forming factors, namely, climate, relief, organisms, parent material, and time. Using the designations s for soil, cl for climate, o for organisms, r for relief, p for parent material, and t for time, Jenny wrote

$$s = f(cl, o, r, p, t, \dots),$$

where the dots stand for additional soil-forming factors not yet identified. Jenny updated this concept of the soil in 1961. Looking at the role of soil-forming factors in an open system, he reduced the five factors to three "state factors", applicable to both soil systems and ecosystems. Jenny's revised proposal may be written

$$l, s, v, a = f(L_0, P_x, t),$$

where l is ecosystem properties, s is soil properties, v is vegetation properties, a is animal properties, and the terms on the right-hand side are the three state factors of soil and ecosystem formation. The first state factor, L_0, is the "initial state of the system, or its assemblage of properties at time zero when genesis starts", and may include parent material, relief, and some organic material. Climatic and organic factors are grouped as the second state factor, P_x; these control the supply and loss of energy to and from the system and Jenny termed them the "external flux potentials". The third state factor is time, t.

The influence of a single state factor on a soil or ecosystem can be examined by seeing how a soil or ecosystem property varies as all other state factors are held constant. Soil nitrogen content varies with climate in a range of situations where parent material, topography, and biotic factors are similar. The basic equation in this case becomes

$$s = f(cl)_{o, r, p, t},$$

which Jenny (1961) called a climofunction. Similarly, where the effect of relief on a soil or ecosystem property is singled out the relationship is described by a topofunction:

$$s = f(r)_{cl, o, p, t}.$$

Lithofunctions, biofunctions, and chronofunctions are defined using the same logic. The purpose of this kind of model is to see how a soil or ecosystem property is influenced by a single state factor, to establish univariant functions. The approach is called functional analysis and is carried beyond the conceptual level by applying regression and correlation techniques to field data. In a few cases, the clorpt equation, as it is nicknamed, has been solved using a multivariate model in which a soil or ecosystem property is seen as a function of more than one state factor (Jenny 1980, pp. 362–363).

Runge (1973) modified Jenny's equation to produce "a more workable model of soil development". In Runge's conception, soil development, s, is a function or organic matter production, o, the amount of water available for leaching (defined as precipitation less evaporation), w, and time, t:

$$s = f(o, w, t).$$

As this formulation is more explicit than Jenny's it should be easier to implement. However, as it excludes, or at least unrealistically diminishes, the effect of the parent material factor, Yaalon (1975) suggests that it is applicable only to soils developed in loess, till, and other unconsolidated deposits where leaching is by far the main process of soil horizon differentiation.

A conception of soil development which focusses on soil processes was put forward by Simonson in 1959 as an "outline of a generalized theory of soil genesis" and extended in 1978 as a "multi-process model of soil genesis". Both models identify four groups of physical, chemical, and biological processes that are common to all soils: additions of organic and mineral matter as solids, liquids, and gases; their removal; their transfer or translocation; and their transformation. The changing balances of these processes from place to place (and time to time) differentiates one soil from another. For example, mineralization and humification of plant litter engage more or less the same processes of transformation in all environments but different process rates may lead to different end products. Simonson's approach to soil development was extended by Yaalon (1971) to include the time dimension and the size of the soil system in the model.

The catena concept recognizes the linkage of soils along slopes (Milne 1935a, b). Morison et al. (1948) proposed that a soil catena, in analogy with an individual soil profile, contains eluvial, colluvial, and illuvial sections, which are linked to one another by subsurface, downslope movement of water. This notion has been reaffirmed by Blume (1968) and Glazovskaya (1968). The downslope concatenation of soils is produced by selective transport of soil materials by throughflow, in the same way that a soil profile is produced, in part, by selective transport of soil materials in vertically moving water. This has prompted Blume and Schlichting (1965) to observe that the valley soils of a landscape can be thought of as the *B* horizons of hill soils. Although the catena concept is useful, it fails to give full prominence to the three-dimensional character of many soil processes involving transfer. To overcome this weakness in the catena concept, the valley basin or erosional drainage basin, which accommodates catenary relations as well as three-dimensional aspects of soil processes, was proposed independently by Huggett (1973, 1975) and Vreeken (1973) as a conceptual frame in which to study soil development.

3.3.3 Concepts of Soil-Landscape Development

The development of soils cannot be divorced from the development of landscapes. The development of a soil-landscape, or soilscape, a contracted form coined by Walker et al. (1968a), is seldom solely the formation of a skin of soil at the land surface: the landscape in which soils from also changes. Soil processes and geomorphological processes go hand in hand within a unitary soil-landscape system (Huggett 1975, 1982).

Although early workers had realized that geomorphology and soils were in some way related, and indeed Neustuev (cited by Rode 1961) had written of the direct link between the Davisian cycle of erosion and soil development, it was Milne (1935a, b) who recognized how landscape development can affect soils and how different parts of the landscape might have had different histories. Only since 1940, after extensive field research in several parts of the Earth, a selection of which is reviewed by Daniels et al. (1971) and Gerrard (1981), have conceptual models been built which attempt to explain how soils and landscapes develop as a whole. One such model was proposed by Butler (1959) under the name of the *K*-cycle concept. Based on fieldwork in Australia, Butler hypothesized that soil-landscape development is episodic, periods of geomorphological stability alternating with periods of erosion and deposition. During a geomorphologically stable phase soils develop but they may be destroyed during an unstable phase. A complete cycle of stable and unstable phases is a *K*-cycle. The surface formed by erosion and deposition during an unstable phase, and the soils formed in the surface during the subsequent stable phase, is called a ground surface. A similar concept, that of pedomorphic surfaces, was proposed by Dan and Yaalon (1968) to stress the connexion between landscapes and soil profile characteristics.

Similar in spirit to Butler's *K*-cycle concept, if less well known, is Erhart's (1967) concept of alternating periods of ecosystem equilibrium (*biostasie*) and disequilibrium (*rhexistasie*). Based on extensive field work in Africa, Erhart believes that periods of *biostasie* are characterized by maximum development of vegetation, optimal infiltration of water, deep chemical weathering, and little erosion, all of which are conducive to the formation of deep residual soils. Periods of *rhexistasie*, in contrast, represent to Erhart a breakdown of *biostasie* with a loss of vegetation owing to climatic change, increased potential energy, or human interference. They are characterized by erosion and the destruction of soil mantles.

Ruhe and his co-workers, whose fieldwork focussed mainly on the relationship between geomorphological surfaces and soils in Iowa, United States, have made many notable contributions to concepts of soil-landscape development. Ruhe and Walker (1968) and Walker and Ruhe (1968) developed hillslope models of landscape systems, both open systems in which drainage basins are part of a more extensive drainage network, and closed systems in which drainage is in a closed basin. Within low-order open drainage basins they defined a number of geomorphological landscape components (divide, interfluve, sideslope, headslope, noseslope), the sideslopes and headslopes being divided into Ruhe's (1960) slope profile components (summit, shoulder, backslope, footslope, toeslope). On similar lines, Dalrymple et al. (1968) and Conacher and Dalrymple (1977) have proposed a nine-unit, landsurface model as an integrated framework in which to study form and process in soil-landscape systems. In the model, nine distinct slope units are recognized along a slope profile. The units are interfluve, seepage slope, convex creep slope, fall face, transportational slope, colluvial footslope, alluvial toeslope, channel wall, and channel bed. In any particular profile, not all the units will necessarily be present and some may repeat themselves. Each slope unit is, according to Conacher and Dalrymple, associated with a particular set of geomorphological and pedological processes.

Part III   Mathematical Models

CHAPTER 4
Deductive Stochastic Models

The concept of probability is central to some models of Earth surface systems. For instance, in models of hillslopes some of the variables involved are best expressed in probabilistic terms. The uplift which may alter base level and the storms which activate a number of slope processes tend to occur rather erratically; they have a chance-like character and may be represented as stochastic events. By applying rules of stochastic occurrence, the effects of chance can be built explicitly into models, not just of hillslopes, but of most Earth surface systems in which chance seems to play a significant role. Stochastic models, rather than giving an "exact" set of predictions about system change in the manner of deterministic models, focus attention on a set of possible outcomes, each of which has a different probability of occurring in reality. The basis of all stochastic (and statistical) models is probability theory. For this reason, the concepts of probability theory and their simple applications to Earth surface systems will be discussed before more complex and substantial stochastic models, both deductive and inductive, and before statistical models, are considered.

4.1 Introduction to Probability

Probability theory grew out of a correspondence between the French mathematicians Pascal and Fermat. The discussion involved the fairness of various bets in gambling and took place in the second half of the seventeenth century. After a long period of development, the varied theorems of probability theory were brought together in a unified, axiomatic system by Kolmogorov (1933). The axioms of Kolmogorov can be summed up by saying that, for every activity, observation, or whatever in which some random element occurs, a probability space can be set up. The probability space, or triple, is defined as $(\Omega, \mathscr{F}, P)$ where Ω is an abstract sample space, $\mathscr{F}$ is a Borel field containing subsets of Ω (set of events), and P is short for $P(E)$, the probability measure assigned to each set E in $\mathscr{F}$ (see Lindgren 1976). Restating these fundamental precepts in less formal terms, it may be said that the language of probability has at its root a few key concepts: sample space, elementary events, and compound events. The sample space contains all possible outcomes of an experiment. Thus the sample space for casting a six-sided die contains the numbers 1 to 6. Elementary events are the members of a sample space. In the example of a six-sided die each number is an elementary event. A compound event is a subset of the sample space, for in-

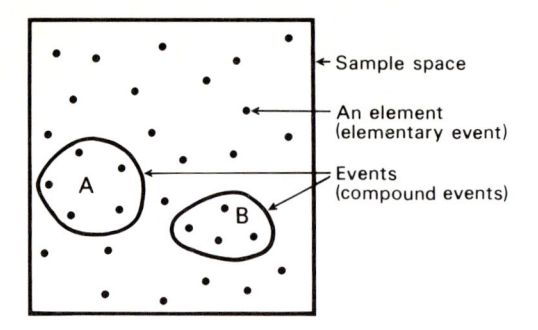

Fig. 4.1. Elements and events in a sample space

stance all the die casts where either a 5 or 6 show. In the case of streamflow, the discharge of a river over a year can be regarded as an experiment. The sample space representing all possible outcomes of the experiment would be the positive numbers describing the discharges. The annual maximum discharges for a set of years can be regarded as the elements or elementary events in the sample space. All discharges over say 5000 cumecs would represent a compound event (normally referred to as an event). These ideas are conveniently summarized as a Venn diagram (Fig. 4.1).

4.1.1 The Classical View of Probability

There are two different views of probability which are of concern to Earth scientists — the classical view and the relative frequency view. The first comprehensive statement on classical probability was given by Laplace (republished in 1951) and the subject is also called Laplacian probability. The classical view of probability rests on the belief that all events in a sample space are equally likely to occur in any one trial on an experiment. Thus the probability of a six-sided die showing a 1 is 1/6. Formally, the classical or a priori definition of probability is:

"If a random event can occur n equally likely and mutually exclusive ways, and if n_a of these ways have an attribute A, then the probability of the occurrence of the event having attribute A is n_a/n and is written prob$(A) = n_a/n$". (Haan 1977, p. 8)

This definition assumes that all equally likely and mutually exclusive ways in which an event can occur and all the ways that an event A can occur can be determined without recourse to observation or experiment. The classical view of probability does not play a large role in the study of Earth surface systems; the relative frequency view of probability does.

4.1.2 The Relative Frequency View of Probability

The relative frequency view of probability rests on the belief that there is some ratio between the actual number of times a particular outcome of an event is recorded and the total number of events. In other words, it assumes that probabilities can only be measured from observational evidence. The frequency of an

Introduction to Probability

event, freq (E_i), is defined as the number of times an event E_i is observed as the outcome in r trials on an experiment. The absolute frequency, freq (E_i), is converted to a probability by dividing its value by the number of trials, r. So if a 2 were observed 7 times in 36 throws of a die, then the probability of the die showing a 2 is prob (E_i) or $p(E_i)$ = freq $(E_i)/r = 7/36$. It has been found that, as the number of trials, r, increases, so the relative frequency comes closer to the true probability. The relative frequency estimate decreases rapidly as the number of trials increases to that relative frequency estimates made from relatively few observations can provide good estimates of the unknown, true probabilities.

A relative frequency definition of probability is:

"If a random event occurs a large number of times n and the event has an attribute A in n_a of these occurrences, then the probability of the occurrence of the event having attribute A is prob(A) = $\lim_{n \to \infty} n_a/n$". (Haan 1977, p. 9)

4.1.3 Axioms of Probability Theory

Axioms are the basic rules for manipulating probabilities. Whether the probabilities are defined on an a priori basis or by measurement, the axioms are the same. Four axioms are particularly useful in the study of Earth surface systems: the addition axiom for mutually exclusive events, the multiplication axiom for independent events, the multiplication axiom for dependent events, and the total probability axiom (see J. R. Gray 1967 or Lindgren 1976).

The addition axiom for mutually exclusive events states that if E_1 and E_2 are two mutually exclusive events, then the probability that either E_1 or E_2 happens is the sum of their individual probabilities. Expressed symbolically, the axiom reads

$$p(E_1 \cup E_2) = p(E_1) + p(E_2),$$

where $\cup$, the union, is read as either/or. An example of mutually exclusive events is the outcome of casting a six-sided die. The probability of casting any one number is 1/6. The probability of casting a 6 or 1 is given by the addition axiom:

$$p(E_6 \cup E_1) = \tfrac{1}{6} + \tfrac{1}{6} = \tfrac{1}{3}.$$

Two simple results follow from this axiom. Firstly, in a single trial on an experiment, the probability that any one of all the n events in the sample space occurs is equal to one. Symbolically,

$$p(E_1 \cup E_2 \cup E_3, \ldots, E_n) = \sum_{i=1}^{n} p(E_i) = 1.$$

In practice, this means that in any single trial one of the elementary events must occur as the outcome. Secondly, the addition axiom can be used to define the idea of a complementary event. When a specified event, E_i, does not occur, the complementary event, $\bar{E}_i$, is said to have occurred. The complementary event is the sum of all events other than the specified event. In any year, a river may either flood, E_F, or not flood, $\bar{E}_F$. Flooding and not flooding are mutually exclusive events and, using the addition axiom,

$$p(E_F \cup \bar{E}_F) = p(E_F) + p(\bar{E}_F) .$$

Because the union of flooding with not flooding is certain to occur, the sum of the individual probabilities must equal one,

$$p(E_F) + p(\bar{E}_F) = 1 .$$

Rearranging this expression gives a definition of the complementary event

$$p(\bar{E}_F) = 1 - p(E_F) ,$$

or, more generally for the ith event

$$p(\bar{E}_i) = 1 - p(E_i) .$$

The multiplication axiom for independent events is used to evaluate the probability that a sequence of specified events occurs as a result of r trials on an experiment. It states that if E_1 and E_2 are two independent events, then the probability that both E_1 and E_2 happen is the product of their individual probabilities. Symbolically,

$$p(E_1 \cap E_2) = p(E_1) \cdot p(E_2) ,$$

where $\cap$, the intersection, is read as and/both. More generally, for a specified sequence of length r, the probability of occurrence is given by

$$p(E_1 \cap E_2, \ldots, \cap E_r) = \prod_{i=1}^{r} p(E_i) .$$

For instance, the probability of throwing three sixes in succession using a six-sided die is

$$p(E_6 \cap E_6 \cap E_6) = p(E_6) \cdot p(E_6) \cdot p(E_6)$$
$$= \frac{1}{6} \times \frac{1}{6} \times \frac{1}{6} = \frac{1}{216} .$$

By the same argument, if a drought occurs 5 times in a 100-year record, and assuming that droughts are independent events, then the probability of a drought occurring in any one year is $p(E_i) = 5/100 = 0.05$. The multiplication axiom could then be used to find the probability of, say, a drought occurring in two successive years. The answer is

$$p(E_1 \cap E_2) = 0.05 \times 0.05 = 0.0025 .$$

The multiplication axiom for dependent (general) events is used for events which are not mutually exclusive or independent. Formally, two events are said to be dependent if the probability of one of the events occurring depends on the occurrence of the other event. Dependence is thus the antithesis of independence and arises when the multiplication axiom for independent events is seen to be incorrect, that is, when

$$p(E_1 \cap E_2) \neq p(E_1) \cdot p(E_2) .$$

The multiplication axiom for dependent events is linked with the idea of conditional probability which is denoted by $p(E_1|E_2)$ or $p(E_2|E_1)$. The expression $p(E_1|E_2)$ is read as the probability of event E_1 occurring when it is assumed that event E_2 has happened; the expression $p(E_2|E_1)$ reads vice versa. Using this notation, the multiplication axiom for dependent events is written

$$p(E_1 \cap E_2) = p(E_1) \cdot p(E_2|E_1)$$

and reads, if E_1 and E_2 are dependent events, the probability that they both occur is the product of the probability of E_1 and the conditional probability of E_1 when E_2 happens. For instance, a study of daily rainfall at Ashland, Kentucky (Haan 1977, p. 13) has shown that, in July, the probability of a rainy day following a rainy day is 0.444, the probability of a dry day following a dry day is 0.724, the probability of a rainy day following a dry day is 0.276, and the probability of a dry day following a rainy day is 0.556. This information, in conjunction with the multiplication axiom for dependent events, can be used to calculate the probability of, say, two rainy days following a rainy day. Letting E_1 be rainy day 1 and E_2 be rainy day 2 following the initial rainy day, then

$$p(E_1) = 0.444,$$

since this is the probability of a rainy day following a rainy day. Now

$$p(E_2|E_1) = 0.444$$

since this is also the probability of a rainy day following a rainy day. The answer is thus

$$p(E_2 \cap E_2) = 0.444 \times 0.444 = 0.197.$$

Many models can be deduced from the definitions and axioms of probability theory. Some of these, such as Bayes's theorem, important though they are in some fields of study, are little used in Earth sciences (but see Krzysztofowicz 1983a, b). The chief models associated with the study of Earth surface systems are probabilistic models, in which events are entirely independent of time, and stochastic models, in which the order of occurrence of events is important in determining the outcome of an experiment (Sumner 1978, p. 125). Independent-events models include the application of binomial and Poisson processes to phenomena observed in both temporal and spatial frameworks. Stochastic models include models of random walks, Markov chains, and entropy.

4.2 Independent Events in Time

4.2.1 Binomial Processes

The binomial theorem is one of the oldest and most basic theorems of probability theory. It provides a theoretical vehicle for summarizing the long-term behaviour of sequences of independent events. It is applied to a discrete sequence of events which are classed as occurrences or non-occurrences, for instance, a run of years in which floods either occur or do not occur. The probability of an event, the occurrence of a flood in the example, is the same at all points in the sequence and defined as

$$p(E) = p.$$

The probability of non-occurrences, of no flood, which is the complementary event to E, is defined as

$$p(\bar{E}) = 1 - p = q.$$

The assumption is that the occurrence of an event is independent of the history of occurrences and non-occurrences in the sequence of events. Any process with these properties is called a Bernoulli process. The definitions given above enable statements to be made concerning the chance of an occurrence or a run of occurrences in a given sequence. To illustrate this point, the example of flooding will be extended. The flood events are independent from year to year so that the probability of a flood in year 3, but not in years 1 and 2, is qqp. The probability of one flood in any 3-year period is $pqq + qpq + qqp$ since the flood could occur in any one of the 3 years. These results may be generalized to find the probability of x floods (events) in r years (trials). The general formula, known as the binomial distribution, is

$$P_x = \binom{r}{x} p^x q^{r-x} \begin{cases} x = 0, 1, 2, \ldots, r \\ r \text{ is any positive integer} \end{cases}$$

and gives probabilities for all the values of x between 0 and r which represent all the possible outcomes of a binomial experiment. The binomial formula consists of two parts. The first part is the binomial coefficient

$$\binom{r}{x} = \frac{r!}{x!(r-x)!}$$

and defines the number of different sequences of occurrences and non-occurrences (outcomes of the experiment). The second part, the $p^x q^{r-x}$ term, defines the probability of any particular combination of x occurrences in a sequence of r trials. The binomial formula can be used, for example, to find the probability of three floods in five years given that the probability of a flood in any one year is $p = 0.2$. The solution is

$$P_3 = \binom{5}{3} (0.2)^3 (0.8)^2 = 0.0512.$$

Further details on the binomial distribution in relation to hydrological sequences may be found in Haan (1977, pp. 70–75).

The geometric distribution enables the probability of the first occurrence happening at the xth time (trial) to be found. The geometric probability distribution is defined as

$$P_x = pq^{x-1} \quad x = 1, 2, 3, \ldots$$

and has a mean $p(x) = 1/p$ and a variance $\sigma^2(x) = q/p^2$. Since the mean is given by $1/p$ it can be interpreted by saying that, on average, a T-year event occurs on

Independent Events in Time

the Tth year, which is the same as the concept of the return period. Using the geometric distribution, the probability of, say, a 5-year flood ($p = 0.2$) occurring in the third year ($x = 3$) of a sequence of years is calculated as

$$P_3 = (0.2)(0.8)^2 = 0.128 \ .$$

The negative binomial distribution defines the probability that the kth occurrence is at the xth time. It is written

$$P_x = \binom{x-1}{k-1} p^k q^{x-k} \quad x = k, k+1, \ldots$$

and has a mean $\mu(x) = k/p$ and a variance $\sigma^2(x) = kq/p^2$. To find the probability of the third occurrence ($k = 3$) of a 5-year flood ($p = 0.2$) in the tenth year ($x = 10$), the answer is

$$P_{10} = \binom{9}{2} (0.2)^3 (0.8)^7 = 0.06 \ .$$

The hypergeometric distribution may be used to find the probability of obtaining x occurrences in a sample of n events from a total sequence of N events which contains k occurrences. The formula is

$$P_x = \binom{k}{x} \binom{N-k}{n-x} \bigg/ \binom{N}{n}$$

where $\binom{N}{n}$ is the total number of ways of selecting a sample of size n from N objects and $\binom{k}{x}\binom{N-k}{n-x}$ is the number of ways of selecting x occurrences and $n-x$ non-occurrences from a population containing k occurrences and $N-k$ non-occurrences. The formula has been applied by Shreve (1966) to find the probability of randomly selecting a link of a given magnitude, from all the links in topologically distinct channel networks of a given magnitude.

4.2.2 Poisson Processes

The Poisson distribution is a limiting case of the binomial distribution when the probability of an event is very small and the total number of trials is large. In these circumstances, the Poisson distribution may normally be used to obtain an approximation for the independent binomial probabilities. The Poisson distribution is defined by the equation

$$P_x = \frac{e^{-\lambda} \lambda^x}{x!} \quad x = 0, 1, 2, \ldots; \lambda > 0$$

where $\lambda = rp$ is the mean rate of occurrence and x is the number of events occurring in a sequence. The mean and variance of the Poisson distribution are both equal to λ.

The Poisson distribution has varied applications. In hydrology, it enables answers to be given to questions such as: what is the probability that a drought with a return period of 100 years ($p = 0.01$) will occur twice ($x = 2$) in a 10-year period ($r = 10$)? An exact answer could be obtained, rather laboriously, using the binomial distribution. The approximate answer provided by the Poisson distribution with $\lambda = rp = 10 \times 0.01$ is

$$P_2 = \frac{e^{-0.1}(0.1)^2}{2!} = 0.0045 .$$

Raup (1981, pp. 8–10), developing the work of Gretener (1967), has shown that the probability of occurrence of a rare event at least once in n years is given by $1-(1-p)^n$. He shows that, for large values of n, this expression may be approximated by the Poisson limit and written in the general form

$$P_{n,1} = 1 - e^{-np} ,$$

where $P_{n,1}$ is the probability of a least one occurrence in n time intervals. Further generalization leads to an expression which includes multiple occurrences

$$P_{n,x} = 1 - e^{-np}\left\{1 + np + \frac{(np)^2}{2!} + \frac{(np)^3}{3!} + \ldots + \frac{(np)^{x-1}}{(x-1)!}\right\} ,$$

where $P_{n,x}$ is the probability that an event will occur at least x times in n years. Raup (1981, pp. 9–10) explores the possibilities of this technique for geological time scales, albeit, as he points out, in a primitive and incomplete manner, by asking the question "If a mass extinction is a random event, why has there not been a mass extinction in the last 65 million years?" To answer this question, he takes the observed fact that in the last 600 million years there have been about five mass extinctions (the number varies according to how a mass extinction is defined). This gives $p = 8.3 \times 10^{-9}$ per year. From here, the probability of at least one mass extinction in 65×10^6 years is

$$P_{65,1} = 1 - e^{-(65 \times 10^6)(8.3 \times 10^{-9})} = 0.42 .$$

This is a large probability indicating that the lack of a mass extinction since the Cretaceous does not challenge the view that mass extinctions during the Phanerozoic are randomly distributed.

Poisson processes have been widely used in the study of time sequences of Earth surface phenomena. In sedimentology, Schwarzacher (1976) developed a simple model of sedimentation in a basin in which a constant rate of sedimentation is interrupted by random "storms" of constant length represented by a Poisson process. The Poisson process model can be extended to incorporate renewal processes (D. R. Cox 1962). Renewal theory has been used to extend the interrupted sedimentation model and also to study volcanic earthquakes (Reyment 1969, 1976).

There are many other distributions used in studying Earth surface systems. Extreme value distributions consist of sets of extreme values taken from some parent population. They were developed mainly by Gumbel (1958) and have proved helpful in describing hydrological, geomorphological, and climatological events of extreme rareness. Many phenomena at the Earth's surface are treated statistically as rare or extreme events. The list includes floods and rainstorms (Gumbel 1958; Dalrymple 1960), earthquakes (Shakal and Toksöz 1977; W. D. Smith 1978; McClellan 1984), faunal interchanges (Simpson 1952), meteoric impacts (Öpik 1958, 1973). Other studies look at the effect of extreme events on river and hillslope processes (Wolman and Miller 1960; Wolman and Gerson 1978; Starkel 1976, 1979) and coastal processes (Cambers 1976). Other distributions include the normal distribution, the exponential distribution, the gamma distribution, the beta distribution, the lognormal distribution, and the circular distribution. Discussion of these is beyond the scope of this book and the reader is referred to the following sources: Elderton and Johnson (1969), Haan (1977), Krumbein and Graybill (1965), and Raudkivi (1982).

4.3 Independent Events in Space

4.3.1 Point Patterns

The binomial distribution may be used to specify the probabilities of points occurring in a specified cell of a spatial grid. The method is best understood by example. Given a 3×3 grid ($n = 9$ equally sized spatial cells) and 5 points to locate independently within it, what is the probability of finding one point in a specified cell? Since there are nine cells of equal size, the probability of a point being located in any one cell is

$$p(E_i) = p = 1/n = 1/9 ,$$

and the probability of the complementary event occurring, of the point being in any of the other cells, is

$$p(\bar{E}_i) = q = 8/9 .$$

Substituting the appropriate values in the binomial formula gives

$$P_{x=1} = \binom{5}{1} (1/9)^1 (8/9)^4 = 0.347 .$$

Calculating P_x for all possible values of x gives a probability distribution which summarizes all possible outcome to this experiment (Fig. 4.2). The first and second moments of the distribution are useful summary measures, especially where r is large. The mean is defined as

$$\mu(x) = rp = r/n = \lambda$$

and the variance is defined as

$$\sigma^2(x) = rpq .$$

When the spatial grid contains a large number of equally sized cells, each of minute area, the probability of a single point being independently located in any specified cell becomes very small. In formal terms, as the number of cells tends to

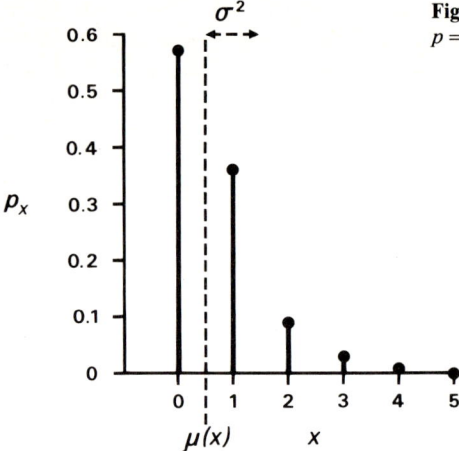

Fig. 4.2. The binomial distribution for $r = 5$ and $p = 1/9$. (Thomas and Huggett 1980)

infinity ($n \to \infty$), so p, the probability of a specified cell receiving an independently located point, will tend towards zero ($p \to 0$). As p tends to zero, so the probability generated by the binomial distribution becomes more and more closely approximated by the Poisson distribution

$$P_x = \frac{e^{-\lambda} \lambda^x}{x!} \quad x = 0, 1, 2, \ldots$$

where, in a spatial context, $\lambda = r/n$ and is the average number of points per cell. As a rule of thumb, the Poisson distribution is used in preference to the binomial distribution if $n \geqslant 50$ and $n < 5$.

A double Poisson model (Schilling 1947) has been used by McConnell and Horn (1972) to describe the frequency distribution of karst depressions on the Mitchell plain, Indiana. The model may be written

$$P_x = a_1 \frac{e^{-\lambda_1} \lambda_1^x}{x!} + a_2 \frac{e^{-\lambda_2} \lambda_2^x}{x!} \quad x = 0, 1, \ldots$$

McConnell and Horn justify the use of this model on the grounds that karst depressions can be divided into two types: dolines, which are small features formed above a water table along zones of weakness in limestone and by ponding of surface runoff; and collapse sinks, which are formed by the collapse of cavern roofs caused by surface drainage. They argue that both types of depression should be randomly distributed but at different densities. The model, they claim, predicts closely the observed frequency distribution of the karst depressions.

A conflicting explanation of the frequency distribution of karst depressions has been offered by LaValle (1967), who believes that, independently of time, a random occurrence of a karst depression leads to an increased probability of other depressions forming nearby because local erosion would be exacerbated by increased diversion of runoff into the underground drainage system. The process envisioned by LaValle leads, not to a double Poisson model, but to a negative binomial model. Using McConnell and Horn's (1972) data for the karst depressions in the Mitchell plain, Thomas (1977) has tested the fit of both double Poisson and negative binomial distributions. He found that both models fitted the data set and he explained that this is because in both cases the model parameters are estimated from the data. Both models make assumptions about the location of points in time and, naturally, the model parameters are deduced from these assumptions. Further information on the development of karst depressions is needed before the apparent conflict between the two theories can be resolved.

4.3.2 Line Patterns

An interesting extension of point pattern analysis is the analysis of linear and areal features (for instance, Crain 1976, 1978; Miles 1964a, b, 1970, 1971; Unwin 1981). The analysis of linear features involves a consideration of random lines in a plane and on a sphere. Generally, a line of infinite extent in a plane (x, y) may be defined by two parameters, ϱ and θ in the following way

$$\varrho = x \cos \theta + y \sin \theta \quad (-\infty < \varrho < \infty; 0 < \theta < \pi),$$

where ϱ can be thought of as a distance from an arbitrary origin. There is a correspondence between random lines generated by this relation and random points in a plane. However, an important result is that the mutual intersection angles of lines has the probability density function $\sin \theta \, (0 \leqslant \theta \leqslant \pi/2)$ which means that intersection is more probable near 90° than 0° (Miles 1964a). Also, the number of lines intersecting a convex figure of perimeter s has a Poisson density with mean $\tau s/\pi$ where τ is the mean number of lines per unit length. These results can be applied to random great circles on a sphere.

Linear features of the Earth's surface are line segments and have a third measurable parameter length. Crain (1976, p. 7) argues that intuitively long or short lineaments will be uncommon, none will be of negative length, and the mean length will exceed the modal length. Accordingly, the resulting positively skewed distribution may be described by a gamma distribution.

The theoretical work of Crain and Miles may prove useful for the rash of studies emerging from the renaissance of interest in tectonics and landforms (for example, Fairbridge 1981). The study of lineaments has a long history and includes Hills's (1956) work in Australia. Recent studies include the work of Ai and Scheidegger (1982), who investigated the principal neotectonic directions in north China as obtained from geological observations, in site measurements made in cores, fault-plane solutions of earthquakes, valley trends, earthquake faults, fracture propagation, and joint orientation. The stress pattern, which indicates the prevalence of a north east – south west compression, results from the collision between Eurasian, Indian, Pacific, and Philippine tectonic plates. Other work in the field of lineaments includes a study of valley trends in Tibet (Ai and Scheidegger 1981), studies of valley trends and neotectonic stress fields in south-east China (Ai et al. 1982), and the Alpine-Mediterranean area (Scheidegger 1981), and a study of neotectonic conditions in peninsular India (Scheidegger and Padale 1982).

Fig. 4.3. Examples of topologically distinguishable channel networks comprising collection Λ_n for $1 < n < 4$. (Dacey 1976)

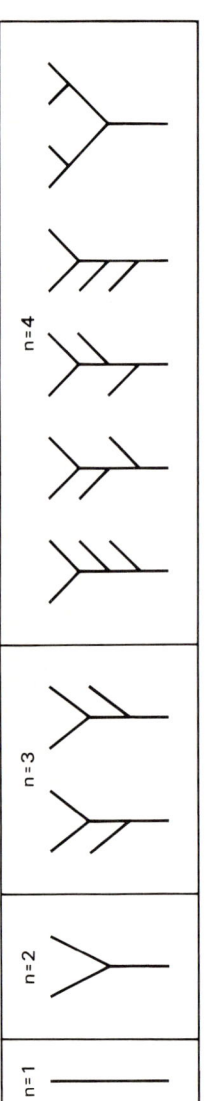

Fig. 4.4. Examples of topologically distinct network patterns comprising collection Ω_n for $1 < n < 4$. (Dacey 1976)

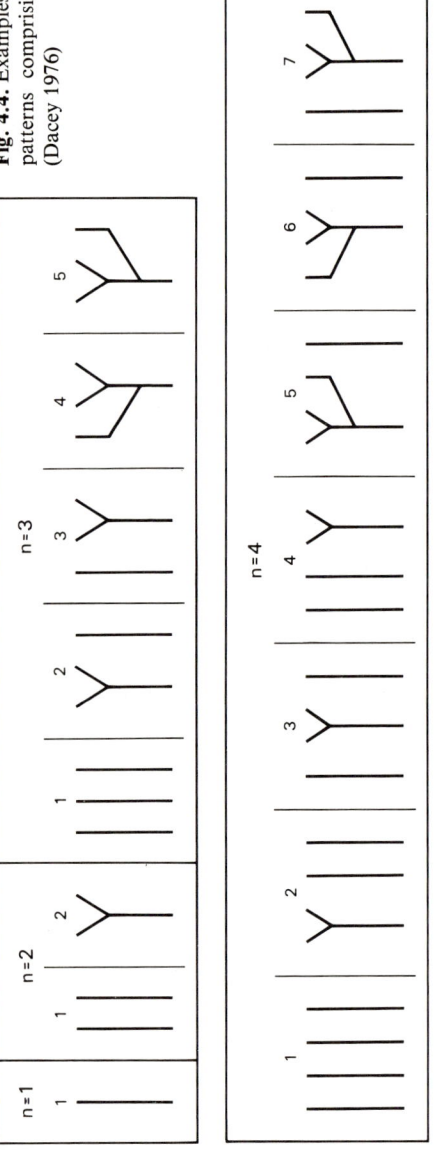

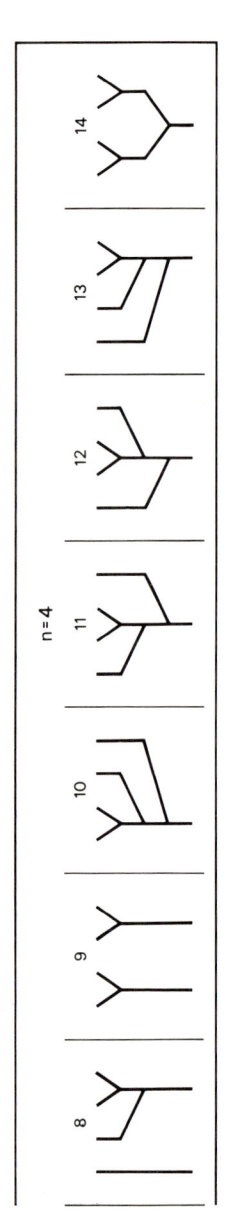

Stream channel networks are a special class of linear feature which, from a topological viewpoint, are rooted trees. Early attempts to quantify drainage networks were made by Horton (1945) and Strahler (1952b). However, it was Shreve who, in his concept of link magnitude (Shreve 1966), provided a system of ordering which is amenable to formal analysis, and who, in his concept of topologically random channel networks (Shreve 1966, 1967, 1969), provided a firm basis for the statistical analysis of drainage systems. Shreve's work was extended by Werner (1971) to the more general case of topologically random network patterns. Examples of topologically distinct channel networks and topologically distinct network patterns for networks of magnitudes up to 4 are shown in Figs. 4.3 and 4.4. The basic properties of channel networks and network patterns have been brought together by Dacey (1976) and are summarized in Table 4.1. Further developments of the statistical study of stream networks include Dacey and Krumbein's (1976, 1979, 1981) work on network growth and particle-size models.

Table 4.1. Properties of topologically distinct channel networks and network patterns. (After Dacey 1976)

Property definition	Property description	Source
$N(n) = \dfrac{(2n-2)!}{(n-1)!\,n!}$	The number of channel networks in Λ_n[a]	Shreve (1966)
$M(n) = \dbinom{2n-1}{n}$	The number of links in Λ_n	Shreve (1967)
$M(k\,n) = \dfrac{(2k-2)!}{(k-1)!\,k!}\dbinom{2n-2k}{n-k}$	The number of links of magnitude k in Λ_n	Shreve (1967)
$R(n) = \dbinom{2n+1}{n-1}$	The number of links in Ω_n[b]	Dacey (1972)
$R(k\,n) = \dfrac{(2k-2)!}{(k-1)!\,k!}\dbinom{2n-2k+2}{n-k}$	The number of links of magnitude k in Ω_n	Dacey (1972)
$S(n) = \dfrac{3(2n)!}{(n-1)!\,(n-2)!}$	The number of channel networks in Ω_n	Werner (1971)
$S(k\,n) = \dfrac{(2n-2k+2)!}{(n-k+1)!\,(n-k+2)!} \cdot \dfrac{(2k-2)!}{(k-1)!\,k!}$	The number of channel networks of magnitude k in Ω_n	Werner (1971)
$T(n) = \dfrac{(2n)!}{n!\,(n+1)!}$	The number of network patterns in Ω_n	Werner (1971)
$T(k-) = \dfrac{k(2n-k-1)}{(n-k)!\,n!}$	The number of network patterns in Ω_n that consist of k channel networks	Dacey (1971)

[a] Λ_n is the collection of topologically distinct channel networks of magnitude n.[c]
[b] Ω_n is the collection of topologically distinct network patterns of magnitude n.
[c] n is, unless otherwise stated, a positive integer and $1 \leq k \leq n$.
The "property description" column does not appear in the original table.

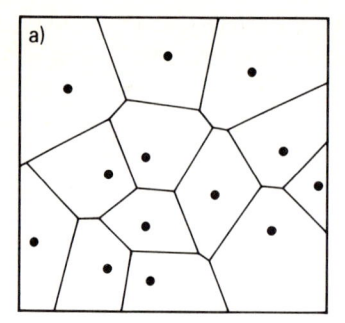

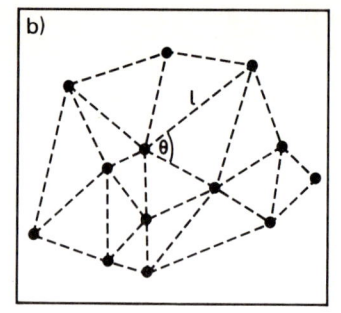

Fig. 4.5. a Voronoi polygons. **b** Delaunay triangles of the same pattern. (Thomas 1981)

4.3.3 Area Patterns

Analysis of random patterns can be extended to the polygons formed by random lines. Of interest are the probability distributions of the number of sides, n, the length of the sides, l, the perimeter, s, the incircle (the inscribed circle of largest diameter), d, and the area, a, of the polygons. Miles (1964a, 1971) has established the probability distributions for these on a plane and on a sphere. Interestingly, the distributions are invariant even if parts of the lines are missing: the polygons observed to be complete will have the same statistical properties as the entire set.

A different class of random polygons is derived by considering growth around random centres or the contraction cracking of a surface. These cell models were derived by Meijering (1953) in the context of work on the structure of crystal aggregates. The polygons are generated by, firstly, locating points randomly in a plane by a Poisson process; and, secondly constructing a convex polygon around each point such that the area within each polygon is nearer to the enclosed point than to any other point. Polygons constructed in this way are called Voronoi polygons, Thiessen polygons, Dirichlet regions, or Wigner-Sietz regions (Boots and Murdoch 1983). Figure 4.5a shows some typical tessellations. Voronoi polygons have been used in studying columnar jointing (Smalley 1966), ice-wedge polygons (Lachenbruch 1962), and cracking in general (Crain 1972, 1976). Associated with Voronoi polygons are a second set of areas known as simplicial graphs or Delaunay triangles. They are obtained by connecting all pairs of points whose Voronoi polygons share a common edge (Fig. 4.5b) and have been found useful by Vincent et al. (1976).

4.4 Random-Walk Models

Classically, the idea of a random walk is associated with a drunkard randomly staggering backwards of forwards one step at a time. The drunkard's progress after a specified time is the sum of the outcomes, backward and forward steps, from the time he set out. Another classical model concerned with random walks is the gambler's ruin problem in which the stake money of one of two players is

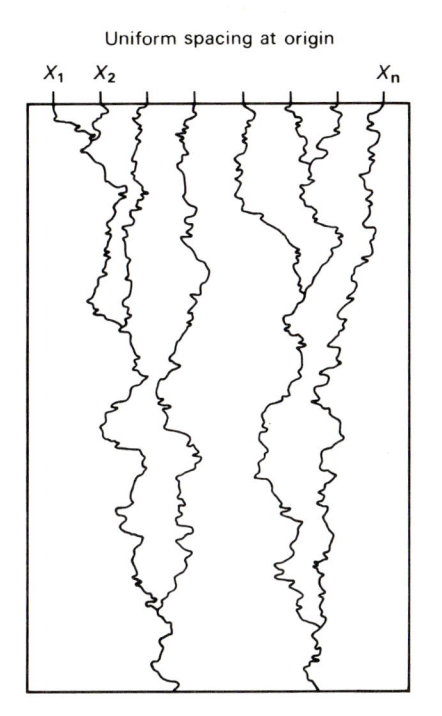

Fig. 4.6. Portion of a random-walk model of a rill or stream network. (Leopold et al. 1964)

exhausted in a long sequence of games in which each player has a fixed probability of winning. Applications of random-walk models to Earth surface phenomena have been mainly to meander patterns (Chap. 5.3.2) and stream channel networks, although a random-walk model of alluvial fan formation has been devised by W. E. Price (1976).

4.4.1 Stream Networks

Leopold and Langbein (1962) produced a random-walk model of a stream network by taking uniformly spaced points at one end of a piece of paper, each of which is the start of a random walk. The walks (streams) may move left, right, or forwards at any angle but never backwards. The several walks are continued until junctions, representing stream confluences, are obtained, after which the walk of just one of the two tributaries is continued. The pattern produced (Fig. 4.6) is evidently reminiscent of a natural stream network. The average distance any one random walk proceeds before joining another may be defined by the gambler's ruin model which predicts the most probable number of plays in a game before one player loses all his capital to his adversary. By strict analogy with the gambler's ruin model, and providing the starting points of the walk are equally spaced, the number of steps taken before adjacent walks meet should depend on the square of the initial separation distance. However, as the stream walks may move to the left and right as well as forwards, they are less constrained than walks in a gambler's ruin model and the number of steps adjacent

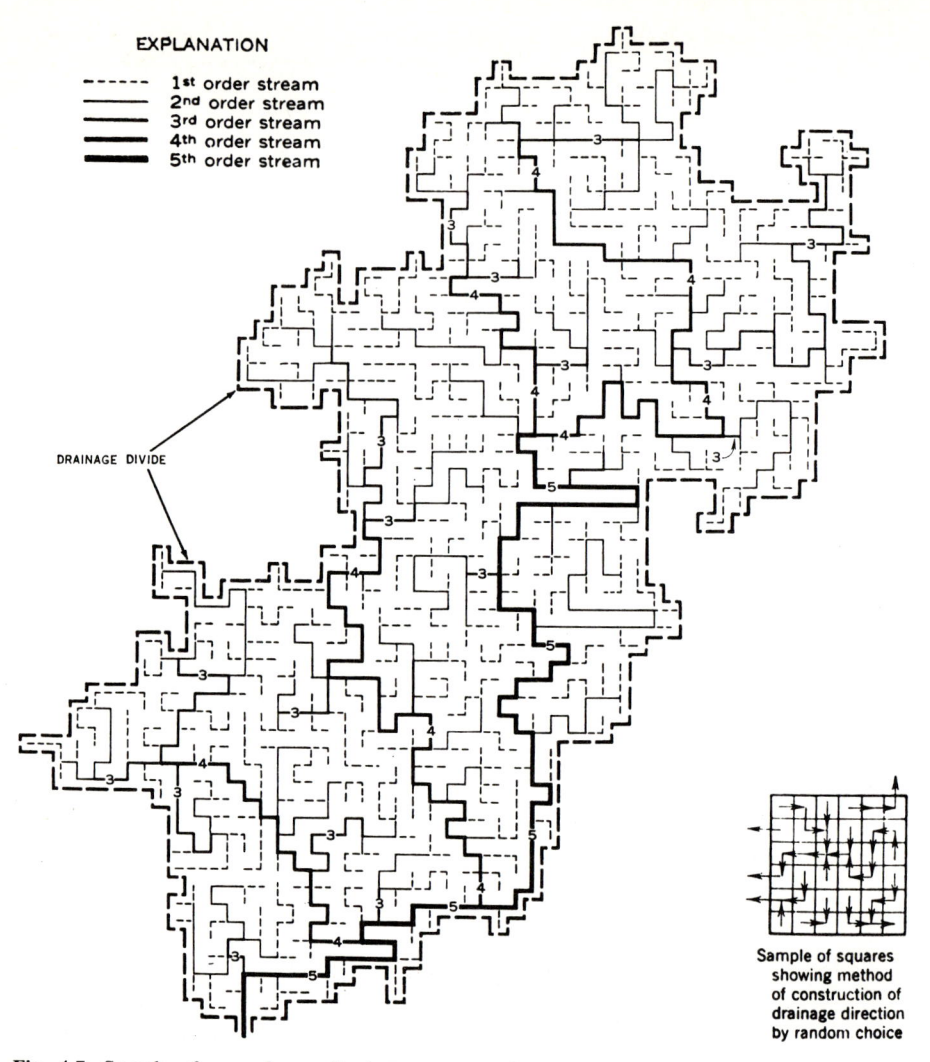

Fig. 4.7. Sample of a random-walk drainage network developed on rectangular graph paper. (Leopold et al. 1964)

walks can be expected to take before meeting is proportional to the first power of the initial distance between starting points (Leopold and Langbein 1962, p. A14). It can also be shown that in the random-walk model the logarithms of stream length vary linearly with stream order, a relationship commonly observed in natural stream networks.

Leopold and Langbein (1962) also developed a random walk model of a stream network on rectangular cross-section paper (Fig. 4.7). Each square on the paper represents a unit area in a developing drainage network and each square is to be drained. The direction of drainage from the centre of each square is in any of the four cardinal directions, the actual direction taken being selected at

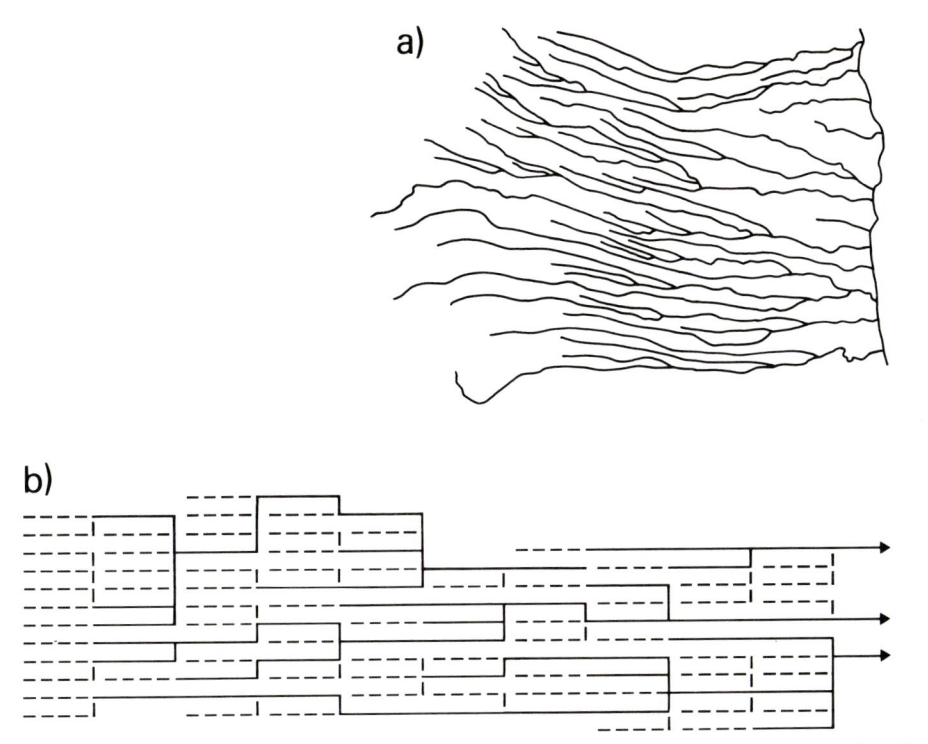

Fig. 4.8a, b. Drainage systems on the Ontonagon Plain, Michigan, USA. **a** Streams between Mineral and Cranberry Rivers. **b** Simulated drainage pattern. (Hack 1965b)

random with an equal chance of drainage in any direction subject to the constraint that no internal sinks can form. The direction of flow from the centre of each square can be depicted by an arrow (Fig. 4.7 inset). The arrows connect and a stream network is generated (Fig. 4.7) which has properties statistically identical to the properties of natural stream networks.

Hack (1965b) applied Leopold and Langbein's first model to simulate the development of postglacial drainage on the progressively exposed bottom of Lake Duluth, Michigan. The observed streams are much attenuated owing to more or less equally spaced glacial grooves running at right angles to the lake shore (Fig. 4.8). The observed patterns show little evidence of capture, migration of divides, or headwards cutting, but valleys have deepened by an amount roughly proportional to the contributing drainage area. Hack constructed a random-walk model assuming that the spacing of the original channels was controlled by the glacial grooves which are roughly 0.1 miles apart; that, as the shoreline of the Ontonagon plain receded, the streams were required to flow further and further to reach the lake, but all the time being subject to chance coalescence with adjacent streams; and that the length of first-order streams was one mile and the drainage density ten. The model predicted that stream length and drainage area are related by the function $L = 4.0 A^{0.67}$, which accords well with field observations.

Table 4.2. Simulated and observed stream numbers. (After Schenck)

Number of streams (dimensionless[a])	Stream order			
	First	Second	Third	Fourth
Simulation model[b]	0.780	0.171	0.036	0.012
United States	0.777	0.174	0.040	0.009
Santa Fe	0.716	0.208	0.058	0.018

[a] Dimensionless numbers of streams are defined as the number of streams of any given order divided by the total number of streams in the network.
[b] Mean of eight runs.

A similar kind of random-walk model was developed by Scheidegger (1967a) to simulate the drainage pattern on an intramontane trench. Starting points of the random walk are again equally spaced but each step of a walk can move randomly forward to the left or to be the right. To simulate this process, Scheidegger used a Monte Carlo method to generate 20 rows of randomly selected zeros and ones. The rows were then staggered so that a drainage network could be constructed by starting at the top of the grid and taking zero to mean drainage to the left and one to mean drainage to the right. The pattern of drainage areas produced by the random-walk model matches the pattern of drainage areas observed in the Rhône valley.

The generation of stream networks by random-walk models is admirably well suited to the computer. The first attempt to use a computer for this purpose was made by Schenck (1963) who devised a programme to generate a fourth-order drainage basin on a 20 × 30 matrix. A fair match between simulated patterns and selected natural patterns was achieved (Table 4.2). Smart et al. (1967) developed a programme which was based on a modified version of Leopold and Langbein's (1962) second model. The modifications allowed bias in the probabilities of drainage lines forming in the cardinal directions and allowed the setting up of a series of rules to prevent the occurrence of source juntions, triple junctions, loopings, and trappings. Drainage networks were generated under assumptions of random and biased stream direction.

A major difficulty in using random-walk simulations to explore the changes of a drainage system with time is that the model may have to be so simplified and its development may proceed through such unreal and artificial steps that its only value lies, not in its representation of network development, but in its ability to approach and to exemplify some state of equilibrium (Chorley and Kennedy 1971, p. 283). To overcome this problem, partly at least, more realistic constraints can be added to the models. A good example of this is A. D. Howard's (1971a, b) study of capture models. A key component of these models is that the probability of capture of a given stream segment by a nearby stream is expressed as a function of the upstream drainage areas of the two streams and the gradient across the site of potential capture. If the probability is such that capture is permitted, then the drainage area of the captured stream is regraded according to the formula:

stream gradient = contributing area$^{0.6}$,

a relationship derived from field data. Repeated runs of various versions of the model suggest that network changes increase rapidly to a peak and then decline to a roughly constant value, and that, when ordered according to the Strahler system, the networks produced accord well with natural stream networks. Indeed, Howard believes that the properties of natural stream networks are better reproduced by capture models than by models not involving capture. He concludes therefore, with certain reservations, that stream capture is an important process in the development of stream networks. His reservations are that the slightly better results of the capture models may be due to chance; that factors other than stream capture may produce similar results; and that the slightly better results may be an artefact of the model itself resulting from, for example, the spacing of the spatial cells.

4.4.2 Alluvial Fans

W. E. Price (1976) has built an interesting random-walk simulation model of alluvial-fan deposition which contains a large number of realistic boundary conditions and constraints. Figure 4.9 is a plan view of the model grid system and the boundary conditions used in some of the model runs. In the top centre of the grid is a canyon mouth from which debouches the flow upon the depositional basin in which the fan is built. The boundary along the top of the grid is a mountain front which is bordered by a fault along which uplift of the mountain block takes place. The sides of the grid represent other fans or mountain areas. The boundary along the bottom edge of the grid may be a playa or a flowing stream. The main dynamic elements in the model are relative uplift of the mountain area containing the fan source basin, a storm which produces flow on the fan, change in the thickness of weathered material in the source basin, a random walk of storm flow into the fan, deposition of sediment on the fan, and erosion both in the source basin and on the fan.

Relative uplift of the mountain area and storms that produce flow on the fan are regarded as independent stochastic events. Relative mountain uplift is described by equations defining the time of uplift and the magnitude of the earthquake assumed to have caused the uplift. The time of uplift follows a Poisson process with exponentially distributed inter-occurrence intervals:

$$f(t_u; \lambda_u) = \lambda_u e^{-\lambda_u t_u},$$

where t_u = a period of time and λ_u is the mean rate of occurrence of uplifts. The magnitude of the earthquake is also modelled by a negative exponential distribution:

$$f(M_e; \beta) = \beta e^{-\beta(M_e - M_0)},$$

where β is a parameter related to b in the formula of Gutenberg and Richter (1954, p. 17), M_0 is the minimum magnitude of earthquake events, and M_e is the magnitude of a particular earthquake event. Storms producing flow on the

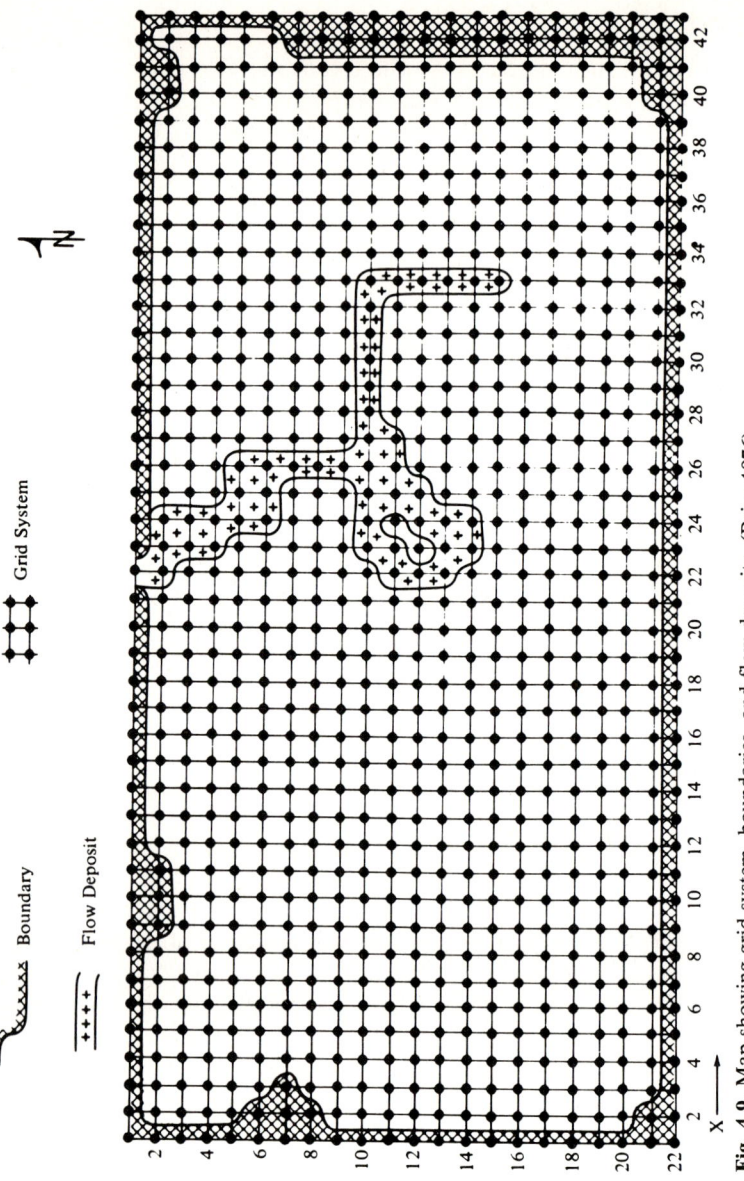

Fig. 4.9. Map showing grid system, boundaries, and flow deposits. (Price 1976)

Random-Walk Models

fan are modelled by exponential distributions expressing the time and magnitude of the flow. Time of flow is given by

$$f(t_f; \lambda_f) = \lambda_f \exp(-\lambda_f t_f),$$

where t_f is a time period and λ_f is the mean rate of occurrence of flow events. Magnitude of flow (Shane and Lynn 1964, pp. 9–12) is modelled by

$$f(y; \gamma) = \frac{1}{\gamma} \exp(-y/\gamma),$$

where γ is the mean peak flow rate and y is the magnitude of the peak flow rate.

The amount and type of material deposited on the fan during a flow event are assumed to be independent of the magnitude of the flow but to be dependent upon the volume of weathered material that is immediately available for erosion form the source basin. Throughout a model run, the areal extent of the material is taken as constant but its thickness increases with time according to the relation

$$y_s = m_s(1 - e^{-\eta t})$$

where y_s is the thickness of the weathered layer, m_s is the maximum thickness of the weathered layer, η is a constant related to the rate of development of the weathered layer, and t is a time period (Fig. 4.10).

Storms occurring over the source basin generate runoff which picks up readily erodible material and carries it to the canyon mouth where the random walk starts on the model grid. Walks may move in any of the four cardinal directions, the probability of moving in any particular direction, P_d, being set as proportional to the gradient, s, in that direction:

$$P_d = 0.25 - 0.75 s.$$

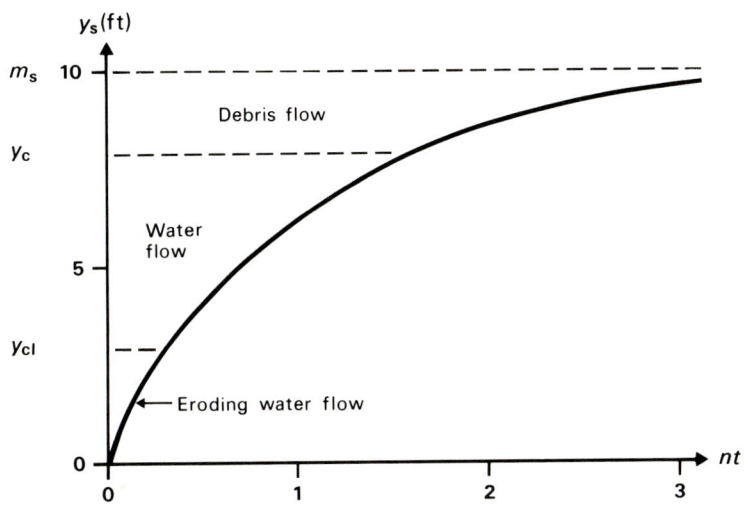

Fig. 4.10. Graph showing rate of increase of weathered layer in basin and critical values for flow events in the model. (Price 1976)

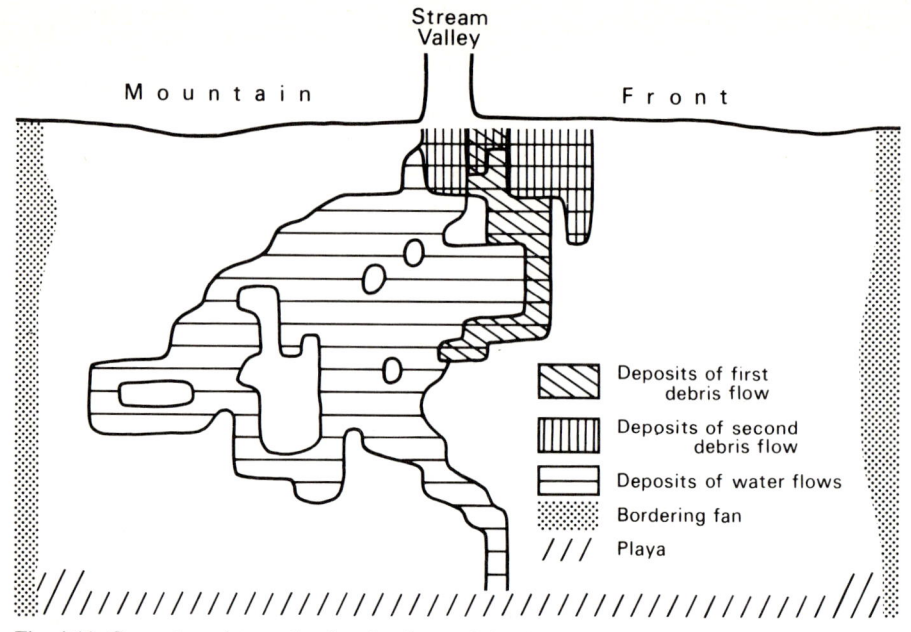

Fig. 4.11. Computer printout showing the shape of simulated debris-flow and water-flow deposits. (Price 1976)

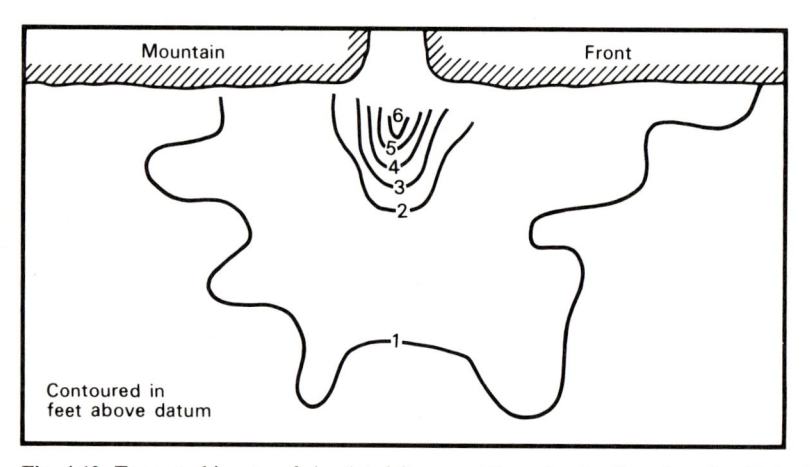

Fig. 4.12. Topographic map of simulated fan consisting of water-flow deposits. (Price 1976)

The probabilities calculated by this formula are weighted by a momentum coefficient which takes account of the fact that once a stream of water is headed in a particular direction, it will tend to continue in that direction.

Two types of deposit were simulated: debris flows and water flows. Debris flows consist of poorly sorted material ranging in size from silt and clay to boulders. Debris flows are assumed to occur when the weathered layer in the

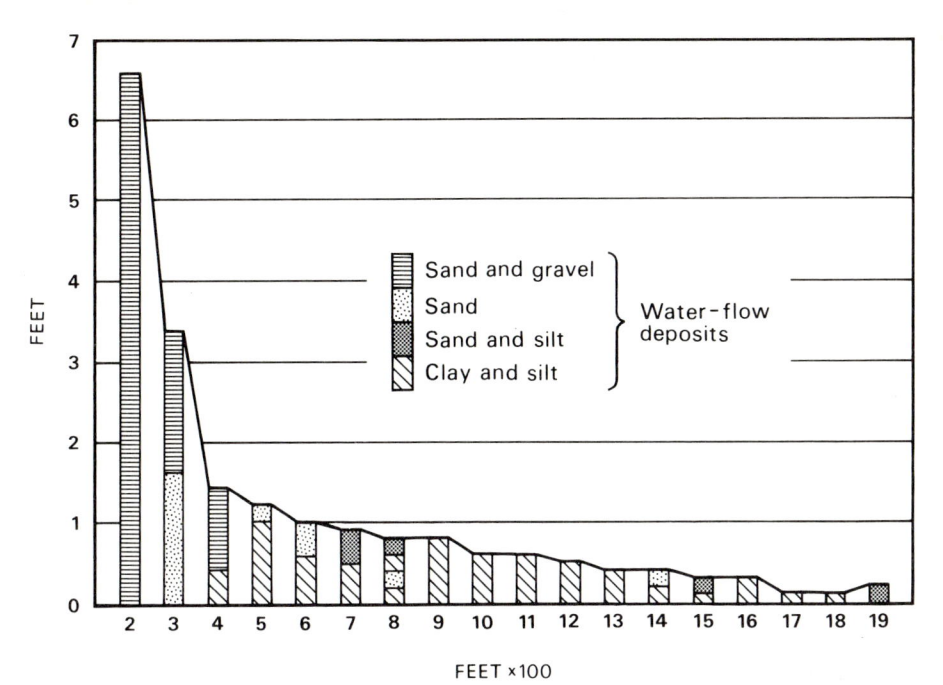

Fig. 4.13. Radial section through a simulated fan consisting of water-flow deposits. Section perpendicular to mountain front. (Price 1976)

source basin attains a critical, arbitrary thickness, y_c (see Fig. 4.10) and a storm hits the basin. Below this critical thickness, if a storm occurs, water flows will deposit moderately well-sorted silty to gravelly sand. If the weathered thickness is equal to or less than another critical thickness, y_{c_1}, erosion will take place. The grain size of water-flow deposits is assumed to be related to slope in the following way:

$$d = c_f s,$$

where d is median grain size, c_f is a coefficient, and s is slope. Deposition is assumed to occur instantaneously along the entire length of a channel. Deposits may develop irregular shapes or may branch. Their thickness tapers in the direction of flow.

Two kinds of erosion are modelled. The first kind is expressed by

$$h = h_0 e^{-kt},$$

where h is the elevation (in feet) of the rock stream channel above base level at a particular time t, h_0 is the elevation of the rock stream channel immediately following uplift at time t_0, and k is a dimensionless parameter expressing the rate of decline of the stream channel. This erosion model leads to downcutting of the main stream into its rock channel above the fan. The second kind of erosion involves the erosion of sediments from the fan and is caused by water flows virtually unladen with basin sediment; this kind of erosion is treated as negative deposition.

Examples of simulated flow events and fans are given in Figs. 4.11 to 4.13. The general form of the simulated deposits is consistent with actual fan deposits, the pattern of simulated flows matches real flows, and the simulated fan facies show the expected concentration of debris flow near the apex and a particle-size decrease of water-flow deposits down the fan. These results indicate that the digital model based on a random walk can be used with profit in the study of alluvial-fan deposition.

4.5 Markov Chains

The idea of Markov chains was first posited by the Russian mathematician Markov. A Markov chain can be thought of as a series of transitions between different states. The transitions are expressed as probabilities and depend only on preceding states. If the present state depends only on the state at the immediately preceding time, then the Markov chain extends back over one time step and is a first-order Markov chain. If the dependence of state extends back to even earlier states, then higher-order Markov chains are formed (see R. A. Howard 1971).

4.5.1 Transition Probabilities

In a Markov chain, the dependence of one state upon another is represented by a transition probability. Take the hypothetical example of daily wheather change given by Huggett (1980). The weather for each day in a year is divided into three types or states — fine, fair, and foul. The record of state changes for one year's observations is listed in Table 4.3. The data in this table can be converted into a set of transition probabilities (Table 4.4). For instance, a day of fine weather follows a day of fair weather 66 times out of the 165 times that a day of fair weather was recorded during the year; the transition probability from fair to fine is thus $66/165 = 0.4$. The set of transition probabilities may be expressed as a matrix P, where

$$P = \begin{array}{c} \\ Fine \\ Fair \\ Foul \end{array} \begin{array}{c} Fine\ Fair\ Foul \\ \left[\begin{array}{ccc} 0.3 & 0.2 & 0.5 \\ 0.4 & 0.4 & 0.2 \\ 0.3 & 0.6 & 0.1 \end{array} \right] \end{array}$$

and represented as a diagram (Fig. 4.14). In the general case, for a three state system,

$$P = \begin{array}{c} \\ x_1 \\ x_2 \\ x_3 \end{array} \begin{array}{c} x_1\ \ x_2\ \ x_3 \\ \left[\begin{array}{ccc} p_{11} & p_{12} & p_{13} \\ p_{21} & p_{22} & p_{23} \\ p_{31} & p_{32} & p_{33} \end{array} \right] \end{array},$$

where the p's are transition probabilities between states x_1, x_2, and x_3. Note that a transition probability of 1.0 is an absorbing state — once entered, the state

Markov Chains

Table 4.3. Daily weather changes over 1 year. (Huggett 1980)

		to			
		Fine	Fair	Foul	Row total
from	Fine	48	32	80	160
	Fair	66	66	33	165
	Foul	12	24	4	40
					$\Sigma = 365$

Table 4.4. Transition probabilities between weather types. (Huggett 1980)

		to			
		Fine	Fair	Foul	Row total
from	Fine	0.3	0.2	0.5	1.0
	Fair	0.4	0.4	0.2	1.0
	Foul	0.3	0.6	0.1	1.0

will not change. The information in the matrix P can be used to discover the probability of changing between any two states over one or two days given a starting state. Take, for instance, a fine day as a starting point. A tree diagram may be constructed which shows the transitions to other states after one and two days (Fig. 4.15). This shows that there are three different ways of changing from fine weather to foul weather in two days. The probability of changing from fine to fair weather in two days, p_{13}^2, is defined by

$$p_{13}^2 = (p_{11}p_{13} + p_{12}p_{23} + p_{13}p_{33})$$
$$= (0.3 \times 0.5) + (0.2 \times 0.2) + (0.5 \times 0.1)$$
$$= 0.24 .$$

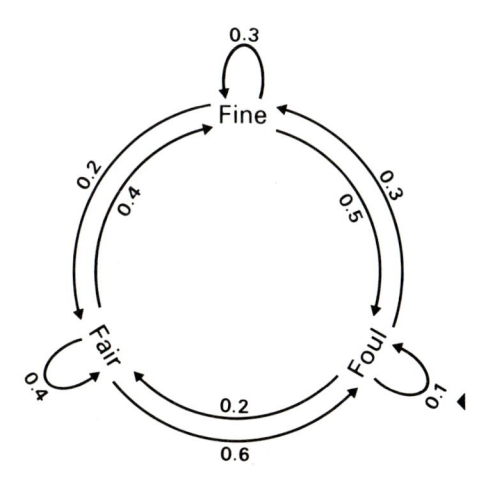

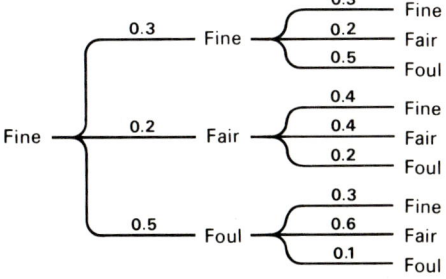

Fig. 4.14. State changes as transition probabilities. (Huggett 1980)

Fig. 4.15. A tree diagram showing transitions to other states over two days starting with a fine day. (Huggett 1980)

The probabilities of changing between all pairs of states over two days may be evaluated in like manner. The general formula for probability of change over n days (time steps) is

$$p_{ij}^n = \sum_{i=1}^{n} p_{ik} p_{kj},$$

where k is the number of different paths which may be taken to pass from one state to another in n days. A more efficient method of calculating these probabilities is to raise the transition probability matrix, P, to the power n. This matrix, P^2, will give the probabilities of passing between pairs of states over one time step, matrix P^3 will give the probabilities of passing between pairs of states over two time steps, and so forth. Eventually, successive powering of matrix P will lead to a state which does not change on being raised to a higher power. In the weather example, this happens at the fourth power:

$$P^4 = \begin{pmatrix} 0.3387 & 0.3871 & 0.2742 \\ 0.3387 & 0.3872 & 0.2742 \\ 0.3387 & 0.3871 & 0.2742 \end{pmatrix}.$$

The rows in this matrix give the equilibrium proportions of the three states.

4.5.2 Sedimentary Sequences

Krumbein (1976, p. 40) has noted that a Markovian matrix of transitional probabilities is an excellent device for summarizing stratigraphic sequences, giving a good estimate of the relative probabilities with which lithological units follow one another, and allowing major and minor deposition cycles to be identified, in a stratigraphic section. Such information is valuable for reconstructing the succession of environments in sedimentary basins which in turn may suggest physical and chemical conditions that succeed one another. Many studies have been made along these lines, starting with the paper by Vistelius (1949), a Russian geologist who counted transitions of bedding types in a Cretaceous flysch sequence in the Causacus and calculated a transition matrix. The technique made no headway until the mid-1960's when, belatedly following Vistelius's lead, papers started to appear (Allègre 1964; Carr et al. 1966; Krumbein 1967; Potter and Blakely 1968; to name but a few).

The Markov chain model has been little used in geomorphology though it seems to have potential applications in that field (Thornes and Ferguson 1981, p. 165). An example of its use in geomorphology is the work of Miall (1973, 1977), who employed Markov chains to describe and to analyse successions of sedimentary facies in the alluvium of braided streams. Thornes (1973) has experimented with applying Markov chains to slope series. Curl (1959) has applied the Markov model to limestone cavern development and Melton (1962) has used it to distinguish between regional and local random variations in river profiles in Arizona. The application of Markov chains to certain fields of hydrology has proved a success, particularly in the modelling of rainfall (Gabriel and Neumann 1962; Haan et al. 1976) and understanding reservoir behaviour (Lloyd 1967).

4.5.3 Volcanic Activity

Wickman (1976) has built six Markov models to represent idealized patterns of volcanic activity. A number of states of volcanic activity, consistent with established volcanological classification, are defined. These are the eruption state, E_e, the persistent activity state, E_p, and one to several repose states R_0, R_1, to R_n, the exact number depending on how many different time scales are necessary to describe the patterns of activity. Transitions between states are defined by time-independent parameters, λ_i, calibrated in units of per month. Wickman (1976) constructs six different models each representing a specific physical situation, though the parameters in each case admit of many physical interpretations. The six cases are: a simple, two-stage volcano of the Fuji type without persistent activity; a volcano of the Hekla type with a loading time and dormancies; a volcano of the Vesuvius type with a loading time and persistent activity; a volcano of the Kilauea type with lava-lake activity; a volcano of the Asama type with accelerating cycles of eruption; and a volcano with both excitation and a loading time. With appropriate rate parameters fitted to these models, Monte Carlo simulations were made to show periods of repose and activity (see for instance Fig. 4.16). The results indicate that simple Markov models can describe complex patterns of volcanic behaviour, in a qualitative way at least. Wickman (1976) declined to test the results against observed records of volcanoes because of the great inadequacies in the historical data.

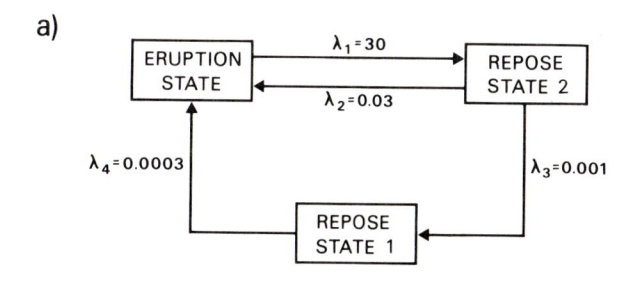

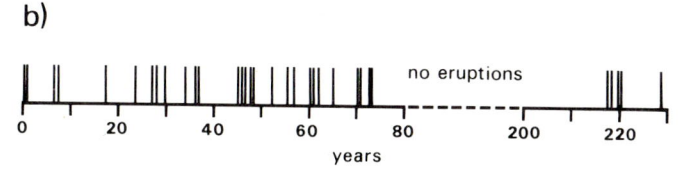

Fig. 4.16. a A model of a two-stage volcano without persistent activity (Fuji type). The λ's are rate constants for transitions. **b** A Monte Carlo simulation for the same model. There are two active periods separated by a long repose period. (After Wickman 1976)

4.6 Entropy Models

4.6.1 Entropy Maximization

In geomorphology, Leopold and Langbein (1962) set up an analogy between landscape variables and thermodynamic variables. A two-dimensional thermodynamic field is described by temperature, $T(x, y)$, and the quantity of heat, Q. The landscape field is described by the height of the land surface above a datum level, $h(x, y)$, and mass, M. Leopold and Langbein (1962) then proposed these heuristic analogies: $T \leftrightarrow h$, $dQ \leftrightarrow dM$. Scheidegger (1964a) justified this analogy on the grounds that the statistical basis for both systems is the same and that there is an established thermodynamic analogy for all transport processes (Scheidegger 1961a, 1967a, b). Thus, in a system which is a linear combination of fluctuating components, and in that system a certain quantity (heat or mass in the examples) is a non-negative definite constant of the motion, it can be shown that, under equilibrium conditions and assuming linear regression of the fluctuations and microscopic reversibility, the quantity in question is subject to a diffusivity equation with a symmetrical diffusivity tensor. On this basis, it is possible to define analogous entropies S in thermodynamics and landscape evolution:

$$dS = dQ/T \leftrightarrow dM/h$$

as well as other thermodynamic quantities. For instance, the quantity of heat introduced into a system by a given substance is

$$dQ = \gamma dT,$$

where γ is a heat capacity coefficient. In the landscape, the analogy is

$$dM = \gamma dh,$$

where γ is an analogue of the heat-capacity coefficient. In a regular thermodynamic system, the first law states that

$$dU = dQ + dW,$$

where U is internal energy, Q is a quantity of heat introduced from outside, and W is worked performed externally on the system. In a landscape, there is a similar relation, according to the thermodynamic analogy

$$dU = dM + dW,$$

where U signifies some potential, M the mass that was introduced, and W some, as yet undefined, work. For an ideal gas,

$$W = -\int_V p dV,$$

where V is the geometrical domain in which the variables vary. According to the ideal gas law,

$$p = RT/V = \text{const } T/V.$$

Entropy Models

From this relation, Scheidegger (1970, p. 277) set up, for the equilibrium case, an analogy to "pressure" in landscapes

$$P_{\text{landscape}} = \text{constant } h/A$$

where A is the area of the geometrical domain over which the variables vary. For a profile the analogue becomes

$$p_l = \alpha h/L$$

and the analogue for work may be written as

$$W = -\int p \, dV = -\int \alpha(h/L) \, dL$$

where L is the geometrical domain of the profile.

Leopold and Langbein (1962) and Leopold (1962), applying the thermodynamic analogy, have likened a river system to a series of perfect heat engines which operate between heat sources and heat sinks, each engine yielding a quantity of work. They obtain equations to express change of height, the landscape analogue of temperature, with horizontal distance in terms of entropy change. The river long-profile corresponds to a state of maximum entropy.

4.6.2 Entropy Minimization

Scheidegger and Langbein (1966) have shown that for a steady state the rate of entropy production, σ, must be a minimum. In a one-dimensional system

$$\sigma = -\int \frac{dT/dx}{T^2} J \, dx = \text{minimum},$$

where J is the heat flux per unit time. In the landscape, the analogous condition is

$$\sigma = -\int \frac{dh/dx}{h^2} J \, dx = \text{minimum},$$

where J is now the mass flux per unit time. Scheidegger and Langbein (1966) apply this to a river profile, though it in fact applies to any geomorphological profile. For a steady state,

$$\delta\sigma = -\delta\int \frac{dh/dx}{h^2} J \, dx = 0,$$

where J, being proportional to mass flow, represents sediment transport along a river. Intuitively defining $J = J(dh/dx)$ as a process equation (the flow is proportional to slope) gives the following Euler-Lagrange equation for the minimization:

$$\frac{d^2h}{dx^2}\left[2\frac{\frac{dJ}{dx} \cdot \frac{dh}{dx}}{h^2} + \frac{dh}{dx} \cdot \frac{\frac{d^2J}{dx^2} \cdot \frac{dh}{dx}}{h^2}\right] - 2\frac{\frac{dJ}{dx} \cdot \frac{dh}{dx}}{h^3}\left(\frac{dh}{dx}\right)^2 = 0.$$

Choosing specifically $J = c\,dh/dx$ (that is, flow is proportional by a constant, c, to slope) yields the minimization equation

$$(d^2h/dx^2)/dh/dx - (dh/dx)/h = 0,$$

which when solved yields

$$h = c_1 e^{-c_2 x},$$

where c_1 and c_2 are constants of integration. This equation is analogous to that obtained by a deterministic model.

Scheidegger (1980, p. 277) has modified this model by choosing as a process equation

$$J = \left(-q\,\frac{dh}{dx}\right)^a,$$

where q is the discharge volume per unit river width and a is a constant. Discharge, q, may be assumed to be proportional to the distance from the source of the river, x, because drainage area is roughly proportional to x^2, the river width proportional to x, hence q to x; this yields

$$J \sim \left(x \cdot \frac{dh}{dx}\right)^a.$$

Minimizing the appropriate Euler-Lagrange equation gives

$$h = \left[c\,\frac{a-1}{a+1}\log_c kx\right]^{\frac{a-1}{a+1}}$$

where c and k are constants of integration. Thus

$$\frac{dh}{dx} = \frac{2}{ch^{1+a}/x}.$$

A regression analysis on a large number of streams yields $a = 1.5$ which, according to Scheidegger (1970, p. 277), is excellent confirmation of the theory.

4.6.3 Developments of the Thermodynamic Approach

Yang (1972) has developed the thermodynamic approach and proposed two basic laws of fluvial geomorphology: firstly, the law of average stream fall – under conditions of dynamic equilibrium, the ratio of the average fall between any two different order streams in the same river basin is unity; and secondly, the law of least-rate energy expenditure – during the approach to equilibrium conditions, a natural stream chooses its course of flow in such a way that the potential energy expenditure per unit mass of water along the course is a minimum.

Davidson (1978, p. 74) notes that the behaviour of a drainage network can, in part, be explained by these laws. As well as seeking to minimize the expenditure of energy, rivers also adjust so as to minimize discontinuities in energy loss along their lengths. The alternation of pools and riffles in a straight channel introduces discontinuities and the river's response is to develop meanders which increase the loss of energy in pool sections. Thus meanders develop, according to this explanation, because the river seeks a smooth grade line (Leopold et al. 1964).

On a global scale, plate tectonics has been interpreted as a gigantic, stationary, irreversible process which minimizes the production of entropy produced by thermal convection in the upper mantle (Elsasser 1971; Miyashiro et al. 1982, p. 195). A number of models of plate tectonics have been built according to this assumption (Forsyth and Uyeda 1975; Solomon et al. 1975, 1977).

In a recent re-evaluation of the thermodynamic approach to the study of landscapes, Davy and Davies (1979) conclude that the application of specific entropy concepts to stream systems is illegitimate because the character of the system lies outside the proper domain of the principles and assumptions governing entropy behaviour. In an equally condemnatory mood, E. Culling (1981) questions Scheidegger's analogy between elevation and temperature, arguing that a closer analogy is between the thermal vibration of molecules in a liquid or amorphous solid and "vibrational" random activity of soil particles. Karcz (1980) is more sympathetic towards the thermodynamic approach, however, and gives a clear account of the premises upon which it is built. His reservation about minimum entropy formulations is that they apply only to systems near equilibrium. When, as is commonly the case in Earth surface systems, imposed forces drive the system away from equilibrium, minimum entropy formulations are no longer valid. The thermodynamics of systems far removed from equilibrium will be discussed in Chap. 9.

CHAPTER 5
Inductive Stochastic Models

The raw data for inductive stochastic models are observed time and distance series. In theory, the series may be continuous or discrete. In practice, it is usual to sample a continuous series, such as a hillslope profile or river discharge, at discrete distance or time intervals to give a set of observations $\{x_i\}$ or $\{x_t\}$ where i and t denote distance and time increments respectively. Distance series of interest to Earth scientists include elevation series along hillside slopes and along rivers, and series of river channel form variables. Time series of interest include average annual values of river flow, temperature, precipitation, tree-ring indices, and wind-varve thickness and seasonal series of temperature, precipitation, and streamflow.

Observed series may consist of four parts: a trend, a cycle, a persistent stochastic element, and a random element (Fig. 5.1a). If two series are compared, then relationships may be sought between each of the four components (Fig. 5.1b). An example of a time series which consists of cyclical and stochastic parts is given in Fig. 5.2 which depicts the mean monthly rainfall and runoff data collected for the Lagan drainage basin, Kenya by Blackie (1972).

Not all time series are well behaved or regular. Some display distinctly irregular behaviour involving steps (jumps), ramps, and transient changes, or a combination of all three (Fig. 5.3a). Rao (1980) illustrates a number of misbehaved hydrological time series (Fig. 5.3b). The river Nile may be divided into two sets of flows. The first set is the flows prior to the building of the Aswan Dam in 1902 which may be regarded as decreasing from 1871 to 1902. The second set of data is the generally lower, regular flows occurring since 1902. The same kind of ramp function as in the pre-Aswan Dam Nile river flows, though not as pronounced, can be seen in the annual rainfall data for Dunedin, New Zealand. The flows of the Niger river are variable, consisting of rising and falling ramps. The Krishna river, India, steps on a seasonal basis from low flows to high, monsoon runoff, and back again.

5.1 Box and Jenkins's Models: an Introduction

Natural time and distance series may be analysed using a family of stochastic models developed by Box and Jenkins (1976), whose work was a culmination of the research of many prominent statisticians starting with Yule (1927). The Box-Jenkins models have been extended and refined, the better to suit the study of spatial systems (Bennett and Chorley 1978; R. J. Bennett 1979) and geophysical systems (Hipel and McLeod 1981a, b).

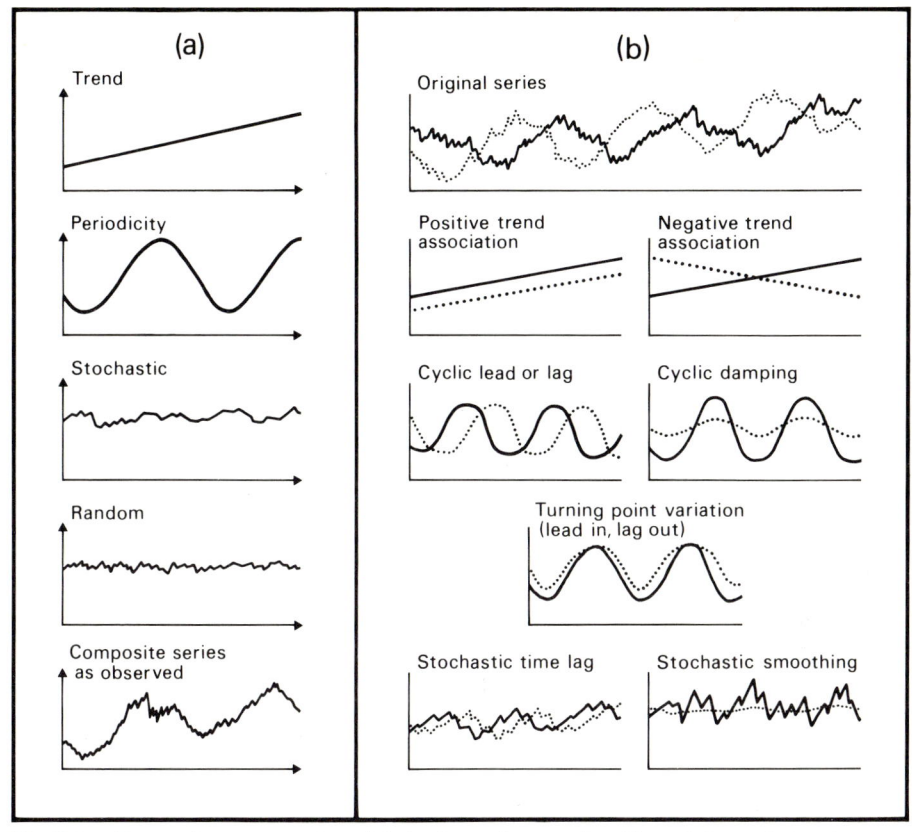

Fig. 5.1. a Time-series components and their observed composite effects. **b** Interrelations between components of two time series. (Bennett 1979)

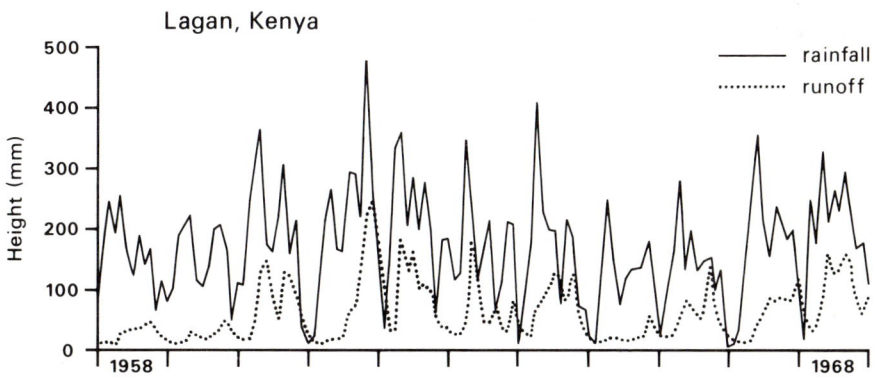

Fig. 5.2. Monthly rainfall and runoff in the Lagan drainage basin, Kenya. (Data from tables in Blackie 1972)

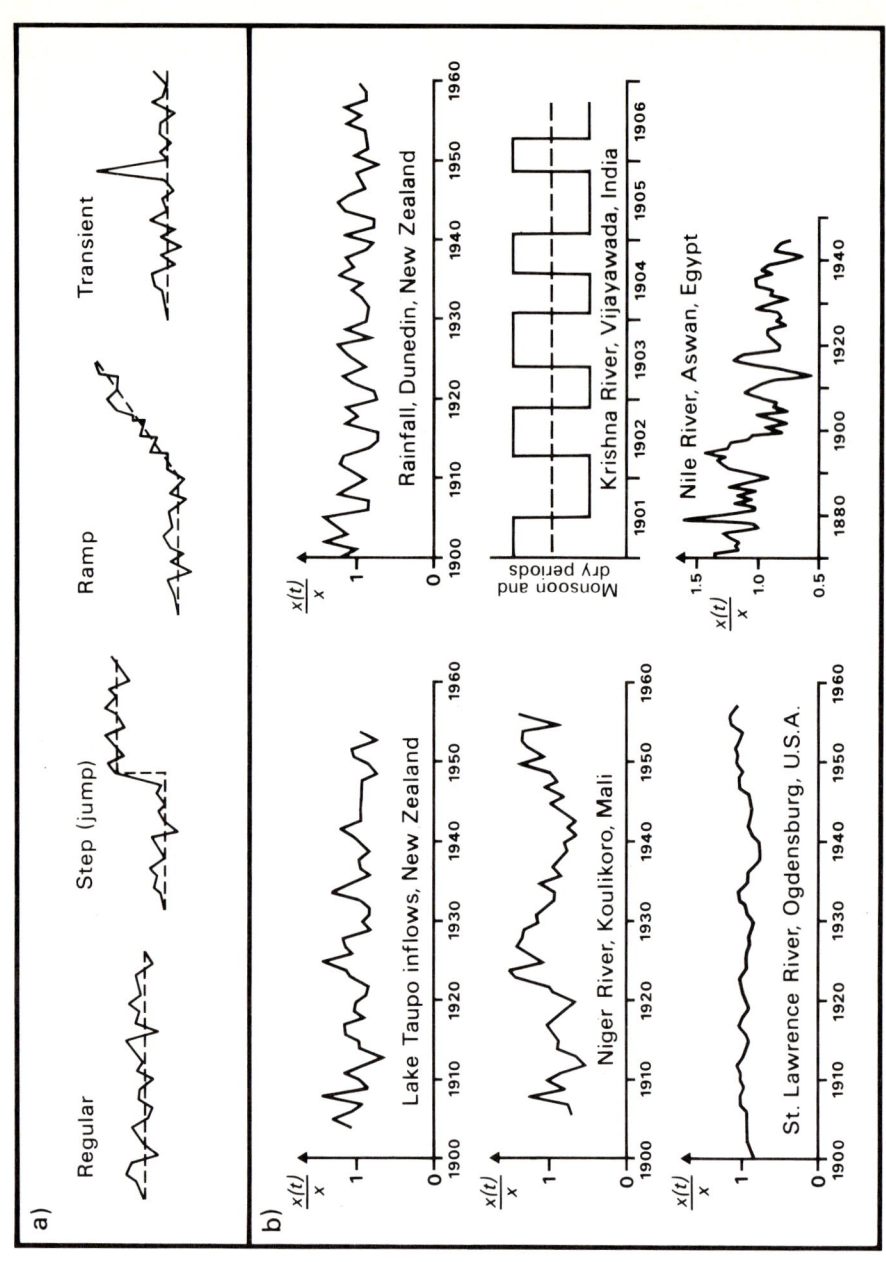

Fig. 5.3. a Some types of behaviour of time series. **b** Some hydrological time series. (Rao 1980)

5.1.1 System Definition

The basis of the Box-Jenkins models is that a system may be defined by an input variable, an output variable, and a translation operator linking the two (Fig. 5.4). The relation between input, X, and output, Y, in the time domain may be written

$$Y_t = SX_t,$$

which indicates that the input variations are transformed into the outputs by an operator, S, termed the transfer function of the system. For discrete systems, three important operations are the lead, the lag, and the summation or integration. These operations may be written for general cases as

$$SX_t = \text{parameters} \times X_{t-L}$$

for a lag process,

$$SX_t = \text{parameters} \times X_{t+L}$$

for a lead process, and

$$SX_t = \text{parameters} \times (X_{t-L} + X_{t-L+1} + \ldots + X_{t-1})$$

for a summation process. In each case, the time base is shifted forwards or backwards by changing the subscripting variable. The form of the shift produces the character of the transfer function, S. In this definition, the transfer function *is* the system in the sense that it completely describes the process or changes induced by the operation of the system modulating system inputs to produce given system outputs (Bennett and Chorley 1978, p. 26).

In a drainage basin, rainfall inputs of various forms and magnitude are translated into runoff outputs. The process can be described, to a first approximation, by a response curve commonly referred to as a basin hydrograph. This curve has a characteristic shape with a rapid rise to peak discharge and

Fig. 5.4. The structure of a simple system

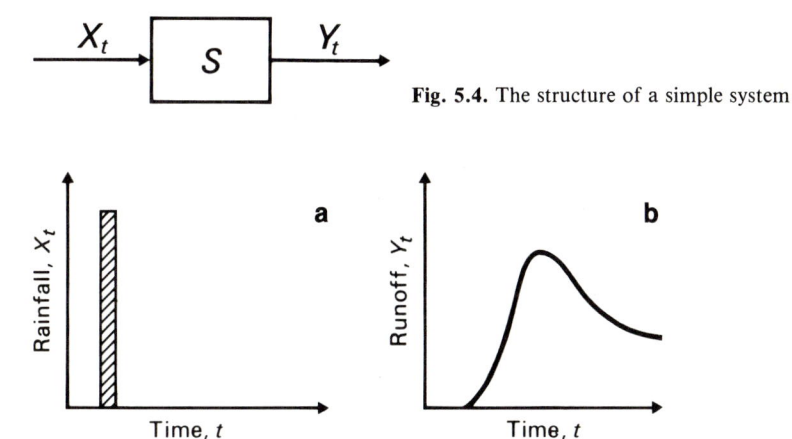

Fig. 5.5 a, b. Drainage basin inputs and outputs. **a** Rainfall input, X_t. **b** Resulting runoff output, Y_t

an attenuated recession curve (Fig. 5.5). The diagram shows how an isolated input of rain is translated, via the transfer function, into the runoff outputs. The response curve of all drainage basins will be similar but differences will reflect the characteristics of individual drainage basins. Different drainage basin characteristics may produce similar response curves (hydrographs) so that the transfer function for each basin is not unique. Nonetheless, as R. J. Bennett (1979, p. 6) states, the transfer function can be very useful in characterizing the properties of a system.

5.1.2 Stages in Systems Analysis

The Box-Jenkins models involve five stages of analysis (R. J. Bennett 1979, pp. 20–23) (Fig. 5.6). The first stage involves specifying the variables, interactions, and linkages thought to be of importance to the structure and function of the system being modelled. The second stage is concerned with identifying the nature of the system operation (transfer function) which governs the relationship between inputs and outputs. In other words, it is concerned with defining the structure of a model which best represents the observed dynamics of a system. If the input, output, and transfer function of a system are all known, then system identification presents no problem. However, it is usually the case when dealing with Earth surface systems that one or more of the system elements is unknown. In these cases, where information is incomplete, the observations must be interpreted statistically to permit the induction of system structure. The methods for diagnosing the unknown system elements differ according to which of the

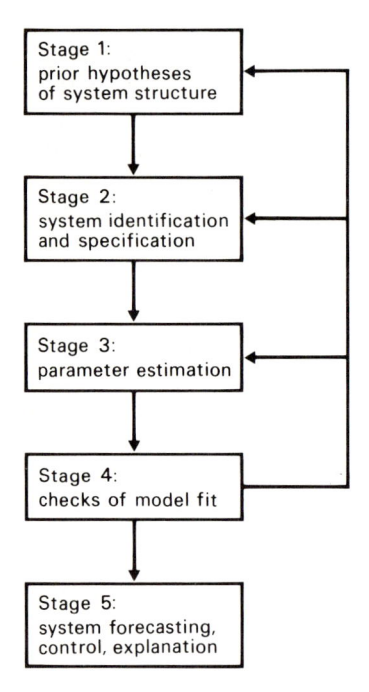

Fig. 5.6. Stages in the analysis of spatial (and temporal) systems. (R. J. Bennett 1979)

Table 5.1. Time and frequency domain diagnostics used in system specification. (R. J. Bennett 1979)

Specification problem	System unknowns	Diagnostics of system unknown	
		Time domain	Frequency domain
1 Identification and estimation	Transfer function	Impulse response (cross-correlation function)	Frequency response (cross-spectrum: gain and phase)
2 Convolution and forecasting	Output	Output autocorrelation function	Output spectrum
3 Deconvolution and back-forecasting	Input	Input autocorrelation function	Input spectrum
4 ARMA identification and estimation	Input *and* transfer function	Impulse response (autocorrelation function)	Frequency response (spectrum)
5 Simulation	Input *and* output	Simulated output autocorrelation	Simulated output spectrum

elements is unknown. Five distinct cases of system identification can be made in both the time and frequency domain (Table 5.1). The third stage of Box-Jenkins modelling involves estimating the values of parameters in the system model identified in stage two. Least-squares regression is a common method for estimating parameters. The fourth stage involves testing the worth of the model as an explanatory device and the fifth and last stage, only carried out when the model seems successful, involves making forecasts, simulations, explanations, or control solutions for the system under study.

In applications in the Earth sciences, just two of the cases of system identification listed in Table 5.1 are commonly used. Firstly, where only a single series is known, an autoregressive moving-average, ARMA, model is employed, which is normally tested by comparing observed autocorrelation and partial autocorrelation functions of the original series with autocorrelation and partial autocorrelation functions predicted by the model. Secondly, where both input and output series are known but the transfer function is not, input-output relationships are used to identify a transfer function, TF, model which can be tested by comparing the cross-correlation function of the original series with the cross-correlation function predicted by the transfer function model. It is to these two types of model that discussion will now turn.

5.2 Autoregressive Moving-Average Models of Time Series

5.2.1 Model Formulation

Autoregressive moving-average models are a family of stochastic process models of great use in Earth surface sciences. For a set of values of Y spaced at equal intervals in time, a general model for Y_t is expressed as

$$C(z) Y_t = D(z) e_t,$$

where

Y_t designates the output series, usually given as residuals from the mean;
t is an index for the discrete time periods;
z is a backwards shift operator defined by

$$z Y_t = Y_{t-1} \quad \text{and} \quad z^k Y_t = Y_{t-k};$$

e_t is a stochastic noise process;
$C(z) (= 1 - c_1 z - c_2 z^2 - \ldots - c_p z^p)$ is a nonseasonal autoregressive, AR, operator or polynomial of order p and the c_i ($i = 1, 2, \ldots, p$) are the nonseasonal autoregressive parameters; and
$D(z) (= 1 - d_1 z - d_2 z^2 - \ldots - d_q z^q)$ is a nonseasonal moving average, MA, operator or polynomial of order q and the d_i ($i = 1, 2, \ldots, q$) are the nonseasonal moving average parameters.

The notation ARMA (p, q) is used to denote an ARMA model with an autoregressive operator of order p and a moving-average operator of order q. If the process is represented by an autoregressive model for order p with no moving-average components, it is usually denoted by AR(p), and not by ARMA$(p, 0)$. Similarly, if the process is represented by a moving-average model only, it is usually denoted by MA(q) in preference to ARMA$(0, q)$.

If a time series is nonstationary, it is customary to remove the nonstationarity and then fit a model to the resulting stationary data. Box and Jenkins (1976) recommend that nonstationary data should be differenced just enough times to remove "homogeneous nonstationarity". In the case of a single series, the difference operator is denoted as

$$\nabla Y_t = Y_t - Y_{t-1} = (1 - z) Y_t,$$

where z is the backwards shift operator. For second-order differencing this becomes

$$\nabla^2 Y_t = (1 - z) \nabla Y_t$$
$$= Y_t - 2 Y_{t-1} + Y_{t-2}.$$

For many applications in Earth sciences a normal difference operator is combined with a seasonal difference operator to render the series stationary and remove any trend. The operations are written as

$$\nabla \nabla^d Y_t = (1 - z)(1 - z^d) Y_t.$$

If $d = 12$, then monthly data with an annual cycle and a linear trend may be operated on by

$$\nabla \nabla^{12} Y_t = (1 - z)(1 - z^{12}) Y_t$$
$$= (1 - z - z^{12} - z^{13}) Y_t.$$

In general, the term $(1 - z)^d$ is a difference operator of order d. A Box-Jenkins model which includes a difference operator is called an autoregressive integrated moving-average, ARIMA, model. In these models, the notation ARIMA (p, d, q)

is used to indicate the orders of the operators in the autoregressive-integrated-moving-average process. An ARIMA $(p, 0, q)$ model is equivalent to a stationary ARMA (p, q) model.

Irregular or nonstationary data can be handled by stochastic modelling techniques. The statistics involved in calculating serial correlation coefficients for nonstationary series has been worked out by Yevjevich and Jeng (1969). Box and Jenkins (1976) developed an autoregressive integrated moving-average, ARIMA, model for seasonal data. An example is given by Rao (1980), who fits the model to annual precipitation data for La Porte, Indiana, from 1915 to 1968. McLeod and Hipel (1978) have presented a family of stochastic models that are designed exclusively for modelling monthly geophysical time series.

Differencing operations may or may not be a suitable method for rendering a series stationary. If a change in the mean level of a process occurs because of a known external cause, then the method of intervention analysis should be employed (Box and Tiao 1975). Hipel et al. (1975) use intervention analysis to model the drop in the mean level of the river Nile at Aswan owing to the completion of the dam in 1902. The different mean levels before and after the construction of the dam are automatically accounted for in the intervention model. The need to exercise caution when differencing data is also indicating by the study made by Pickup and Chewings (1983) of monthly mean discharges of the Purari river, Papua New Guinea. Seasonal differencing of the discharge series was found to enhance rather than remove the seasonal component of flows. More satisfactory results were obtained by standardizing the data by subtracting the mean of all the values for each month from the individual observations and dividing by the standard deviation.

5.2.2 Modelling Procedures

Box and Jenkins (1976) recommend a procedure involving stages of identification, estimation, and diagnostic-checking to establish the type of ARMA model to fit a given data set. Essentially, the method involves comparing autocorrelation, partial autocorrelation, and spectral functions for the time series with theoretical forms of these functions derived from various ARMA models (Fig. 5.7). The correspondence between observed and predicted graphs can be determined by eye, a procedure sometimes called "eyeballing", or by statistical means. For low-order models, that is those with a small number of lags, the following set of generalizations holds. For an autoregressive process of order p, the autocorrelation function tails off exponentially or with exponentially damping oscillations, while the partial autocorrelation is cut off after the lag p. A moving-average process of order q has autocorrelation cut off after lag q, and a partial autocorrelation tailing off. For the mixed autoregressive moving-average process, of order (p, q), both the autocorrelation and partial autocorrelation functions tail off with a mixture of exponentials and damped sinusoids after the first $q - p$ lags for the autocorrelations, and after the first $p - q$ lags for the partial autocorrelations.

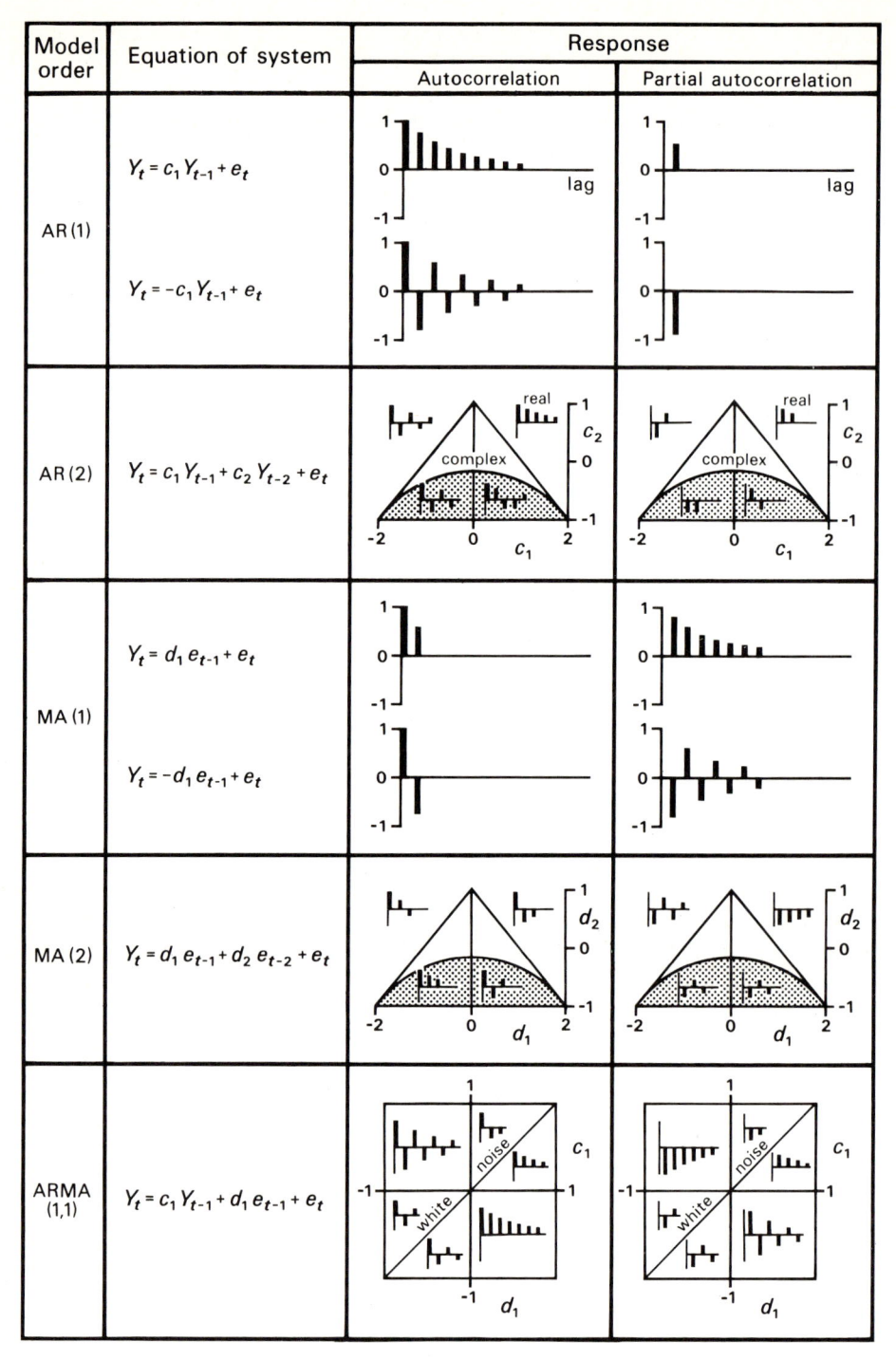

Model order	Equation of system	Response	
		Autocorrelation	Partial autocorrelation
AR(1)	$Y_t = c_1 Y_{t-1} + e_t$		
	$Y_t = -c_1 Y_{t-1} + e_t$		
AR(2)	$Y_t = c_1 Y_{t-1} + c_2 Y_{t-2} + e_t$		
MA(1)	$Y_t = d_1 e_{t-1} + e_t$		
	$Y_t = -d_1 e_{t-1} + e_t$		
MA(2)	$Y_t = d_1 e_{t-1} + d_2 e_{t-2} + e_t$		
ARMA (1,1)	$Y_t = c_1 Y_{t-1} + d_1 e_{t-1} + e_t$		

Fig. 5.7. The autocorrelation and partial autocorrelation functions for various low-order autoregressive moving-average systems. (After Box and Jenkins 1976, R. J. Bennett 1974, 1979)

5.2.3 The Lagan Rainfall Series

As an example of an autoregressive moving-average model, consider the Lagan rainfall series. The autocorrelation functions and partial autocorrelation functions for this series are shown in Fig. 5.8 for various levels of differencing. An important seasonal component at a lag of 12 months is evident. To render the series stationary, R. J. Bennett (1979, p. 157) applied seasonal differencing to the data, the results of which, as Fig. 5.8b shows, had two effects. Firstly, it failed to remove lags of periods greater than 12, so leaving significant correlation in the rainfall series at 2- and 3-year intervals. Secondly, a nonstationary trend component was revealed in the autocorrelations up to lag four. The trend was removed by differencing the seasonally differenced series (Fig. 5.8c) but the higher-order lags still remained. If higher-order lags are ignored, the partial autocorrelation function (Fig. 5.8c) suggests the identification of an AR(3) process as a suitable model to describe the data. In matrix form, the model may be written

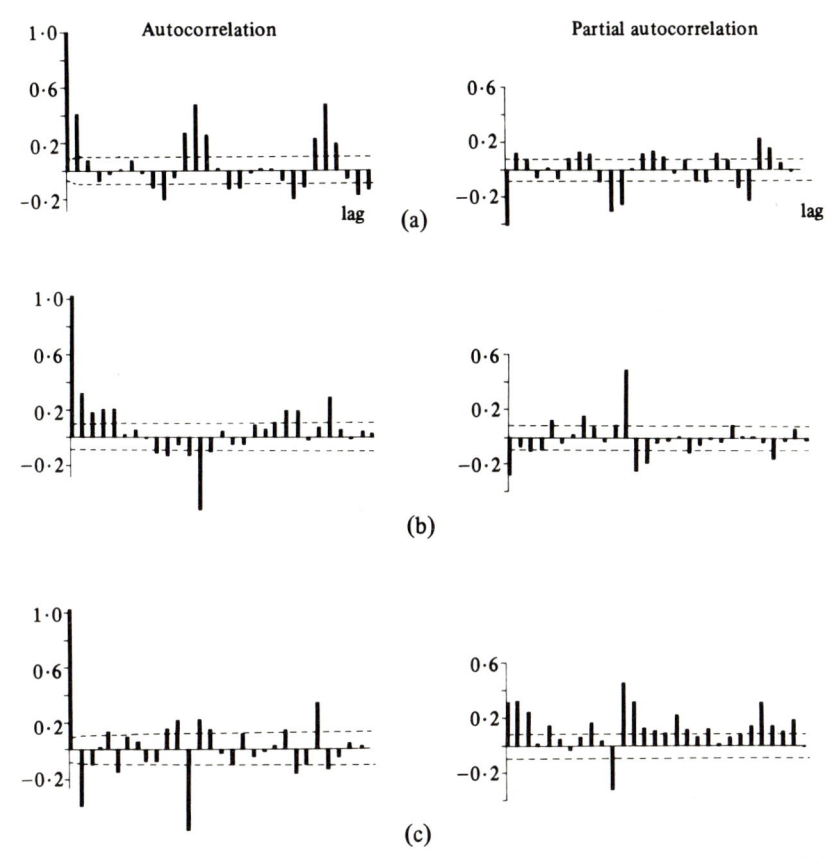

Fig. 5.8a–c. Autocorrelation and partial autocorrelation functions for Lagan rainfall series. **a** Original series. **b** Seasonally differenced, ∇^{12}, series. **c** Seasonally and first-order differenced, $\nabla\nabla^{12}$, series. (R. J. Bennett 1979)

$$Y_t = [-Y_{t-1}, -Y_{t-2}, -Y_{t-3}] \begin{bmatrix} a_1 \\ a_2 \\ a_3 \end{bmatrix} + e_t.$$

R. J. Bennett (1979, p. 237) used ordinary least-squares regression to estimate the parameters a_1, a_2, and a_3. The resulting equation is

$$\nabla\nabla^{12} Y_t = -0.62\, Y_{t-1} - 0.46\, Y_{t-2} - 0.25\, Y_{t-3},$$
$$\qquad\quad (0.135)\qquad (0.068)\qquad (0.036)$$

where the terms in brackets are the standard errors of the parameters which may be used to test each parameter for statistical significance. The overall model and individual parameters are all significant at $p = 0.05$.

Many other examples of ARMA models fitted to time series can be found in the literature. Pickup and Chewings (1983), on the basis of autocorrelation and partial autocorrelations of standardized mean monthly discharges of the Purari river, QP', identify an AR(1) process which, when calibrated, gives the following stochastic model:

$$QP'_t = 0.472\, QP'_{t-1} + a_t,$$

in which the series a_t has a mean of 0.0091 and a standard deviation of 0.5728. McLeod et al. (1977) and Hipel and McLeod (1981b) have determined autoregressive moving-average models for a variety of time series including river flow, sunspots, tree rings, precipitation, and temperature.

5.3 Autoregressive Moving-Average Models of Distance Series

5.3.1 Model Formulation

The formulation of an autoregressive moving-average model for distance series is identical to the formulation for time series except that values of a variable Y are taken at equal intervals in space rather than in time. A general model for Y_i is expressed as

$$C(z)\, Y_i = D(z)\, e_i,$$

where

Y_i designates the distance series;
i is an index of discrete distance increments;
z is a backwards shift operator defined by

$$z\, Y_i = z\, Y_{i-1} \quad \text{and} \quad z^k Y_i = Y_{i-k};$$

e_t is a stochastic process;
$C(z)\,(=1-c_1 z - c_2 z^2 - \ldots - c_p z^p)$ is an autoregressive, AR, operator of order p and the $c_j\,(j = 1, 2, \ldots, p)$ are the autoregressive parameters; and
$D(z)\,(=1-d_1 z - d_2 z^2 - \ldots - d_q z^q)$ is a moving-average, MA, operator of order q and the $d_j\,(j = 1, 2, \ldots, q)$ are the moving average parameters.

The variable Y is usually a series of elevations along a hillslope or river bed or a series of directions along a river. Most form variables measured at regular distance intervals could in fact be analysed with an autoregressive moving-average model.

5.3.2 River Meanders

A spatial series is obtained for river meanders by measuring the direction of the river, $\theta(=Y)$, at various distances, s, along its course. The directional series $\{\theta\}$ can be differenced to produce a direction change series $\{\nabla\theta\}$ which approximates changes in river curvature. The representation of meander patterns by direction was first done by Leighly (1936) for semi-arid gullies. Speight (1965) and Langbein and Leopold (1966) rediscovered the idea of using direction series to study meander patterns some 30 years later (Ferguson 1979, p. 230). The main achievement of Langbein and Leopold (1966) was to demonstrate that sequences of two or three meander bends, ranging in size from the Mississippi to small laboratory streams, show a more or less sinusoidal variation of direction with distance downstream, the relation being written

$$\theta = \omega \sin ks,$$

where channel direction, θ, is measured about its mean and oscillates between the limits ω and $-\omega$ with a period $\lambda = 2\pi/k$. The oscillation described by this equation is not a sine curve but what Langbein and Leopold (1966) term a sine-generated curve. Although the sine-generated curve is widely accepted as a simple but satisfactory description of individual meander bends, it does not satisfactorily describe complete meander patterns which seldom, if ever, consist of a string of identical bends (Ferguson 1979, p. 231). The irregularities exhibited by meanders consisting of more than a few bends are best studied statistically by spatial series analysis, a technique first used in this context by Speight (1965) on the Angabunga river in Papua New Guinea. A perfectly regular meander pattern would have a spectrum with a single sharp peak at a frequency of $1/\lambda$ whereas Speight's (1965) computed spectra have multiple peaks. Speight interpreted the multiple peaks as a set of superimposed harmonics that persist downstream and over time. Ferguson (1979, p. 232) points out, however, that none of Speight's subsidiary spectral peaks is significant at the 5 per cent level. Thus, the evidence can be better interpreted as showing an irregular periodic pattern rather than stable, superimposed wavelengths. The same interpretation can be given to Trenhaile's (1979) power spectra of detrended river thalwegs in the Canadian Cordillera (Scheidegger 1983, p. 4).

Langbein and Leopold (1966), in the same paper in which they argue in favour of a sine-generated curve to describe meander bends, were the first workers to suggest that meander patterns may be random. Although it may seem paradoxical to propose that meander patterns at once may be described by a sine-generated curve and are random, the logic behind the proposal is that the apparent regularity of meanders is the expected outcome of an entropy-maximizing or variance-minimizing tendency in rivers. Thakur and Scheidegger (1968) supported a random-walk model but both they (Thakur and Scheidegger 1970) and Surkan and van Kan (1969) found that meander direction series have autocorrelation functions which tail off rapidly, a fact in contradiction to the random-walk model in which the mean is nonstationary, the expected variance is infinite, and the expected autocorrelation is undefined. Computer simulations of randomly walking meanders produce patterns that wander all over the place, cross one another, and produce correlograms that tail off very slowly (Ferguson 1976).

Taking account of the fact that sine-generated curves fit individual meander bends well but require different parameter values for different bends, Ferguson (1976) proposed a disturbed periodic model of meander patterns. This model is thus consistent with the idea that meandering is essentially a deterministic process which takes place in a heterogeneous environment. It is a space domain version of the time domain model proposed by Yule (1927) to describe the sunspot cycle based on the analogy of the motion of a pendulum damped by friction being bombarded by small boys armed with peashooters. In Ferguson's (1976) model, trees, clay-filled oxbow lakes, and other obstructions take the place of small boys as disruptors of the regular oscillation. It may be expressed as a differential equation:

$$\theta + \frac{2l}{k}\frac{d\theta}{ds} + \frac{1}{k^2}\frac{d^2\theta}{ds^2} = \xi$$

with $0 < l < 1$ and where ξ is a randomly varying function which disturbs the centreline of oscillation. It may also be written in difference form as a second-order autoregressive process:

$$\theta_j = \beta_1 \theta_{j-1} + \beta_2 \theta_{j-2} + e_j.$$

The parameters in this equation are related to those in the differential equation in the following way

$$\beta_1 = 2\exp(-kl)\cos\{k(1-l^2)^{1/2}\}$$
$$\beta_2 = -\exp(-2kl).$$

If actual meander patterns are generated by a disturbed periodic process, then their direction series should have the approximate structure of the second-order autoregressive model and the estimates of the β parameters should lie within the oscillatory region of the parameter space (Fig. 5.9). To test his model, Ferguson

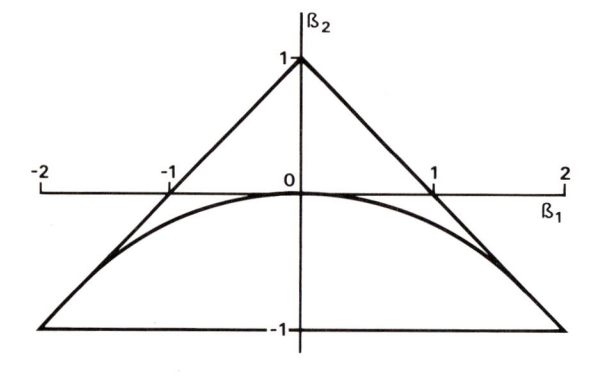

Fig. 5.9. AR(2) parameter space showing stationary region (*within triangle*) and oscillatory subregion (*below curved line*). (Ferguson 1979)

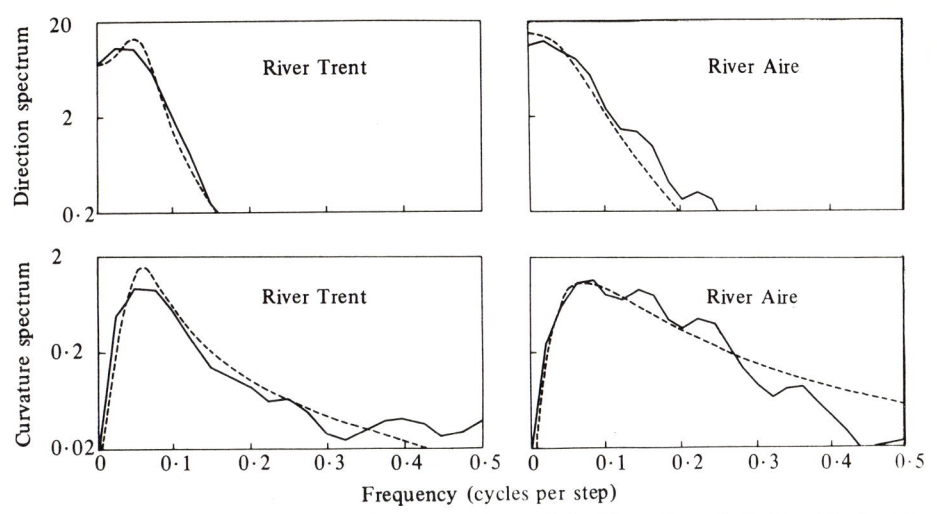

Fig. 5.10. Spectra of direction (*top*) and curvature (*bottom*) for Rivers Trent (*left*, $\nabla s = 60$ m) and Aire (*right*, $\nabla s = 70$ m). *Solid lines* are observed spectra, *broken lines* are theoretical spectra for $\lambda = 1000$ m, $l = 0.4$ (Trent) and 0.8 (Aire), direction variance = 0.85 (Trent) and 1.0 (Aire). (Ferguson 1979)

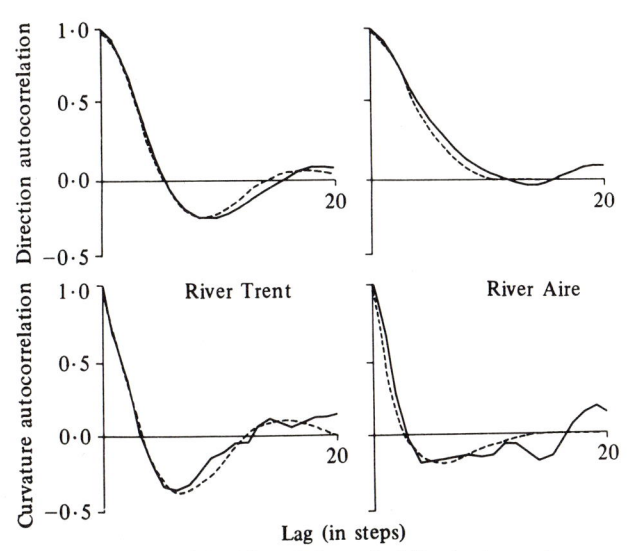

Fig. 5.11. Observed (*solid*) and theoretical (*broken*) correlograms of direction (*top*) and curvature (*bottom*) for Rivers Trent (*left*) and Aire (*right*). Parameters as in Fig. 5.10. (Ferguson 1979)

(1976) applied it to 19 reaches of English and Scottish rivers. Correlograms and spectra of the 19 direction series were very varied with some spectra having a peak at a finite wavelength and others not, some correlograms tailing off with, others without, pronounced oscillation (Figs. 5.10 and 5.11). Correlograms and spectra of the curvature series were less varied, all spectra having single significant peaks and all correlograms dropping to a significant negative value at a

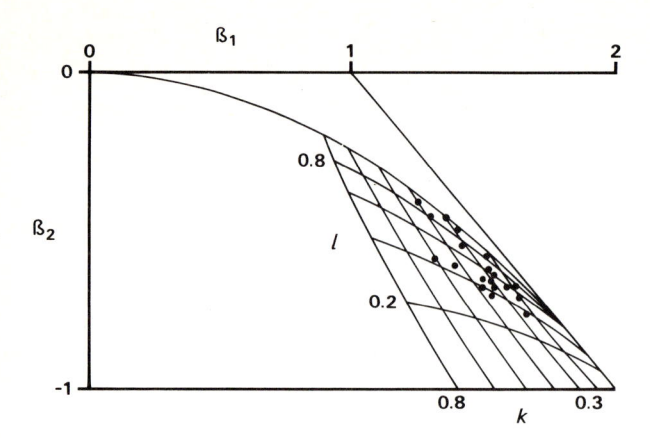

Fig. 5.12. Contours of the disturbed-periodic parameters k and l in AR(2) parameter space of Fig. 5.9, with points for 19 British rivers. (Ferguson 1979)

low lag before tailing off (Figs. 5.10 and 5.11). These findings refute the random-walk model and accord with the kind of behaviour expected to result from a disturbed periodic model. The match between observed and theoretical spectra and correlograms is close (Figs. 5.10 and 5.11). As an additional test, Ferguson (1979) fitted the second-order autoregressive approximation to the 19 direction series by a least-squares technique. In every case, the parameters are significantly different from zero and lie within, or at the edge of, the oscillatory region of the parameter space (Fig. 5.12).

5.3.3 Landforms

Craig (1982b) has used autocorrelation functions and partial autocorrelation functions of a series of elevations along a landform transect to identify a Box-Jenkins model that satisfactorily describes the surface processes which are shaping the landform. In general, he finds that the models belong to the autoregressive integrated moving-average, ARIMA (p, d, q) class having the general form

$$E_i^d = \phi_1 E_{i-1}^d + \ldots + \phi_p E_{i-p}^d + e_i - \theta_1 e_{i-1} - \ldots - \theta_q e_{i-q},$$

where E_i^d is the ith value of the dth difference of the elevation series, E^0 is the original elevation, E^1 is the slope series, and E^2 is the change in slope or curvature series so that

$$E_i^1 = E_i^0 - E_{i-1}^0$$

and in the general case

$$E_i^d = E_i^{d-1} - E_{i-1}^{d-1}.$$

Craig determines the values of p, d, and q by using autocorrelation and partial autocorrelation functions for a particular elevation series as a diagnostic tool. The first step is to determine the order of differences, d. If the chosen d is too small, then the autocorrelation function will decay extremely slowly from a value of $r \sim 1.0$. Should this behaviour be found in the autocorrelation function or the

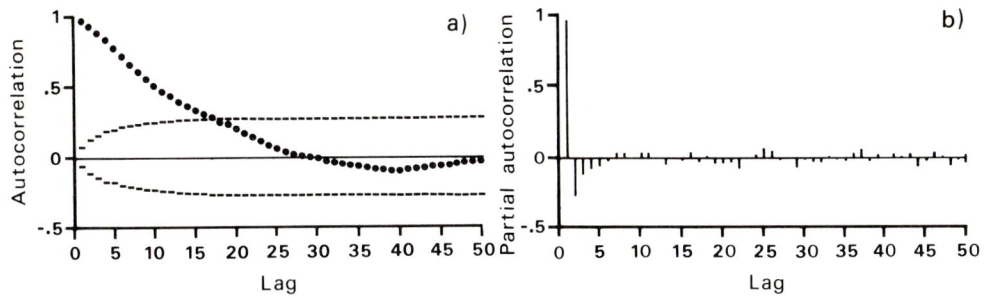

Fig. 5.13. a Autocorrelation and **b** partial autocorrelation functions of a nonstationary series following an ARIMA (1,1,0) model. Functions were computed from a traverse of 200 observations at a spacing of 48 m in the Snow Shoe, SE Pa $7\frac{1}{2}'$ quadrangle. Note the extremely slow decay of the autocorrelation function and the large value at lag one in the partial autocorrelation function. (Craig 1982b)

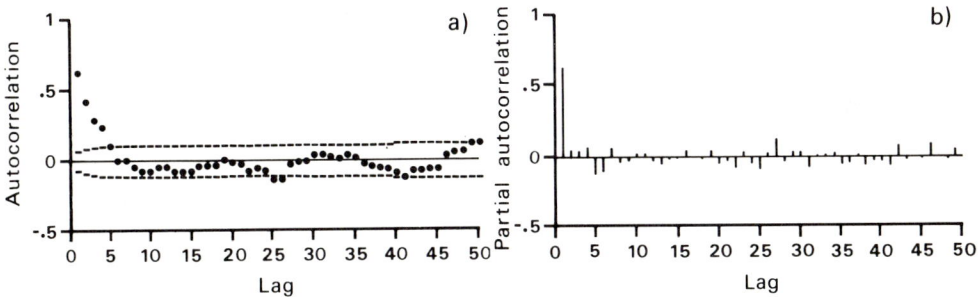

Fig. 5.14. a Autocorrelation and **b** partial autocorrelation functions of the first difference of the series of Fig. 5.13. Note the rapid decay of the autocorrelation function and the single "spike" at lag one of the partial autocorrelation function. This value (0.63) at lag one is a good estimate of ϕ_1 for the ARIMA (1,1,0) model. (Craig 1982b)

partial autocorrelation function then the series should be differenced until slow decay is replaced by rapid decay or by a plot containing isolated spikes. Figure 5.13 shows the autocorrelation and partial autocorrelation functions of an elevation series that requires differencing while Fig. 5.14 shows the equivalent functions for the difference series with $d = 1$. The exponential decay of the autocorrelation function and the single spike at lag one in the partial autocorrelation function indicate that the series has $p = 1$ and $q = 0$. The model that fits the series is thus an ARIMA (1,1,0):

$$E_i^1 = \phi_1 E_{i-1}^1 + e_i.$$

The coefficient ϕ_1 has been shown by Craig to be related to one of the coefficients used in deterministic models of slope development. This will be discussed later (Chap. 10.1) but, briefly, ϕ_1 is a "recessional coefficient" or "coefficient of retreat" and the equation applies to an elevation series produced by direct removal of material including surface wash and solution.

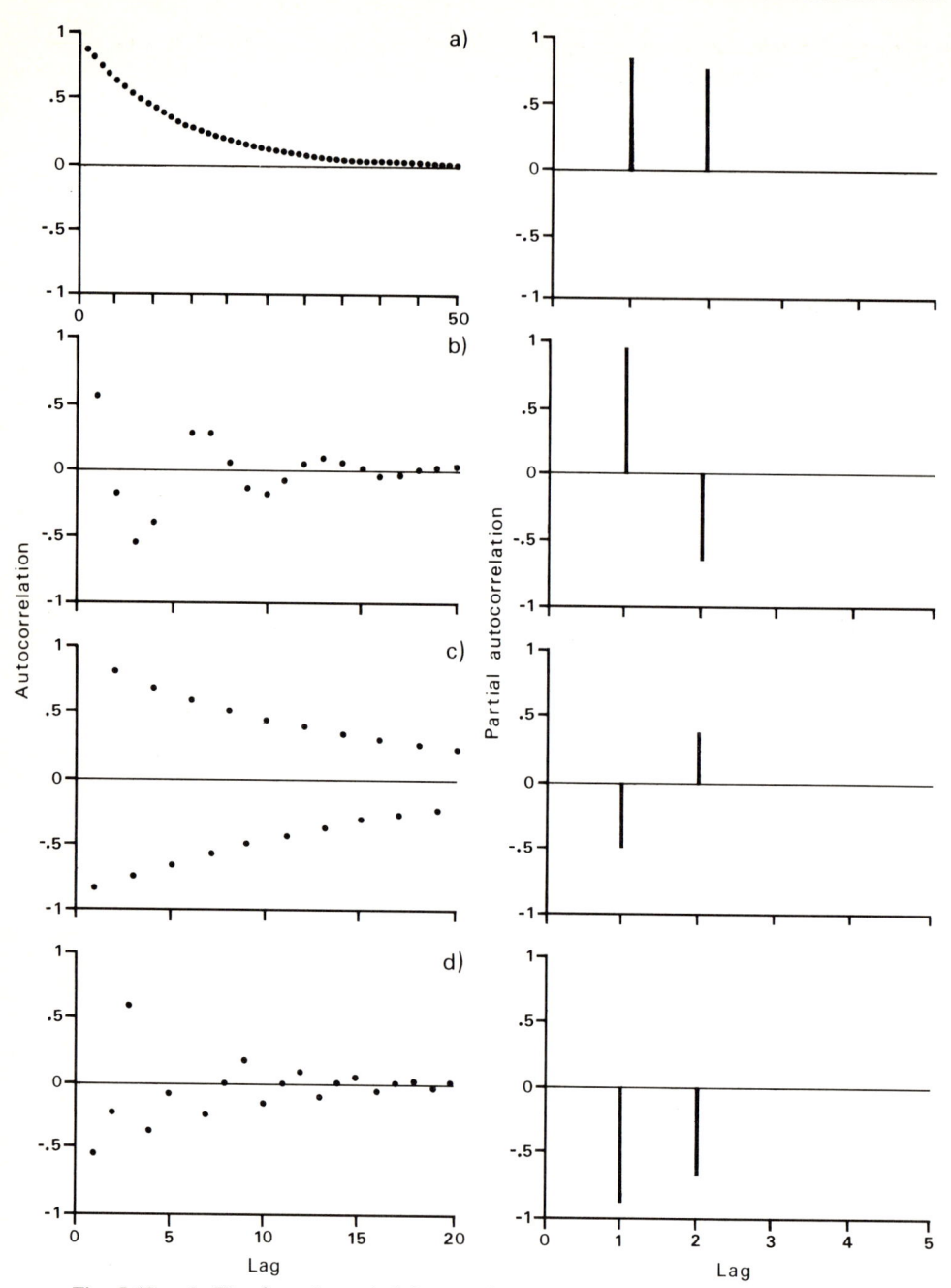

Fig. 5.15a–d. The four theoretical forms of the autocorrelation and partial autocorrelation functions of an ARIMA (2,1,0) model. **a** Both parameters positive. **b** $\phi_1 > 0$, $\phi_2 < 0$. **c** $\phi_1 < 0$, $\phi_2 > 0$. **d** Both parameters negative. (Craig 1982b)

Transfer Function Models

Another model discussed by Craig (1982b, p. 92) is an ARIMA (2,1,0) model which takes the form

$$E_i^1 = \phi_1 E_{i-1}^1 + \phi_2 E_{i-2}^1 + e_i.$$

For this equation, four basic forms of autocorrelation and partial autocorrelation functions are possible (Fig. 5.15). Craig demonstrates that this equation is, in effect, a stochastic partial differential equation describing the development of a landform whose temporal changes are controlled by a combination of slope wash (which leads to recession) and creep (which leads to smoothing). So if an elevation series has autocorrelation and partial autocorrelation functions conforming to those shown in Fig. 5.15, then the development of the series can be described by an ARIMA (2,1,0) model.

5.4 Transfer Function Models

It is commonly the case with Earth surface systems that input series and output series are known but the relationship between them is not. Such systems can be modelled using transfer-function analysis (Box and Jenkins 1976). The basic idea behind transfer-function analysis is that an input variable, X, influences the level of an output variable, Y, through the black box of the transfer function. A change in the level of input will often, because of inertia in the system, produce a delayed response in the level of output which eventually will attain a new equilibrium level. The change in output in the face of a change in the input is called the dynamic response. A transfer-function model describes this response and characterizes the inertia of the system.

5.4.1 Model Formulation

Given an input series X_t and an output series Y_t, then the transfer-function model relating the two may be expressed in general form for the univariate case as

$$A(z) Y_t = B(z) X_t + e_t$$

or

$$Y_t = A^{-1}(z) B(z) X_t + e_t,$$

where

$A(z) = 1 + a_1 z + a_2 z^2 + \ldots + a_r z^r$ and r is the order of the lag polynomial of the output series and determines the extent to which past values of the series influence Y_t;

$B(z) = b_0 + b_1 z + b_2 z^2 + \ldots + b_s z^s$ and s is the order of the lag polynomial for the input series and determines the way in which the input series influences Y_t;

e_t is a stochastic noise model; and

z is a backwards shift operator defined such that

$$z Y_t = Y_{t-1} \quad \text{and} \quad z^n Y_t = Y_{t-n}.$$

Model order	Equation of system	Cross-correlation
	Input	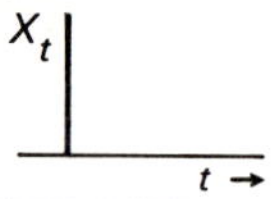 X_t, $t \rightarrow$
TF(1,0)	$Y_t = b_0 X_t$	 Lag
TF(2,0)	$Y_t = b_0 X_t + b_1 X_{t-1}$	
TF(3,0)	$Y_t = b_0 X_t + b_1 X_{t-1} + b_2 X_{t-2}$	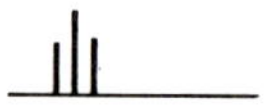
TF(1,1)	$Y_t = b_0 X_t + a_1 Y_{t-1}$	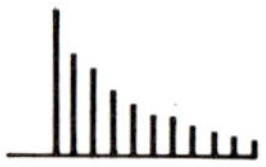
TF(2,1)	$Y_t = b_0 X_t + b_1 X_{t-1} + a_1 Y_{t-1}$	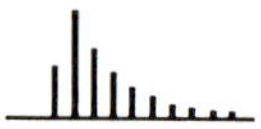
TF(3,1)	$Y_t = b_0 X_t + b_1 X_{t-1} + b_2 X_{t-2} + a_0 Y_t$	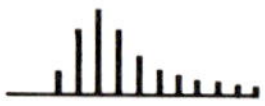
TF(1,2)	$Y_t = b_0 X_t + a_1 Y_{t-1} + a_2 Y_{t-2}$	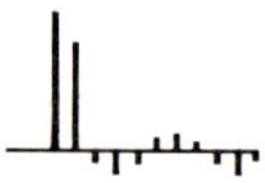
TF(2,2)	$Y_t = b_0 X_t + b_1 X_{t-1} + a_1 Y_{t-1} + a_2 Y_{t-2}$	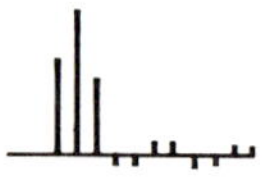
TF(3,2)	$Y_t = b_0 X_t + b_1 X_{t-1} + b_2 X_{t-2} + a_1 Y_{t-1} + a_2 Y_{t-2}$	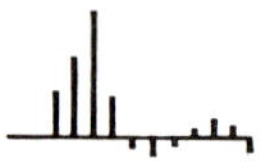
TF(1,3)	$Y_t = b_0 X_t + a_1 Y_{t-1} + a_2 Y_{t-2} + a_3 Y_{t-3}$	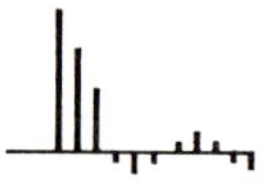

Fig. 5.16. The cross-correlation functions for various low-order transfer-function systems. (After R. J. Bennett 1974, 1979)

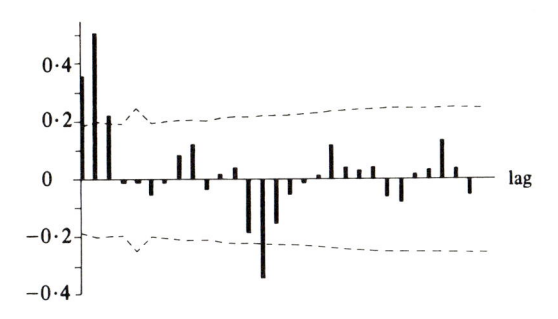

Fig. 5.17. Cross-correlation of rainfall and runoff series for the Lagan drainage basin after first differencing and AR(3) prewhitening. (R. J. Bennett 1979)

The usual practice in Earth surface systems is to identify the order of the transfer function model (the r and s values) by comparing the estimated cross-correlation functions for the two series with cross-correlation functions that are known to characterize a particular transfer-function system (Fig. 5.16). A suitable transfer function model having been identified, the parameters of the model are then estimated, usually by a least-squares regression technique.

5.4.2 Rainfall and Runoff in the Lagan Drainage Basin

The cross-correlation function between rainfall and runoff in the Lagan drainage basin, Kenya, is shown in Fig. 5.17. The series have been differenced to remove the seasonal cycle, differenced again to remove the trend, and prewhitened using an AR(3) model to remove correlation in the rainfall data. The resulting cross-correlation function when compared with the cross-correlation functions in Fig. 5.16 is indicative of a TF (2,2) model. Notice, however, that there is still a significant seasonal contribution. Ignoring the long-lag effects, the model may be written

$$Y_t = a_1 Y_{t-1} + a_2 Y_{t-2} + b_0 X_t + b_1 X_{t-1}.$$

The parameters in this model may be estimated using ordinary least-squares regression. (For other methods of parameter estimation, see R. J. Bennett 1979, Chap. 4.) For the 132 observations available on the system, and taking the difference series, the calibrated model is

$$\nabla \nabla^{12} Y_t = 0.207\, Y_{t-1} + 0.199\, Y_{t-2} + 0.075\, X_t - 0.094\, X_{t-1}.$$
$$\phantom{\nabla \nabla^{12} Y_t = }(0.034)\phantom{Y_{t-1}} (0.035)\phantom{Y_{t-2}} (0.011) (0.015)$$

The values in brackets are the standard errors of the parameter estimates and may be used to test each parameter for statistical significance. The application of Student's t-test reveals that each parameter is significant at $p = 0.05$. The overall fit of the model can be tested using the F statistic and correlation coefficient. Again, significance at $p = 0.05$ is indicated suggesting that the model gives a satisfactory description of the system.

5.4.3 Channel Form in the Afon Elan, Wales

Transfer function models may be applied to distance series. A good example is Anderson and Richards's (1979) study of the Afon Elan in mid-Wales. The river

98   Inductive Stochastic Models

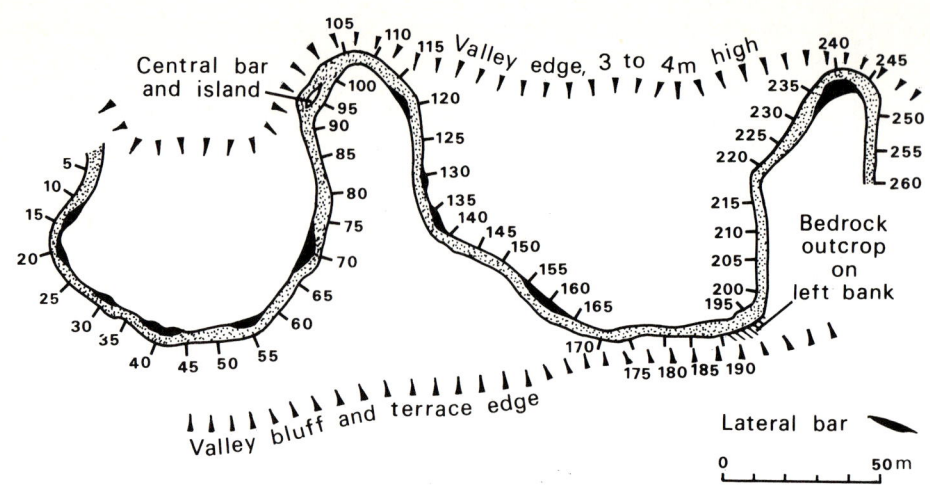

Fig. 5.18. Surveyed reach of the Afon Elan, mid-Wales. (Anderson and Richards 1979)

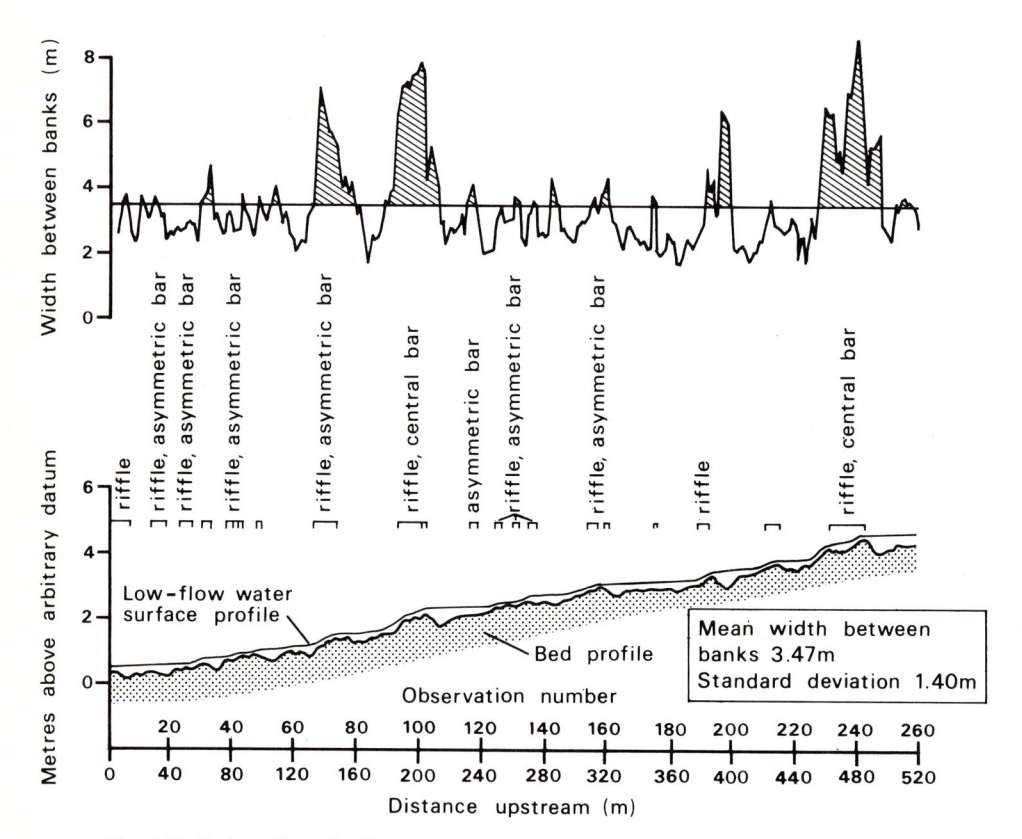

Fig. 5.19. Bed profile and width series of the Afon Elan. (Anderson and Richards 1979)

Transfer Function Models

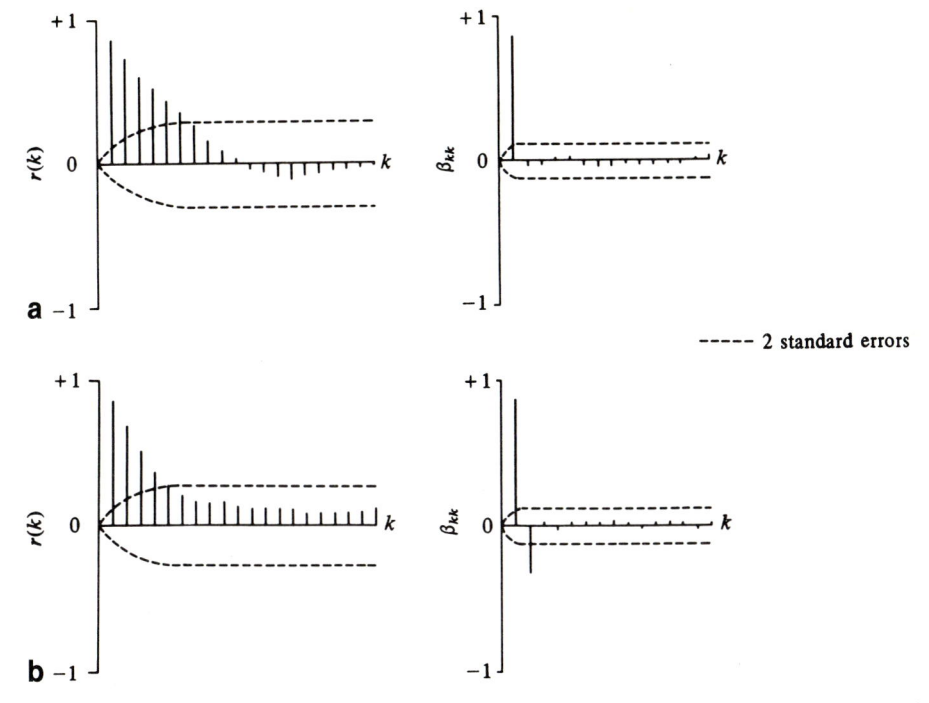

Fig. 5.20a, b. Correlograms and partial autocorrelation functions of **a** width and **b** profile series of the Afon Elan. (Anderson and Richards 1979)

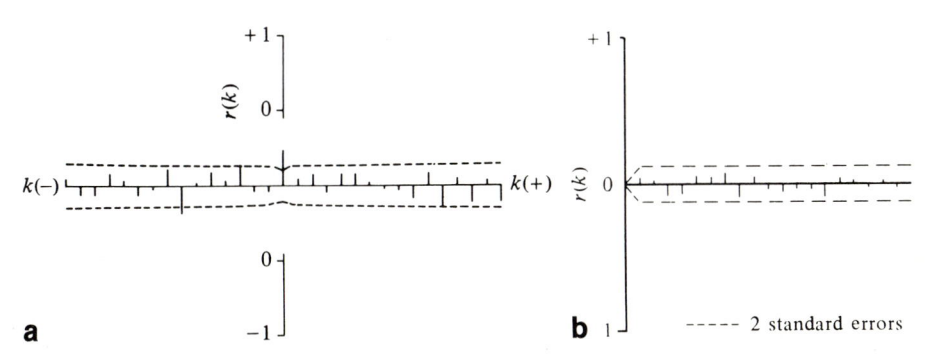

Fig. 5.21. a Cross-correlation function of width and profile series for the Afon Elan. **b** Correlogram of final error series of width variation, Afon Elan. (Anderson and Richards 1979)

was surveyed to provide 260 observations at a spacing of 2 m (Fig. 5.18). It is an underfit stream with large meanders cut into a valley floor between the bluffs of a terrace. Within the mapped reach there are nineteen riffles giving a mean wavelength of 27.4 m and a wavelength-width ratio of 7.89 m, the mean width being 3.47 m. The bed profile and the bank-to-bank widths are plotted on Fig. 5.19. The trends in the profile and width series were removed by a linear regression, though, as Anderson and Richards (1979, p. 222) observe, the residual variance, notably of the width series, may not be constant. The correlograms and partial autocorrelation functions of both series are shown in Fig. 5.20. The slow decline in both series suggests a trend in the residuals from the regression line but Anderson and Richards argue that, since there are only five significant terms in the profile correlogram and six in the width correlogram, and by looking at Fig. 5.21a, there is no justification in using a higher-order polynomial or logarithmic trends. The partial autocorrelation functions show cutoffs after the first and second terms respectively suggesting that first-order and second-order autoregressive models are appropriate in each case. The cross-correlation between the width and profile series displays a single significant term at zero lag (Fig. 5.21a). (The significant cross correlation at a lag of -7 may be ignored because it simply expresses the fact that the negative width residual of a pool is correlated with a positive profile residual (a riffle) seven sample units upstream, the data being analysed from the downstream end.) The noise series, that is the residuals of the width-profile transfer-function model, when cross-correlated with the input series of profile elevation, indicates a lack of cross-correlation and that errors are generated by a white noise process (Fig. 5.21b). A general conclusion of the study, apart from the apparent appropriateness of simple autoregressive models, is that bed profile series tends to lead width series in downstream direction.

5.5 Problems of Inductive Stochastic Modelling

Before leaving inductive stochastic models, it is helpful to make a few cautionary comments about them. First of all, wherever stochastic models are applied to a set of data, certain commonsense modelling guidelines should be followed (Hipel and McLeod 1981a). The prime guideline is that the physical phenomenon being modelled should be soundly understood. The limitations of the stochastic model used to analyse a phenomenon should also be appreciated. For example, for many time series, individual monthly averages may have constant mean values, but the means vary from month to month and the data are, by definition, nonstationary. It might seem appropriate therefore to use a nonstationary model but this must be done cautiously since models which employ differencing operators to remove homogeneous nonstationarity may obscure reality. If the fitted stochastic model does not allow for second-order stationarity within each month, serious problems may arise. Another guideline is that the selected model should be kept as simple as possible. It should, following the principle of parsimony, incorporate the minimum number of parameters necessary to define adequately the data.

A second point of caution is that the inductive transfer-function strategy suffers from a number of limitations, three of which Thornes (1982) regards as severe. The three severe limitations of the strategy are that it reveals little of the nature of system equilibrium and has little to say about the sensitivity of a system to change other than by reference to the amplitude of environmental fluctuations in relation to empirically defined thresholds; that it may in some cases enable a pattern of equilibrium behaviour to be discerned but is of little help in identifying the several different ways in which such patterns may arise; and that it requires near perfect series of data, whereas the data for Earth surface systems are usually short runs of spotty and imperfect information. In addition to these severe limitations, the transfer-function model is bugged by technical problems. For instance, extraneous noise is mixed with the input (R. J. Bennett 1979).

A third point of caution is that the process of "eyeballing" to compare observed and theoretical correlation functions is, in some ways, unsatisfactory. However, improved ways of determining the orders of autoregressive and moving-average parameters are being produced (Hipel et al. 1977; McLeod 1977) as are improved ways of estimating parameters (R. J. Bennett 1979). Lastly, for spatial series at least, there are problems of assessing the autocorrelation. N. J. Cox (1983) has shown that the choice of one of the many available estimators of correlation for a one-dimensional spatial series should be made carefully because variation between estimates for a series which appears only mildly nonstationary may be of the same order as the sampling variation expected under stationarity.

Despite these caveats, it seems likely, given the nature of many Earth surface systems and the ready availability of package programmes, that inductive stochastic modelling will become widely adopted in the near future. Developments will probably involve the study and application of nonlinear models and multivariate models (for example, R. J. Bennett 1979; Lai 1979).

CHAPTER 6
Statistical Models

An outgrowth of the relative frequency view of probability is the development of inductive statistical models. The statistical models used in the Earth sciences are many and varied. They range from simple techniques, such as the chi-square test, Student's *t*-test, and the more versatile analysis of variance, all of which are used in testing the significance of differences between class data; through the more sophisticated techniques of regression and correlation, which are used for establishing relationships between variables; to complex multivariate models, such as multivariate regression, multivariate correlation, canonical correlation, multidimensional scaling, and principal component analysis, which enable relationships to be detected within sets of variables.

The simpler statistical models make regular if inconspicuous appearances in the literature. One of the earliest applications of the chi-square test is in Strahler's (1950) work where he used it to test for normality in his data, a use to which it is commonly put. The chi-square test crops up in a variety of fields including till fabric analysis (P. W. Harrison 1957), coastal change (So 1974), river water quality (Walling and Webb 1975), and soil-slope relationships (Roy et al. 1980). The *t*-test was used in Strahler's (1950) study of maximum valley-side slope angles in the Verdugo Hills, California to see if a difference could be established between slopes which were protected at their base and slopes which were being corraded at their base. Schumm (1956) applied the *t*-test in his classic study of slope development in the Perth Amboy badlands, New Jersey. More recent examples of the *t*-test are found in the work of Mitchell and Willimot (1974) and Mills and Starnes (1983). Analysis of variance was applied to the study of sediments in the middle and late 1940's and soon became widely used by geologists of all kinds (see Krumbein and Graybill 1965). Its adoption in geomorphology is somewhat limited, partly, if Thornes and Ferguson (1981, p. 215) are to be believed, because geomorphological systems do not lend themselves to experimental design, a requisite of analysis of variance tests. Where used however, analysis of variance has proved valuable, as in Melton's (1960) study of the effect of the erosional environment upon slope angle, Kennedy and Melton's (1972) comparison of valley-side slope angles in various situations, and R. B. Bryan's (1974) analysis of soil erodibility.

6.1 Simple Regression and Correlation

Regression is a technique, widely used by Earth surface scientists, that enables a relationship between two variables to be expressed as a mathematical function. The raw data for regression are paired observations of two variables in a system. One is designated the dependent variable, y, and this is thought to be related, usually causally related, to the independent variable, x. The relationship between the dependent and independent variable is easily seen in a scatter-plot of the paired variables x_i, y_i. The assumption in the regression is that x always produces a change in y but not necessarily vice versa. As Haggett (1965, p. 294) remarked, rainfall can be expected to affect wheat yields but wheat yields can hardly be significant in determining rainfall. It is assumed that values of y can be predicted from values of x. In fact, the covariation of x and y, in terms of cause and effect, can be interpreted in several ways (Chorley and Kennedy 1971, p. 24): that x is the cause of y, which conclusion can only be based on a sound understanding of the processes involved; that x and y change in harmony but without any clear cause-and-effect connotation – this is autocorrelation and may arise if x and y are related through a third variable not included in the analysis; and that the relation between x and y has arisen by chance, perhaps owing to some malpractice in the collection of data.

6.1.1 The Regression Line

There are a number of ways in which a regression line can be fitted to sets of data. The most common way is the method of linear least squares. This method is attributed to Gauss, who devised it in measuring the orbit of planets, though other claimants to its inventorship are Legendre and Adrain (Sorenson 1970). In any linear regression, a line of the form $y = a + bx$ is fitted where y is regressed on x. The estimates of the parameters a and b are the regression coefficients. In the method of least squares, the most probable values of the parameters are defined when the sum of squares of the residuals, that is the sum of squares of all departures of y from the regression, is minimized. The sum of squares of the residuals may be written

$$M = \Sigma e_i^2 = \Sigma(y_i - \hat{y}_i)^2 = \Sigma(y_i - a - bx_i)^2,$$

where $M = \Sigma e_i^2$ is the sum of squares of the residuals, y_i is an observed value of y, $\hat{y}_i$ is a predicted value of y, and e_i is the difference between the observed value of y and its predicted value. This expression can be minimized with respect to a and b by taking the partial derivatives of M with respect to a and b and equating to zero:

$$\frac{\partial M}{\partial a} = -2\Sigma(y_i - a - bx_i) = 0$$

$$\frac{\partial M}{\partial b} = -2\Sigma x_i(y_i - a - bx_i) = 0.$$

These expressions can be written as the normal equations

$$\Sigma(y_i - a - bx_i) = 0$$
$$\Sigma x_i(y_i - a - bx_i) = 0,$$

the solution of which in terms of a and b is

$$b = (\Sigma x_i y_i - \Sigma x_i \Sigma y_i/n)/(\Sigma x_i^2 - (\Sigma x_i)^2/n)$$
$$= \frac{\Sigma(x_i - \bar{x})(y_i - \bar{y})}{\Sigma(x_i - \bar{x})^2}$$

$$a = (\Sigma y_i - b\Sigma x_i)/n$$
$$= \bar{y} - b\bar{x}.$$

Because this algorithm involves minimizing Σe^2, the parameters a and b are called the least squares estimates.

A second method of fitting a regression line, the reduced major axis method, was devised by Pearson (1901). This method is appropriate where cause-and-effect relationships and inferences about the goodness-of-fit are not important. Davidson (1977), for example, has used the reduced major axis method to illustrate the different sediment characteristics of upland soils in mid-Wales. The advantage of the least squares method is that inferences about the fit of the regression line can be made.

6.1.2 The Correlation Coefficient

The question of how adequately the data can be described by the regression line may be answered by considering the expression for the total sum of squares corrected for the mean, $\Sigma(y_i - \bar{y})^2$. This expression is made up of two components: the sum of squares of deviations from the regression (also termed the error or residual sum of squares) which is written $\Sigma(y_i - \hat{y}_i)^2$; and the sum of squares due to the regression which is written $\Sigma(\hat{y}_i - \bar{y})^2$. In other words,

$$\Sigma(y_i - \bar{y})^2 = \Sigma(y_i - \hat{y}_i)^2 + \Sigma(\hat{y}_i - \bar{y})^2.$$

The larger the sum of squares due to the regression is in comparison with the residual sum of squares, the better the regression line explains the total sum of squares corrected for the mean. The ratio of the sum of squares due to the regression to the total sum of squares corrected for the mean may be used as a measure of the ability of the regression line to explain variations in the dependent variable (Haan 1977, p. 185). The ratio, denoted by r^2 and called the coefficient of determination, ranges between zero, when the regression explains none of the variation in y, and unity, when the regression explains all the variation in y. It may be written in several ways, including

$$r^2 = b^2 \Sigma(x_i - \bar{x})^2 / \Sigma(y_i - \bar{y})^2$$

or

$$r = b\Sigma(x_i - \bar{x})/\Sigma(y_i - \bar{y}).$$

Simple Regression and Correlation

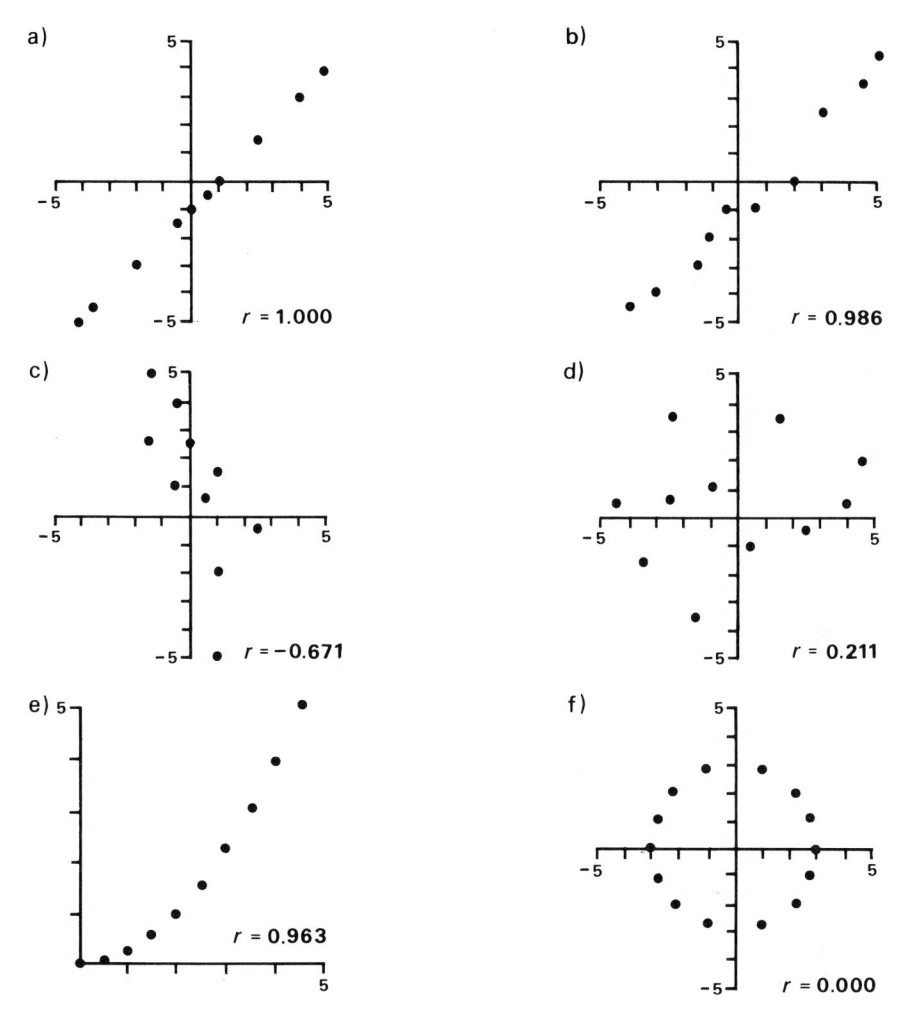

Fig. 6.1a–f. Examples of the coefficient of correlation, r. (Haan 1977)

This last equation may also be written as

$$r = \frac{\Sigma(x_i-\bar{x})(y_i-\bar{y})}{\{\Sigma(x_i-\bar{x})^2 \cdot \Sigma(y_i-\bar{y})^2\}^{1/2}}$$

$$= \frac{\text{covariance } x, y}{\text{variance } x \cdot \text{variance } y}.$$

The value of r, which is a dimensionless ratio called the coefficient of correlation, measures the degree to which the independent variable, x, explains variations in the dependent variable, y. In other words, it measures how good x is as a predictor of y. It is a widely used coefficient in studies of Earth surface systems. A matrix of correlation coefficients showing correlations between all pairs of variables in a system is the basis of a number of multivariate methods. Values of r range from 1.0 for a perfect, direct correlation (y increases as x increases); through 0 for no correlation between x and y; to -1.0 for a perfect inverse correlation (y decreases as x increases) (Fig. 6.1).

6.1.3 Problems of Correlation

Several facets of correlation need further discussion. Firstly, the strength of the correlation is expressed by the value of the correlation coefficient, r, and is the degree to which the data points are coincident with the regression line. However, a high correlation between two variables does not mean that there is necessarily a cause-and-effect relation between them. For instance, the monthly flows of two adjacent small streams may be highly correlated but this does not mean that the monthly flow in one stream causes the monthly flow in the other – more likely both flows are caused by the same external factors operating on both watersheds (Haan 1977, p. 230). Nor does a lack of correlation mean that the variables are in fact unrelated. It may simply indicate that the relation between the variables is not a linear one. Secondly, the direction of the correlation is expressed by the sign attached to r. This sign is always the same as the sign attached to the parameter b in the regression equation. It indicates whether an increase in the magnitude of y is associated with an increase or decrease in the magnitude of x. Thirdly, the sensitivity of the correlation indicates the extent to which a change in the magnitude of one variable leads to a change in the other. In a sensitive relationship, a small change in one variable will produce a large change in the other. Fourthly, and very importantly, it should be stressed that every correlation is based on a number of sampled points and, because of this, the apparent strength and sensitivity of the correlation might have arisen by chance. Even when sampling from uncorrelated populations of variables, it would be rare for the correlation coefficient to be exactly zero. More likely it will deviate from zero owing to chance in sampling. So inferential statistical tests are needed to see whether the deviation of the sample correlation coefficient from zero can, or cannot, be attributed to chance.

Spurious correlation, that is correlation between variables that are in fact uncorrelated, can cloud relationships. It may arise because of clustered data or because ratios, dimensionless terms, or standardized variables have been used (Fig. 6.2). Thornes and Ferguson (1981, p. 287) warn that spurious correlation in geomorphological studies is frequently unheeded or misunderstood, despite good reviews of the problem (Benson 1965). As an example, they note that in hydraulic geometry, discharge is defined as the product of width, depth, and velocity and so cannot fail to have a non-zero, log-log correlation with at least one of those variables.

Correlations should always be given a physical interpretation wherever possible. It is not enough simply to say that y is related to x. The important step is to work out what the relation means in terms of form and process in the landscape. Pitty (1982, p. 53) points out that the interpretation of correlation coefficients, even with the help of inferential statistics, is never easy, sometimes not obvious,

Simple Regression and Correlation

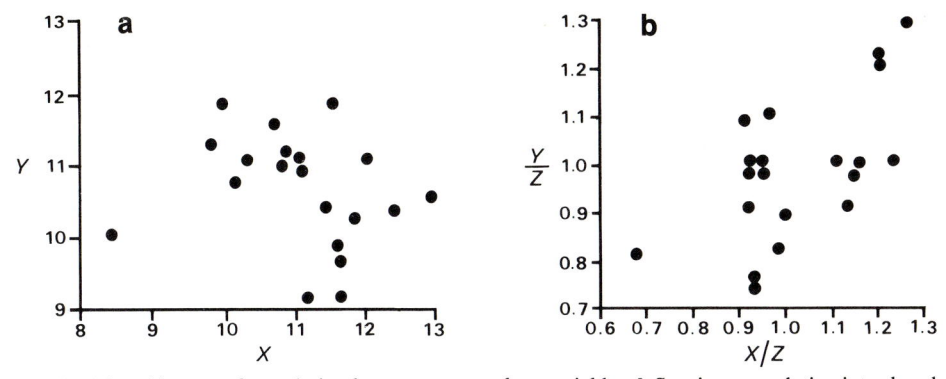

Fig. 6.2. a Absence of correlation between two random variables. **b** Spurious correlation introduced by dividing two random variables by a common third variable. (Haan 1977)

and on occasions produces unexpected results. For instance, in a study of rock creep in Colorado, Tamburi (1974) assumed that rock creep rates would increase at times of higher precipitation. In fact, measured rates of rock creep were higher during dry spells because, Tamburi suggested, under clear skies a greater diurnal temperature than under cloudy skies when rain may fall leads to a greater rate of creep. Problems of this sort lead Pitty to conclude that the complexities of the natural environment make simple and very high correlation unlikely.

6.1.4 Linear Relations

A variety of linear relations have been found among Earth surface phenomena. Jenny (1941, p. 152) fitted a line to data on the clay content (per cent) of soils derived from basic rocks in the eastern United States and temperature (°C):

$$y = -37.4 + 4.49x \quad (r = 0.814; n = 21).$$

In Alaskan soils, Stephens (1969) found that the cation exchange capacity of A_2 horizons (me 100 g^{-1} soil) is linearly related to organic carbon content (per cent):

$$y = 4.90 + 3.45x \quad (r = 0.92; n = 44).$$

In geomorphology, Prior et al. (1971) established a linear relationship between mudflow movement (cm) and rainfall (mm) for sites in northeast Ireland:

$$y = 16.31 + 16.06x \quad (r = 0.71; n = 19).$$

A number of workers have also discovered that linear relations provide a good fit to certain soil properties and various measures of relief. Ruhe and Walker (1968) found a linear relation between grain size indices and slope gradient for the convex slope segments in loess watersheds in Iowa. Likewise, Furley (1968, 1971) found a linear relationship between organic carbon, nitrogen, and pH of the surface soil and slope gradient on both upper and lower slope segments of chalk and calcareous grit in England. Vreeken (1973) too has found that clay and

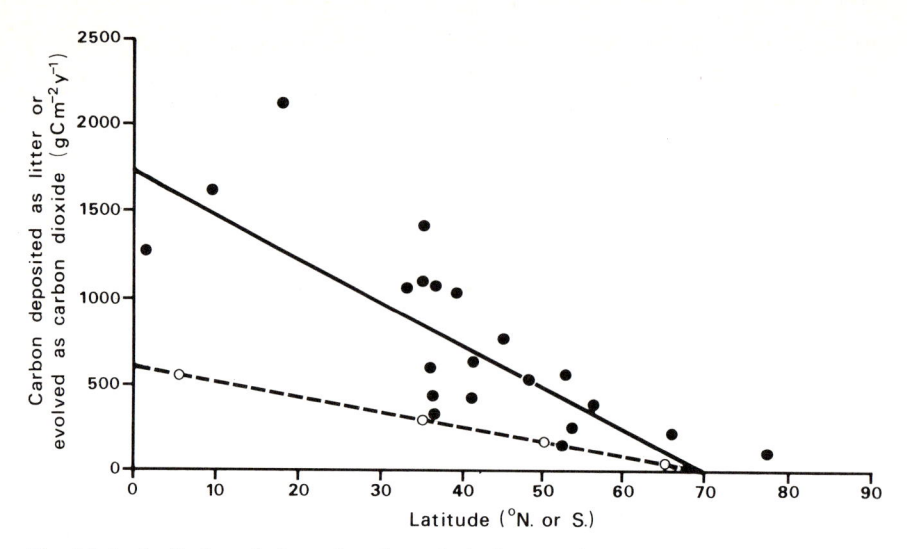

Fig. 6.3. Latitudinal trends for carbon dynamics in forest and woodland soils of the world. *Dashed line* shows the mean annual carbon input to soils in litterfall; *solid line* shows the pattern of carbon loss as CO_2 evolved from soils. See original article for data sources. (Schlesinger 1977)

organic carbon content of soils are linearly related to slope gradient on both upper slopes and backslopes in a loess watershed in Iowa.

Linear relationships have also been discovered on a global scale. Schlesinger (1977) established that the amount of carbon evolved as CO_2 from soils ($g\,cm^{-2}\,yr^{-1}$) is related to latitude north or south in the following way (Fig. 6.3):

$$y = 1721.5 - 24.2x \quad (r = 0.77; n = 40).$$

6.1.5 Linear Versus Nonlinear Relations

Pitty (1982) has noted two problems in seeking linear relations. The first is the problem of thresholds: a relationship between two variables may change suddenly when a threshold is crossed or a new factor emerges at a particular time or beyond a certain distance, leading perhaps to a reversal in the trend of the y values beyond a certain value of x (Pitty 1982, p. 53). The second is the problem of non-monotonic rates of change. For instance, Hickin and Nanson (1975) found that, for a site in eastern British Colombia, the rate of change of channel bend migration with increasing radius/width ratio is low at first, reaches a maximum with a radius/width ratio of around 3.0, and then declines rapidly. The answer to both these problems is simply to recognize a nonlinear relationship when it exists by plotting the scatter of points for every pair of variables. A scattergraph should reveal any discontinuities (Chorley and Kennedy 1971, p. 43) and any nonlinearities within the data. Indeed, it is salutary to recall here the wise words of Reichmann (1961): never accept a regression line unless the data points are also plotted. To this advice might be added: never accept a regression equation unless it is accompanied by a coefficient of correlation or determination and a statement of the range of data. The work of Knighton (1975) is exemplary in this regard.

In general, there is a lack of linearity in nature (Thornes and Ferguson 1981, p. 287). But the problem is not so much recognizing nonlinearity as deciding upon the appropriate nonlinear model to use. The most commonly employed nonlinear functions are logarithmic functions. A semi-logarithmic function takes the general form

$$y = a + b \log x,$$

where the logarithms are either to the base 10 or to the base e. If the logarithms are to the base e the function is called an exponential function. An example of a semi-$\log_{10}$ function is given by Reid et al. (1981) who found that the concentration of chemical species (mg l^{-1}) in the river Dye, northeastern Scotland is related to river discharge (m^3 s^{-1}) in the following manner:

$$y = 2.86 - 2.53 \log_{10} x.$$

(Solute-discharge relationships are normally described by taking logarithms of concentration and discharge but Reid et al. (1981) found that the semi-logarithmic form fitted the data as well as a double-log form.) Soil nitrogen seems to relate to mean annual temperature in an exponential manner as has been found for soils developed on igneous rocks in California (Harradine and Jenny 1958) and for equatorial soils in Colombia (Jenny et al. 1948). A semi-logarithmic function also seems to fit the long profile of rivers. The equation for the river Bollin, in Cheshire, England (Fig. 6.4) is (Knighton 1975):

$$y = 260 - 65 \log_e x.$$

Double-log functions take the general form

$$\log y = \log a + b \log x$$

and relate the log of the dependent variable to the log of the independent variable. This type of equation is commonly written as a power function, the general case being

$$y = a x^b.$$

The area of alluvial fans is related to the area of their source basins in this way (Fig. 6.5). Many relationships in hydrology are well described by log-log functions. A classic example is the relationship between width, depth, and velocity of a river and discharge. The equations for Brandywine Creek, Pennsylvania are

$$w = a Q^{0.04}$$
$$d = c Q^{0.41}$$
$$v = k Q^{0.55}.$$

Other hydrological relationships may be described by log-log functions. In a study of the hydraulic geometry of river channels of South Island, New Zealand,

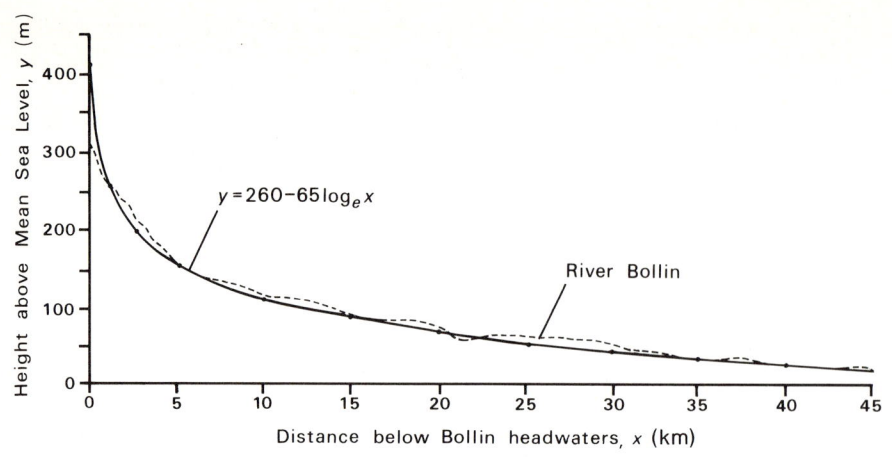

Fig. 6.4. Longitudinal profile of the River Bollin, Cheshire, in relation to profile derived from integration of slope-length relation. (Knighton 1975)

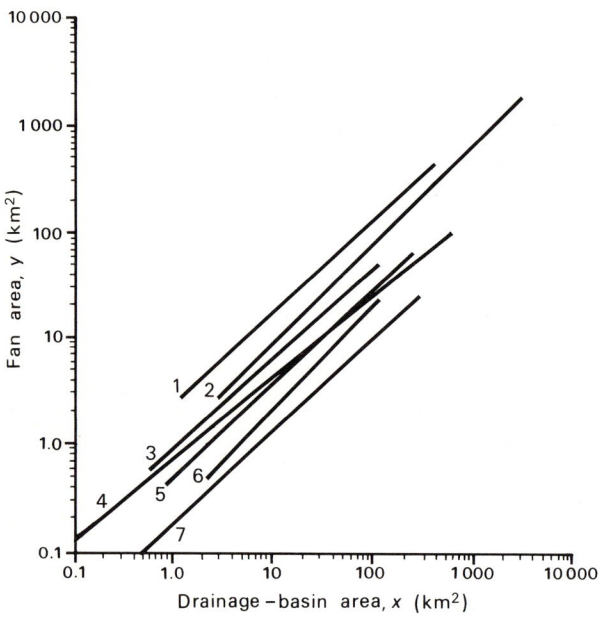

Fig. 6.5. Relations of fan area to drainage basin area for groups of fans in California and Nevada. The equations and sources of data are as follows: $(1,2) y = 2.1 x^{0.91}$, $y = 0.96 x^{0.98}$, least-squares revisions from Bull (1962); $(3) y = 0.7 x^{0.98}$, Hawley and Wilson (1965); $(4) y = 0.5 x^{0.8}$, depositional parts of fans, Denny (1965); $(5,6,7) y = 0.42 x^{0.94}$, $y = 0.24 x^{1.01}$, $y = 0.15 x^{0.90}$, Hooke (1968). (After Hooke 1968 and Bull 1977)

Simple Regression and Correlation

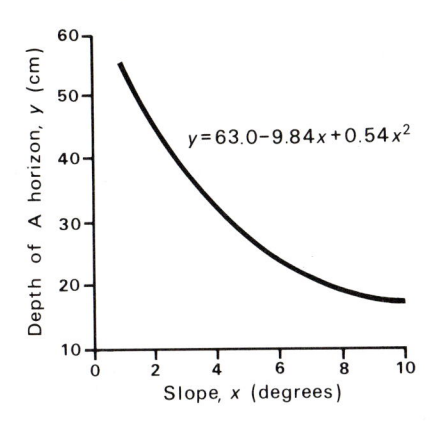

Fig. 6.6. The quadratic relation between the thickness of A horizons and slope angle in loess soils near Kraków, Poland. (Koreleski 1975)

$$y = 63.0 - 9.84x + 0.54x^2$$

Mosley (1981) found the following relationship between dominant meander wavelength and mean annual flood:

$$y = 46x^{0.51} \quad (r = 0.86;\ n = 53).$$

The relationship between concentration of dissolved chemical species in a river and river discharge is most commonly described by a power function. Edwards (1973), for instance, has found that a power function described a variety of solute-discharge relationships in catchments in Norfolk, England. Another case where power functions have been employed is in the study of karst landscapes. A number of workers have found that the logarithm of karst depression width is related to the logarithm of karst depression length. On the western Highland Rim, Tennessee (Kemmerly 1976) the relationship is

$$y = 1.656 x^{1.024} \quad (r = 0.741;\ n = 471);$$

on the eastern Highland Rim (Mills and Starnes 1983) the relationship is

$$y = 2.153 x^{0.944} \quad (r = 0.862;\ n = 127);$$

and for major dolines in parts of Iowa mantled by glacial drift and loess the relationship is (Palmquist 1979)

$$y = 2.85 x^{0.85} \quad (r = 0.30;\ n = 308).$$

Some data are best described by a polynomial function which takes the general form

$$y = a + bx + cx^2 + \ldots + nx^{n-1}.$$

Many relationships between soil properties and measures of relief seem to follow a polynomial form. For instance, Koreleski (1975), working on loess soils near Kraków, Poland found that the thickness of A horizons (cm), is related to slope angle (degrees) by the following quadratic equation (Fig. 6.6):

$$y = 63.0 - 9.84x + 0.54x^2.$$

In a closed bog watershed on drift in Iowa, Walker (1966) found that polynomial functions described the relationships of grain size indices and surficial sediment thickness to distance from summit while Walker and Ruhe (1968) found that

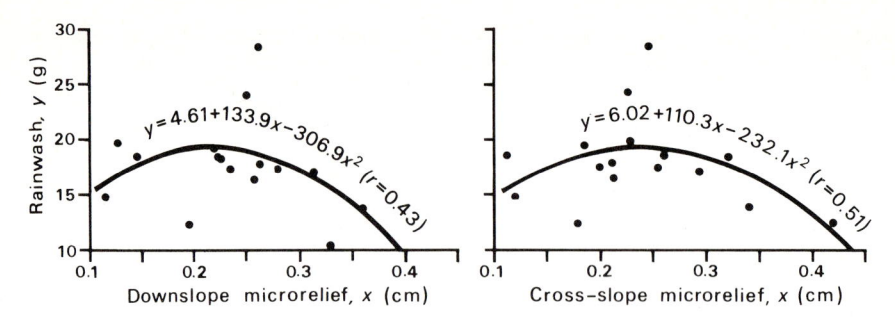

Fig. 6.7. Relationships between rainwash and microrelief indices for the King clay loam (steep phase). (Luk 1982)

organic matter content and depth to carbonates are also related to distance from summit in a polynomial way. Similar relationships have been revealed by Kleiss (1970) working in surficial sediment on till in Iowa. Other polynomial functions have been established between soil bulk density and soil water content (Camp and Gill 1969). Luk (1982) correlated rainwash (g) with downslope microrelief (cm) and, separately, with cross-slope microrelief (cm). He found that in both cases a quadratic relation provided a better fit than a linear equation. The equations were

$$y = 4.61 + 133.9x - 306.9x^2 \quad (r = 0.43; n = 16)$$
$$y = 6.02 + 110.3x - 232.1x^2 \quad (r = 0.51; n = 16)$$

for downslope microrelief and cross-slope microrelief respectively (Fig. 6.7).

6.2 Multiple Regression

The rest of the statistical models considered in this chapter deal with more than two variables. They are therefore multivariate models. The theory of multivariate statistics was begun at the turn of the century and developed during the 1920's, particularly by R. A. Fisher, who devised methods for studying the simultaneous operation of several variables in agricultural surveys. The calculations involved in multivariate methods are enormously time-consuming if carried out by hand. So, until the electronic computer became generally available, multivariate models considered just a few variables (for example Kendall 1939). Highspeed computers have allowed conventional problems to be attacked more vigorously and have speeded up what might otherwise be impossible or unimagined (Chorley 1967b). This is certainly true of multivariate models which, during the 1960's, found their way into all branches of Earth Science. The models dealt with here are multiple regression and the various techniques used to detect patterns in correlation systems – principal component analysis, factor analysis, and canonical correlation.

In most systems at the Earth's surface it is possible to identify one variable which occupies an important position in terms of the direction of causation and

Multiple Regression

it is important to see how far changes in this dependent variable are conditioned by the interaction of the independent variables. Thus on a beach in Virginia, Harrison and Krumbein (1964) selected near-shore bottom slope as the dependent variable y and defined six independent, or quasi-independent, variables x_1 to x_6 (wave height, wave period, angle of wave approach, water depth, shore currents, mean grain size). The kinds of question posed in this case are: How many of the independent variables account for variations in the dependent variable? Are the independent variables interrelated and how do their relevant interactions operate? What is the relative importance of the independent variables? Does the importance of an independent variable vary from one situation to another? (Chorley and Kennedy 1971, p. 37). Questions such as these can be tackled by multiple regression and correlation analyses.

In cases where more than one independent variable is thought to influence a dependent variable, the simple linear regression model can be extended to include several independent variables. In the general case, the model is written

$$y = b_1 x_1 + b_2 x_2 + \ldots + b_k x_k,$$

where y is the dependent variable, the x's are the independent variables, and the b's are the parameters which define the rate of change of y with the independent variables. In practice, n observations would be available on the dependent variable, y, along with n corresponding observations on each of the k independent variables. An equation could be written for each set of the n observations, the full set of equations being written as

$$y_1 = b_1 x_{11} + b_2 x_{12} + b_3 x_{13} + \ldots + b_k x_{1,k}$$
$$y_2 = b_1 x_{21} + b_2 x_{22} + b_3 x_{23} + \ldots + b_k x_{2,k}$$
$$\vdots$$
$$y_n = b_1 x_{n,1} + b_2 x_{n,2} + b_3 x_{n,3} + \ldots + b_k x_{n,k}.$$

The method of solution is to solve the n equations for the k unknown parameters (see Krumbein and Graybill 1965, pp. 283–295, and Haan 1977, pp. 197–203 for details). To solve the equations, the number of observations, n, must equal or exceed the number of independent variables, k. In practice, it is recommended by Haan (1977, p. 198) that n should be three of four times larger than k. Because of this, the sampling design can be gigantic and severely limit the application of the model if too many independent variables are included. Another limitation of the model is that nonsensical results are likely to arise if an independent variable is a perfect linear function of another independent variable or if an independent variable is linearly dependent on any linear function of the other independent variables (Haan 1977, p. 199; Draper and Smith 1966).

6.2.1 "Simple" Multiple Regression

There is a difference between multiple regression between a dependent variable and two or a very few independent variables and multiple regression between a dependent variable and a large set of independent variables. The first case is illustrated by a variety of studies of Earth surface phenomena. Salter and Williams (1967) found a relationship between available water capacity (in ft^{-1}), y, and two independent variables – per cent coarse sand (2 – 0.2 mm), x_1, and per cent silt (0.05 – 0.002 mm), x_2:

$$y = 1.86 - 0.011 x_1 + 0.018 x_2.$$

Studies of soil-slope relationships have produced a number of "simple" multiple regression relationships. Ahnert (1970), working on weathered gneiss in North Carolina, related solum thickness to distance from summit and slope gradient. In drift and loess landscapes in Iowa, Walker et al. (1968b) found depth of A horizon mottling was related to slope gradient, elevation, and aspect while in the Thoms watershed, Iowa, Vreeken (1975) found that depth to carbonates was related to slope gradient and the azimuth of slope direction.

6.2.2 Trend Surface Analysis

Trend surface analysis is, in effect, a special form of multiple regression in which a dependent variable is regressed against two independent variables, usually Cartesian coordinates x and y. Denoting the dependent variable by h, the general expression for a trend surface is

$$h = f(x, y) + e_i,$$

where $f(x, y)$ is the trend surface, equivalent to a regression line in the one-dimensional case, and e_i are the residuals of the data from the trend. The function $f(x, y)$ can take a great variety of forms but polynomial and Fourier functions have been found useful. The lower-order polynomial functions, linear and quadratic, have been most widely employed. The linear trend surface is decribed by the equation

$$h = a_0 + a_1 x + a_2 y + e,$$

whereas the quadratic trend surface is defined by the equation

$$h = b_0 + b_1 x + b_2 y + b_3 x^2 + b_4 xy + b_5 y + e.$$

The addition of further terms produces cubic, quartic, quintic surfaces and so forth but these have not been widely used. A double Fourier series can be used to build up complex harmonic surfaces.

Bassett and Chorley (1971) used trend surface analysis to identify components of the erosional terrain in a region from the Emporium Quadrangle, Pennsylvania (Fig. 6.8). They sought to see if the technique could sort out the regional trend of elevations, probably of geological origin, from the superimposed, branching hierarchy of erosional drainage basins. Elevations in the test region were measured at the intersection of a 15 × 15 regular grid, providing a total of 225 elevations. Polynomial trend surfaces up to order eight were fitted to the data and residuals from the surfaces were mapped (Fig. 6.9). The simple linear surface (Fig. 6.9a) has a strike which closely reflects the regional geological strike and, more interestingly, it has a low dip which is appropriate to a location

Multiple Regression

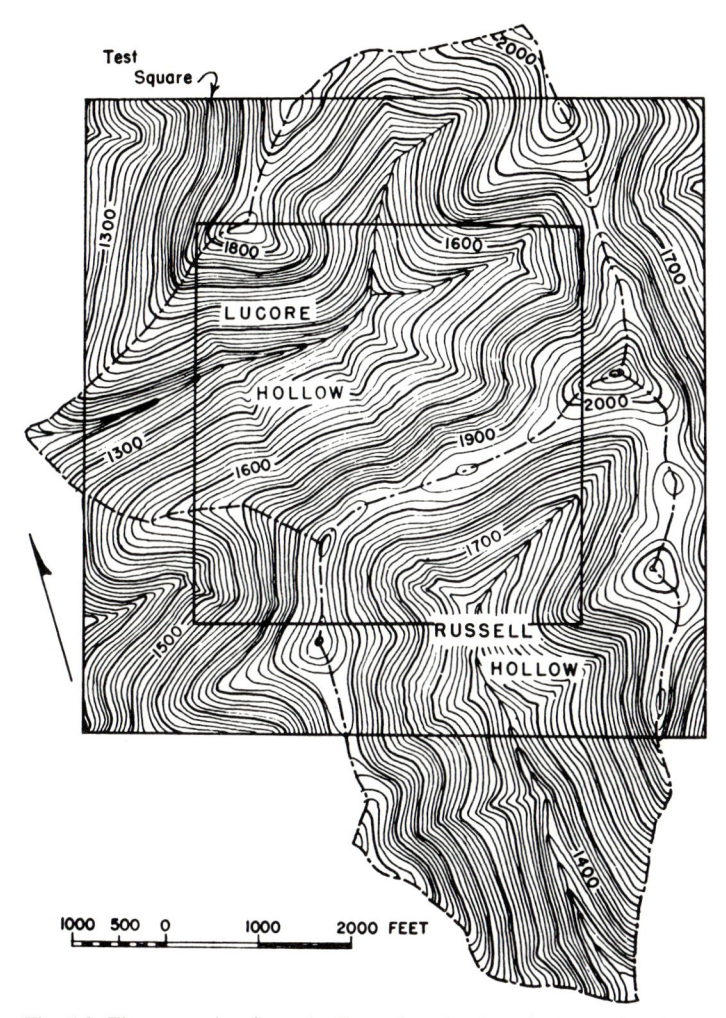

Fig. 6.8. The test region from the Emporium Quadrangle, Pennsylvania. (Strahler 1956)

near a synclinal axis. Because of the high unexplained variance and the undoubted influence of the Lucore Hollow (Fig. 6.8) on the dip direction, the surface cannot be explained unambiguously in terms of the regional dip component. However, the close correspondence between the map residuals from the linear surface (Fig. 6.9a) and the erosional terrain supports the view that the linear surface has picked out a significant regional trend. The successively higher order polynomials describe increasingly complex surfaces which provide better fits to the elevation data. However, two third-order drainage basins are not significantly described by the fitted surfaces until a polynomial of the fifth order is applied (Fig. 6.9b) and second-order drainage basins are not clearly represented until a polynomial of the seventh order is used (Fig. 6.9c). The eighth order polynomial provides only some explanation of the details of topography attributable

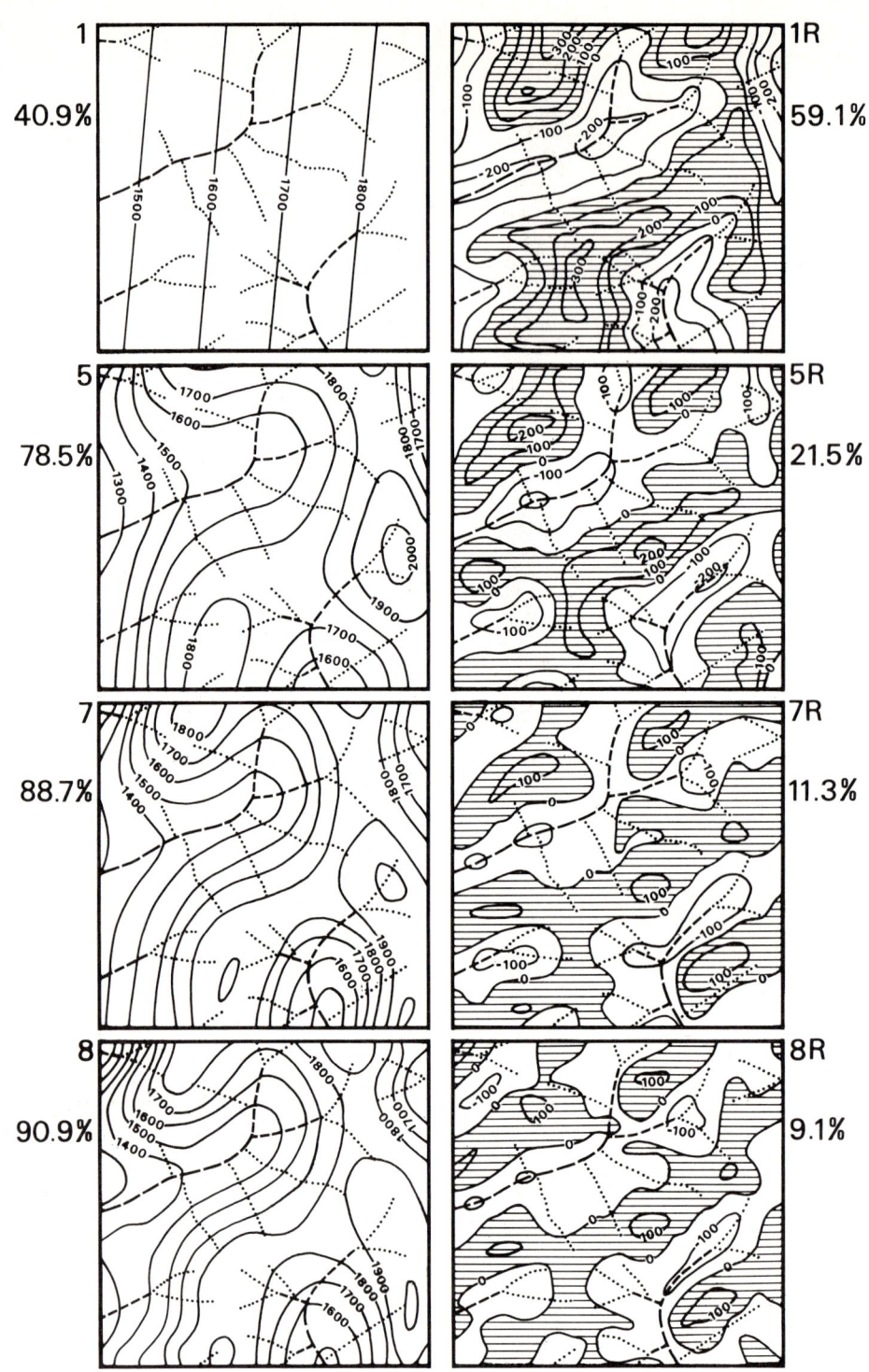

Fig. 6.9. Polynomial trend surfaces (*left*) and the associated residuals (*right*) for first, fifth, seventh, and eighth orders for the test region shown in Fig. 6.8. (Bassett and Chorley 1971)

to first-order drainage basins and at this level of polynomial 9.1 per cent of the variation still remains unaccounted for (Fig. 6.9d).

When first applied to Earth surface systems, trend surface analysis held out much promise as a tool for establishing patterns in spatial data. Examples of usage include soil texture (Chorley et al. 1966), the relationship between elevation and lithology (Chorley 1969b), topography (Thornes and Jones 1969), and corrie elevations (Robinson et al. 1971), as well as Bassett and Chorley (1971). The technique has proved to be a damp squib. There are a number of reasons for this. For one thing, surfaces can be fitted to just one variable at a time and to compare trends between different variables multiple correlation techniques must be employed which introduce a deal of error since few trend surfaces have a 100 per cent fit (Chorley and Kennedy 1971, p. 155). However, an algorithm for quantitative comparison of polynomial trend surfaces has been developed by Goodman (1983). The nature of the fitted surface, whether it should be cubic or whatever, is made by an arbitrary decision which may lack geomorphological significance (Thornes and Ferguson 1981, p. 286). An exception to this drawback is the application to raised beaches (Andrews 1970; J. M. Gray 1978), where the fitting of a quadratic surface seems to make good sense in terms of glacio-eustatic rebound. Pitty (1982, p. 54) is rather scathing about the value of using mathematical functions as descriptors of landforms. He argues that the complexity of a series of hills and valleys is so great that it is not possible to describe them, even approximately, with an equation; and that even if a surface can be fitted, it is difficult to assign meaning to the coefficients, and items of real interest, such as slope gradient, are not readily obtained from them. The charges against using trend surfaces to describe landforms can be answered by pointing out that the approximate form of landforms can be described by equations. Indeed, double Fourier series can, if there are as many harmonic terms included as there are data points, give a 100 per cent fit. Moreover, gradients, curvature and so on can be readily derived for all points on a mathematical surface (see Unwin 1981, p. 169).

Today, trend surfaces are sometimes used purely as a means of smoothing data. E. Bryant (1983), for instance, in an attempt to ascertain any trends towards erosion or accretion on Stanwell Park Beach, New South Wales, Australia, smooths data on mean high tide run-up position, collected at 50-m intervals between 1890 and 1980, with a cubic trend surface.

6.2.3 Stepwise Regression

Where the set of independent variables in a multiple regression is large, analysis usually involves sorting out the usefulness of the independent variables in terms of their value as predictors of the dependent variables. The procedure for doing this is known as stepwise regression and entails evaluating various contributions of the independent variables and the degree to which they explain the dependent variable. An excellent example of the technique is Harrison and Krumbein's (1964) work on a beach in Virginia (this is reported also in Krumbein and Graybill 1965). Other examples include Mosley's (1981) study of the hydraulic geometry or rivers in South Island, New Zealand. The best-fit equations for the dependent variable are shown in Table 6.1. Another example is Arnett's (1974)

Table 6.1. Best-fit equations for the dependent variables in Mosley's analysis of the hydraulic geometry of river channels in South Island, New Zealand. (After Mosley 1981)

Equation	Sample size n	Coefficient of determination R^2	Standard error of estimate
$XS = 0.41\ Q_{maf}^{0.86}\ d^{-0.299}\ SILT^{-0.162}$	72	0.88	0.25
$WD = 0.028\ POW^{-0.576}\ d^{-0.982}\ VI^{0.749}\ SIG^{-0.973}\ SILT^{-0.200}$	65	0.66	0.38
AR = no equations with coefficient significantly different from zero			
$SP = 0.071\ Q_{maf}^{-0.275}\ d_{50}^{0.550}\ SIG^{-0.491}\ Q_{mfbr}^{0.459}\ Q_r^{0.607}$	65	0.75	0.19
$SN = 0.304\ POW^{-0.06}$	61	0.30	0.06
$BI = 0.198\ POW^{0.159}\ d^{-0.133}\ VI^{-0.245}$	65	0.34	0.18
$RN = 0.28\ d_{50}^{-0.892}\ Q_{maf}^{0.249}\ VI^{0.420}\ SIG^{-0.322}$	65	0.78	0.19
$WL = 46\ Q_{maf}^{0.51}$	53	0.75	

Variables: XS = cross-sectional area (m^2); Q_{maf} = mean annual flood (m^3 s^{-1}); d = mean diameter of bed material (m); $SILT$ = silt-clay percentage, bank sediment; WD = width-depth ratio; POW = stream power (N s^{-1}); VI = variability index, flow; SIG = standard deviation of bed sediment (phi units); AR = aspect ratio; SP = slope of reach; Q_{mfbr} = ratio of Q_{maf} to mean discharge; Q_r = range of instantaneous flow; SN = sinuosity; BI = braiding index; d_{50} = mean diameter of bed sediment (m); RN = relative roughness; WL = meander wavelength (m).

Note: Mosley also gives standard errors of regression coefficients but, for clarity, they have been omitted here.

Table 6.2. Stepwise multiple correlation matrix between interflow volumes and the temporal independent variables, Caydale, Yorkshire. (Arnett 1974)

Site	1	2	3	4	5	6	7	8	9	10	11	N
1a	—	—	—	—	—	—	—	—	0.73[2]	0.67[1]	—	30
1b	—	—	—	—	—	—	—	—	0.66[2]	0.59[1]	—	23
1c	—	0.77[1]	—	—	—	—	—	—	—	—	—	9
1d	—	—	—	—	—	—	—	—	0.73[1]	0.84[2]	—	16
1e	0.77[2]	—	—	—	—	—	—	—	0.83[3]	0.67[2]	—	27
1f	0.52[1]	0.83[2]	—	—	—	—	—	—	0.90[3]	—	—	17
1g	—	0.91[1]	—	0.95[2]	—	—	—	—	—	—	—	14
1h	—	—	—	—	0.80[1]	—	—	—	—	—	—	21
2a	—	0.83[2]	0.74[1]	—	—	—	—	—	—	—	—	26
2b	—	—	0.71[1]	—	—	—	—	—	—	—	—	36
2c	—	—	0.76[1]	—	—	—	—	—	—	—	—	31
2d	0.80[3]	0.63[2]	0.51[1]	—	—	—	—	—	—	—	—	30
2e	—	—	—	—	—	—	—	—	—	—	—	12
2f	—	0.77[2]	0.74[1]	—	—	—	—	—	—	—	—	37

Variables: 1 = topsoil permeability; 2 = difference in permeability; 3 = total antecedent weekly rainfall; 4 = rainfall over 0.5 mm h^{-1}; 5 = rainfall over 1.0 mm h^{-1}; 6 = rainfall over 1.5 mm h^{-1}; 7 = rainfall over 2.0 mm h^{-1}; 8 = rainfall less than 0.5 mm h^{-1}; 9 = rainfall 0.5–1.0 mm h^{-1}; 10 = rainfall 1.0–1.5 mm h^{-1}; 11 = rainfall 1.5–2.0 mm h^{-1}; N = number of weekly observations in each regression.

Note: The correlation coefficients listed are those from a stepwise multiple regression analysis. r^1 denotes the correlation coefficient between the dependent variable and the most significant independent variable. R^2 and R^3 represent the multiple correlation coefficients when a second and third independent variable respectively are added to the regression set. Only correlation coefficients significant at the 0.05 probability level are shown.

Table 6.3. Stepwise multiple correlation matrix between interflow and spatial independent variables, Caydale, Yorkshire. (Arnett 1976)

Week	1	2	3	4	5	N
1	–	0.85	–	–	–	10
2	–	0.80	–	–	0.93	11
3	–	0.64	–	–	–	12
4	0.62	0.79	0.88	–	–	12
5	–	–	–	–	–	13
6	–	0.88	–	–	–	12
7	–	0.77	–	–	–	7
8	–	0.84	–	–	–	13
9	–	0.73	–	–	–	12
11	–	0.74	–	–	–	12
12	–	0.81	–	–	0.91	11
13	–	0.69	–	–	–	11
14	–	0.83	–	–	–	11
16	–	0.80	0.94	–	–	12
17	–	–	0.71	–	–	11
19	–	–	–	–	–	9
20	–	–	–	–	–	7
21	–	0.91	–	–	–	9
22	–	0.92	–	0.96	–	10
23	–	0.92	–	–	–	11
24	–	0.71	–	–	–	10
35	–	0.84	–	–	–	11
36	–	0.88	–	–	–	9
37	–	0.95	–	–	–	5
38	–	0.85	–	–	–	12
39	–	0.74	–	–	0.85	11
41	–	–	–	–	–	4
42	–	–	–	–	0.96	12
43	–	–	–	–	–	9
44	–	0.93	–	0.98	–	8
45	–	0.89	–	–	–	9
46	–	–	–	–	–	6
51	–	–	–	–	–	4

Variables: 1 = topsoil permeability; 2 = difference in permeability; 3 = slope angle of site; 4 = surface roughness of site; 5 = distance of site from basal stream; N = the number of sites involved in the regression.

Note: see note for Table 6.2

study of environmental factors affecting the speed and volume of topsoil interflow in Caydale, a small moorland catchment in the Hambleton Hills, northeast Yorkshire. Arnett correlates interflow volumes against temporal independent variables (Table 6.2) and against spatial independent variables (Table 6.3). The results in Table 6.2 show differences between the two plots studied in the catchment (Fig. 6.10). The first plot is dominated by pasture grasses while the second plot is dominated by bracken (*Pteridium aquilinium*) and has greater topsoil permeabilities resulting from a multitude of lateral channels produced, and later

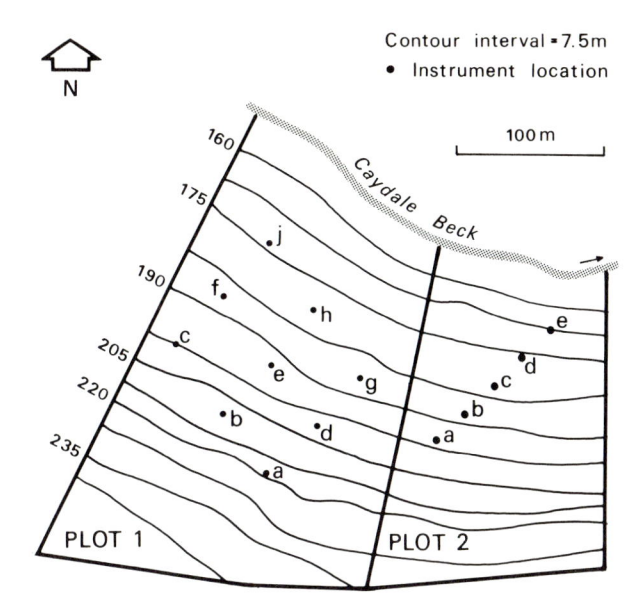

Fig. 6.10. The location of the field sites in Caydale. (Arnett 1974)

abandoned by, colonizing rhizomes of the bracken. Notice that in plot 2, all sites except site 2e show total antecedent weekly rainfall to be a significant independent variable whereas it is not significant in any of the sites in plot 1. Conversely, at many sites in plot 1 rainfall intensities between $0.5-1.0$ mm h^{-1} and between $1.0-1.5$ mm h^{-1} are significant independent variables whereas they are not significant at any site in plot 2. The results in Table 6.3 show the factors affecting interflow volumes for any given week. The importance of "difference in permeability" (that is, the difference between the permeability of the topsoil and subsoil) is revealed, all other independent variables being significant in just a few of the weeks studied.

6.2.4 Problems of Multiple Regression

Haan (1977, p. 218) believes regression analysis should simply be regarded as a tool for exploiting linear tendencies that may exist between a dependent variable and a set of independent variables. He stresses that any regression analysis should be preceded by a great deal of thought about what variables should be included in the analysis; how these variables might influence the dependent variable; the correlations among the dependent variables; and, if practical applications are an important consideration, the ease of using a predictive model based on the selected independent variables. Other points to bear in mind when using regression analysis or reading papers reporting the results of a regression analysis are as follows. In many multiple-correlation systems of Earth surface phenomena there is considerable replication between independent variables. This replication may lead to the chance omission from the analysis of the variable which is genuinely causing the observed variation in the dependent variable and instead give a reasonable explanation of the variation of the dependent variable in terms of other "independent" factors which, like the dependent variable, are under the control of the excluded variable (Chorley and Kennedy 1971, p. 38). Some regression models can be improved upon by including cross-product terms by multiplying two independent variables together to form a new variable or by including ratios and powers of the independent variables as new variables (Haan 1977, p. 218). If these practices are used, however, extreme care is needed to ensure that large correlations between the independent variables do not result and invalidate the regression analysis. Lastly, although Chorley (1967b) is right in saying that high-speed computers have allowed conventional problems to be attacked more vigorously and have speeded up what might otherwise be impossible or unimagined, it is salutary to bear in mind that the ready availability of computers and canned regression programs has led to much thoughtless collection of data and number-crunching in the hope that a model might emerge. As Terjung writes:

"It is tempting to plug data into programs without understanding or considering the deeper implications, assumptions, and meanings of a particular statistical routine. All too often, instead of freeing the scholar from number drudgery so that more time can be devoted to creative thoughts, the computer is used only to crank out more studies of the same kind". (Terjung 1976, p. 205)

6.3 Correlation Systems

Another way of seeking patterns of relationships between large numbers of variables is to calculate correlation coefficients between all pairs of variables and so draw up a correlation coefficient matrix or correlation system. A correlation coefficient matrix may be subjected to a range of statistical techniques, all of which are designed to detect any pattern which may be present. At the simplest level, the signs of the correlation coefficients are used to identify negative and positive feedback loops. An example of a negative feedback loop is shown in Fig. 6.11. An increase in slope erosion causes an increase in the bed load of streams which decreases the erosion by the river channel leading to a lessening of valley-side slopes and thus a reduction in slope erosion and hence decrease in stream bed load. An example of a positive feedback loop is shown in Fig. 6.12. A decrease in infiltration capacity causes an increase in surface runoff, which causes an increase in slope erosion, leading in turn to the removal of the more permeable upper layers of the soil and a further decrease in infiltration capacity.

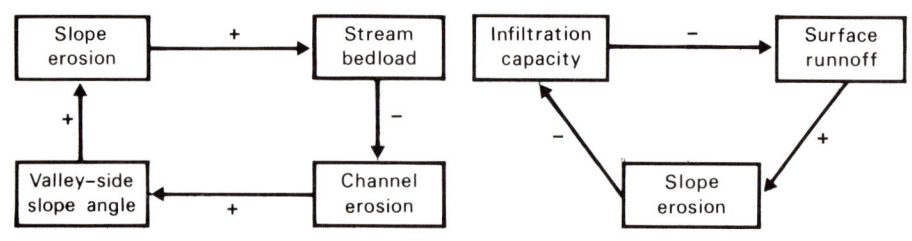

Fig. 6.11. A system exhibiting negative feedback relationships

Fig. 6.12. A system exhibiting positive feedback relationships

Other examples of loops are given in Melton (1958b) and Chorley and Kennedy (1971).

More sophisticated forms of analysis of correlation systems involve algorithms for revealing the so-called causal structure of the matrix. Methods available for doing this include McQuitty's (1957) linkage analysis and Blalock's (1964) causal analysis. None of these methods has proved popular with Earth scientists, though they have been tried by Melton (1958a, b) and Ferguson (1973). Techniques for revealing patterns among sets of intercorrelated variables which have proved popular with Earth Scientists include principal component analysis, principal coordinate analysis, factor analysis, and canonical correlation.

6.3.1 Principal Component Analysis

Principal component analysis was originated by Karl Pearson around the turn of the century and elaborated by Harold Hotelling and others during the 1930's. It is a technique which enables the relationships between a set of p correlated variables to be examined. It does so by transforming the original set of variables into a new set of uncorrelated variables known as principal components. The new variables are linear combinations of the original variables and are derived in decreasing order of importance so that the first principle component accounts for as much of the variation in the original data as possible, the second principal component accounts for as much of the remaining variation in the original data, and so on (Chatfield and Collins 1980, p. 57). The transformation of the original variables involves an orthogonal rotation in p-dimensional space.

There are seven or so steps in the procedure for principal component analysis (Chatfield and Collins 1980, p. 79). The first step is to decide upon the worth of including all the variables recorded in the original data matrix and to decide whether any of the variables need transforming. The second step is to compute the correlation (or the variance-covariance) matrix, bearing in mind that a correlation coefficient should not, in general, be computed for a pair of variables that are related in a nonlinear fashion. The third step is to study the correlation matrix to see if there are any natural groupings of highly correlated variables. If all correlations are small then there is little point in going on with the analysis. The fourth step is to calculate the eigenvalues and eigenvectors of the correlation (or variance-covariance) matrix. The fifth step is to examine the eigenvalues and try to decide how many of them are "large". This should indicate the effective dimensionality of the data. However, because principal component analysis is not based on an underlying statistical model, the sampling behaviour of the eigenvectors and eigenvalues is unknown. A consequence of this is that there is no way of knowing how many of the eigenvalues are "large". A rule of thumb, commonly used, is that all eigenvalues less than unity should be disregarded but this rule has no theoretical justification. A better practice is to seek a "natural" break point within the run of eigenvalues; this can be done by applying the "scree test" (Cattell 1965). Neither of these methods is very satisfactory, however, and the lack of an objective rule for deciding how many components to retain is a serious handicap of the method.

The sixth step is to look at the grouping of variables suggested by the components and to consider whether the components can be interpreted in some meaningful way. Chatfield and Collins (1980, p. 71) caution that the interpretation is difficult and does not always receive the attention it warrants. On the other hand, both they and Kendall (1975, p. 23) take pains to warn of the danger of reading too much meaning into components. There are a number of reasons for this. Firstly, no assumptions are made about the probability distribution of the original variables, although if the observations are assumed to be multivariate normal more meaning to the components can generally be given. Secondly, the components depend upon the scales used to measure the variables. If one variable has a much larger variance than all the other variables, then this variable will dominate the first principle component whatever the correlation structure (Chatfield and Collins 1980, p. 70). If the variables are scaled to have unit variance then the first principal component will be different. The fact is, therefore, that there is little point in carrying out a principal component analysis unless all the variables have "roughly similar" variances. The scaling problem is avoided, though not solved, by using the correlation coefficient matrix in place of the variance-covariance matrix.

The final step in principal component analysis is to use the component scores in subsequent analyses as a way of reducing the dimensionality of the problem. This step can involve a variety of techniques. At a simple level, Gonzales and Winz (1977, p. 313) used principle component analysis to reduce the four channels of data obtained from Landsat imagery to two uncorrelated data sets without any significant loss of information. Anderson and Cox (1978) used it to compare methods of measuring soil creep in the field. More sophisticated techniques involve the use of the components in multiple regression analyses (for example Haan 1977, p. 254).

An example of the use of principal component analysis is the research of Anderson and Furley (1975) who, working on catenas in the Berkshire and Wiltshire chalk downs, England, sought to establish the interrelationships amongst selected surface soil properties and topographical measures (slope gradient and slope length). Analysis revealed a consistent pattern in the distribution of soil properties over five slope transects. The first component of the pattern, which accounted for between 50 and 60 per cent of the total variance in soil properties, was related to organic matter and soluble constituents: properties associated with organic matter (carbon content, nitrogen content, exchangeable potassium, and moisture loss) diminish fairly evenly down slope, whereas properties associated with soluble constituents (pH, carbonates, exchangeable calcium, sodium, and magnesium) increase down slope. The second component, accounting for 13 to 18 per cent of the variance, was interpreted as a particle-size factor, the pattern of which showed an abrupt increase in finer soil material immediately down slope of the maximum gradient in the transect, giving a marked discontinuity in the pattern over the slope. In an earlier paper (Furley 1968), it had been suggested that some slopes could be divided into two sections: an upper, generally convex section where net erosion is greater than net deposition; and a lower, concave section where net deposition predominates. The zone of interaction between the two zones is called the junction. The results for the five chalkland transects showed that, with the exception of fine materials, soil properties change gradually over the slope and the transition from net erosion to net deposition in the surface soil layer is diffuse.

6.3.2 Principal Coordinate Analysis

A technique that should not be confused with principal component analysis is principal coordinate analysis or, to give it its other name, classical scaling. Devised by Gower (1966) from a procedure originally proposed by Torgerson (1952, 1958), this technique is, in essence, an algebraic method for reconstructing the coordination of data points where dissimilarities between data points are expressed as Euclidean distances. An excellent account of the method is given by Sibson (1985). It is interesting to note that, as long as distances calculated from the data matrix are Euclidean, the results of classical scaling are exactly equivalent to the results of principal component analysis (Chatfield and Collins 1980, p. 201), an equivalence which extends to a data matrix consisting solely of zeros and ones with distances between individuals calculated using a simple matching coefficient (Gower 1966).

Principal coordinate analysis was used in pedology by Rayner (1966) to examine relationships among soil profiles in South Wales. Subsequent applications have been mainly used to understand relationships between soil profiles and, by and large, have not really done more than to describe and display quantitatively what an experienced soil surveyor would discover intuitively given sufficient time (Webster 1979, p. 292). However, a few surprising results have emerged from principal coordinate analysis of soils, such as Webster and Butler's (1976) finding for soils at Ginninderra, Australia, where correlations between variables were weak. It may be that the contrast between the Welsh and Australian soils, as revealed by principal coordinate analysis, could result from differences in the age of the landforms and soils in question.

6.3.3 Factor Analysis

The basic ideas of factor analysis were put forward by Francis Galton, Charles Pearson and others at the turn of the century. As a technique, factor analysis involves extracting a set of m variables from an original set of p correlated variables. The number of factors, m, is less than the number of original variables, p, and, so the argument runs, should provide a better understanding of the data.

Principal component analysis and factor analysis, though they have similar aims, should not be confused with one another. Principal component analysis starts with a set of data taken from the field. Components are then sought so as to reduce the dimensions of variation in the data. With luck, the components can be given a physical meaning. The process by which principal component analysis works involves an orthogonal transformation of all p variables and is not based on an underlying statistical model. Factor analysis, on the other hand, starts with a model, a set of factors, and sees how well this model agrees with the data

(Kendall 1957, p. 37). Both techniques have been introduced to most branches of the Earth Sciences including geography (Kendall 1939), soil science (Bidwell and Hole 1964), botany (Goodall 1954), geology (Imbrie 1963; Imbrie and Purdie 1962; Klovan 1966; Solohub and Klovan 1970).

Whereas it is probably true to say that principle component analysis has some degree of respectability among statisticians, factor analysis is something of a bête noire. Mather (1981, p. 144), an advocate of factor analysis, admits that factor analysis would appear to resemble mathematics if, as Bertrand Russell would have it, mathematics is the only science where one never knows what one is talking about, nor whether what is said is true. He also notes that critics place applications of factor analysis in the "Lucky Jim" category, after Kingsley Amis's eponymous hero who defined his research as the casting of pseudo-light on non-problems. Chatfield and Collins (1980, p. 89), in view of the large number of drawbacks with factor analysis, recommend that it should not be used in most practical situations. In the same vein, Blackith and Reyment (1971) argue that principal component analysis is preferable to factor analysis and that, perhaps, factor analysis has lasted as long as it has because it allows its user to impose his or her preconceived ideas on the raw data.

The consensus of opinion against it notwithstanding, factor analysis is still used by Earth scientists, and in some cases used to good effect. For this reason alone, the technique warrants discussion.

The basic idea behind factor analysis is that a number of basic, underlying dimensions or factors are hypothesized and searched for in the data matrix. The mathematical procedure is virtually the same as for principal component analysis except that the resultant matrix will have only m factors as specified in the hypothesis. Additionally, in the model, an individual measurement for the rth case on the jth variable is made up of two parts. The first part is attributable to the effect of common factors while the second part is attributable to specific factors which are unique to the individual variables. Four main assumptions are involved in the model: firstly, that the common factors and specific factors are uncorrelated with one another; secondly, that the specific factors are themselves uncorrelated; thirdly, that the common factors are themselves uncorrelated, though this assumption is relaxed when the factors are rotated; and fourthly, that the common factors and the specific factors each have a multivariate normal distribution.

A study which uses factor analysis to good effect is the work carried out by Mills and Starnes (1983) on sinkhole morphometry in a fluviokarst region in the eastern Highland Rim of Tennessee. Using large-scale maps, the size and shape of 127 sinkholes and their drainage basins were described quantitatively. All the variables, except the map coordinates, were transformed to conform, as far as possible, with a normal distribution and a correlation coefficient matrix between the variables was computed. The relationships between the 23 variables were revealed by an R-mode factor analysis. The results, with five factors assumed and rotated according to the varimax criterion, are shown in Table 6.4. The factors can be interpreted by examining the variables that load highly on them. If several variables load highly on a factor, then those variables are strongly correlated and therefore redundant; just one of the variables needs to be specified

Correlation Systems

Table 6.4. Results of factor analysis on karst-depression and drainage basin variables. (Mills and Starnes 1983)

Variable	Varimax rotated factor matrix				
	F_1	F_2	F_3	F_4	F_5
ARDEP	0.51	0.22	0.80	−0.05	0.05
AREA	0.90	0.32	0.24	−0.01	0.02
AVDIAM	0.89	0.34	0.18	−0.07	0.22
DEPTH	0.58	0.15	−0.69	0.10	0.01
LEN	0.83	0.36	0.16	−0.06	0.38
LENDEP	0.18	0.17	0.89	−0.11	0.35
LENWID	−0.02	−0.26	0.04	0.04	−0.65
PERIM	0.86	0.31	0.19	−0.08	0.32
IRI	0.46	0.24	0.03	−0.18	0.63
MAXSLP	0.12	0.07	−0.51	−0.05	0.19
NSLOP	−0.08	−0.07	−0.98	0.06	−0.05
VOL	0.90	0.30	−0.22	0.02	0.00
WID	0.92	0.25	0.21	−0.07	−0.11
WIDDEP	0.20	0.05	0.95	−0.11	−0.07
AZ	−0.07	−0.10	−0.22	0.00	−0.01
ELB	−0.08	−0.15	−0.05	0.97	−0.08
ELHC	−0.00	−0.14	−0.13	0.97	−0.08
BAREA	0.41	0.82	0.07	−0.17	0.24
BLEN	0.34	0.87	0.12	−0.18	0.23
BWID	0.46	0.76	0.10	−0.14	0.23
BPERIM	0.39	0.84	0.10	−0.16	0.25
DIVSWD	0.25	0.90	0.08	−0.19	0.22
RELIEF	0.13	0.66	0.06	0.08	−0.04
Percent variance explained	55.4	20.5	12.1	7.6	4.5

Definition of variables and transformation (if any): LEN = maximum distance across depression (log); WID = maximum distance perpendicular to LEN (log); AZ = azimuth of depression long axis (linear); ELHC = elevation of highest enclosing contour (log); ELB = elevation of bottom of depression (log); MAXSLP = maximum slope of depression flank (square root); AREA = area within highest enclosing contour (log); PERIM = length of highest enclosing contour (log); DEPTH = ELHC − ELB (square root); LENWID = LENGTH/WIDTH (log); LENDEP = LENGTH/DEPTH (log); WIDDEP = WIDTH/DEPTH (log); AVDIAM = (LENGTH + WIDTH)/2 (log); VOL = volume = πr^2 DEPTH/3, where r is radius of a circle with the same area as the depression (log); NSLOP = nominal slope = DEPTH/r, where r is as above (log); IRI = irregularity index = PERIM/P, where P is the perimeter of a circle with the same area as the depression (log); ARDEP = AREA/DEPTH (log); BAREA = basin area (includes karst-depression area) (log); BLEN = maximum distance across basin (log); BWID = maximum distance perpendicular to BLEN (log); BPERIM = basin perimeter (log); DIVSWD = divide-to-swallet distance = maximum distance from basin divide to lowest point of karst depression (log); RELIEF = elevation of highest point in basin minus ELB (log).

to describe the variance of the system. Variables loading highly on the first factor are area, average diameter, maximum distance across a depression, length of highest enclosing contour, volume, and width. These variables are all related to the horizontal size of the sinkhole and the first factor can be though of as a "horizontal-size" factor. The variables loading highly on the second factor are all drainage basin variables and related to drainage basin size, both horizontal size

Table 6.5. Results of multiple correlation of karst-depression variables with drainage basin variables. (Mills and Starnes 1983)

Variable	R	R^2
Horizontal-size variables:		
AREA	0.734	0.538
AVDIAM	0.778	0.605
LEN	0.788	0.621
PERIM	0.785	0.617
WID	0.688	0.474
Depth/mean-slope variables:		
DEPTH	0.413	0.171
LENDEP	0.454	0.206
NSLOP	0.323	0.104
WIDDEP	0.345	0.118

Variables: See list on Table 6.4.

Note: In each correlation, drainage-basin variables used were BAREA, BLEN, BWID, BPERIM, DIVSWD and RELIEF.

as well as vertical size. The second factor can therefore be thought of as a "drainage-basin size" factor. Variables loading highly on the third factor are all sinkhole variables related to the vertical dimension as measured by depth or mean flank slope. Just two variables load highly on the fourth factor which can be though of as a "sinkhole elevation" factor. This factor is unimportant for it simply reflects the east-to-west decrease in surface, and thus sinkhole, elevation across the area. Just two variables too load highly on the fifth factor and indicate that this factor pertains to the horizontal shape, rather than size, of a sinkhole. However, the fifth factor accounts for a mere 4.5 per cent of the total variance.

The most revealing finding of the factor analysis is that sinkhole variables load highly on one factor, whereas those related to depth or mean flank slope load highly on another, suggesting that these two aspects of sinkhole form are controlled by different mechanisms. Pursuing this finding, Mills and Starnes (1983) carried out a multiple regression analysis of sinkhole variables against six measures of drainage basin size. The results (Table 6.5) show that variables related to sinkhole horizontal size have relatively strong relationships which drainage basin size, whereas variables related to sinkhole depth and mean flank slope have only weak relationships with drainage basin size. As an explanation of these results, Mills and Starnes argue that the planimetric size of sinkholes having larger drainage basins grows more rapidly than the planimetric size of sinkholes having smaller basins, mainly because the larger basins generate greater runoff than the smaller basins. The extra runoff in larger basins not only increases solution by groundwater but also leads to the erosion of sinkhole margins by streams entering the sinkhole, thereby producing lateral growth by headwards erosion of the streams. Erosion of this kind is evidenced by the dissected margins of sinkholes fed by ephemeral streams and the more elongate and irregular shape of sinkhole outlines in larger drainage basins. On the other hand, sinkhole depth and mean flank slope appear to be controlled by local bedrock conditions.

Correlation Systems

6.3.4 Canonical Correlation

Principle component and factor analysis are one way of establishing patterns of relationships between two or more sets of variables. Another method, canonical correlation, imposes an a priori structure on individuals (rows) in the basic data matrix. The variables (columns) of the data matrix are divided into sets with r and q variables in each set, so that p, the total number of variables equals $r+q$. The data matrix is thus written

$$\underset{n \times p}{X} = (\underset{n \times r}{X_1} \quad \underset{n \times q}{X_2}).$$

The variance-covariance matrix or correlation coefficient matrix computed from the data matrix is written in partitioned form as

$$S = \begin{bmatrix} A_{rr} & C_{rq} \\ C'_{qr} & B_{qq} \end{bmatrix},$$

where C_{rq} is a correlation coefficient matrix between left-side (r) variables and right-side (q) variables; A_{rr} is a correlation coefficient matrix between left-side variables (r); and B_{qq} is a correlation coefficient matrix between right-side variables (q). From this matrix, the linear combinations

$$\left. \begin{array}{l} u_i = l'_i X_1 \\ v_i = m'_i X_2 \end{array} \right\} \; i = 1, 2, \ldots, s$$

are sought. They have the property that the correlation of u_1 and v_1 is greatest, the correlation of u_2 and v_2 is greatest among all linear combinations uncorrelated with u_1 and v_1, and so on for all possible s pairs.

There have been a few applications of canonical correlation in pedology. Webster (1977) used canonical correlation to relate soil to environment in the Dee catchment of North Wales. Large and significant correlations were found but he concluded that the results only expressed quantitatively what was common knowledge among British soil scientists. A more revealing analysis was made of soils at Ginninderra in the Australian Capital Territory. The area studied covered some 400 ha of rolling terrain in an environment of largely uniform temperature, rainfall, parent material, and vegetation. The area was sampled at 111 intersections of a rectangular grid at intervals of approximately 180 m. Seventeen soil properties and seven local physiographic properties were measured (Table 6.6). The seven physiographic variables were taken as right-side variables in several analyses with a variety of subsets of soil properties on the left side. Variables were standardized to unit variance so that their contributions to the correlations could be compared and the vectors thereby interpreted. The canonical correlations are shown in Table 6.7 and the first two pairs of canonical vectors, corresponding to the two canonical correlations significant at $p < 0.001$, are listed in Table 6.6. Sand content at 0–10 cm, as well as colour hue at both depths and porosity at the lower depths, make substantial contributions to the first (left-side) soil vector. Colour value, porosity, hardness, and coalescence at 41–51 cm are the largest contributors to the second left-side (soil) vector. Absolute height,

Table 6.6. First two canonical vectors for soil (left side) and physiography (right side) calculated from standardized data for Ginninderra. (Webster 1979)

Variables	Vector 1	Vector 2
Soil (0–10 cm)		
Colour hue	−0.370	−0.134
Colour value	−0.130	−0.077
Colour chroma	0.096	−0.044
Structure grade	−0.103	−0.269
Structure size	−0.217	−0.150
Sand	0.537	−0.263
Silt	0.100	−0.081
Soil (41–51 cm)		
Colour hue	−0.473	0.320
Colour value	0.217	0.561
Colour chroma	−0.040	−0.140
Structure grade	−0.096	0.072
Structure size	0.014	0.066
Porosity	0.494	0.611
Hardness	0.345	0.437
Coalescence	0.304	0.508
Sand	0.042	0.208
Silt	0.122	0.065
Physiography		
Absolute height	0.732	0.273
Dispersal area (log)	0.404	0.665
Slope gradient	−0.085	−0.376
Height above drainage channel[a]	−0.129	−0.555
Horizontal distance to drainage channel[a]	0.053	0.036
Height above mean of dispersal area	0.330	−0.298
Horizontal distance to crest (square root)	0.082	0.343

[a] Or zero gradient (see Webster 1979 for details)

dispersal area, and height above mean dispersal area contribute to the first right-side (physiographic) vector and dispersal area is actually the largest element of the second vector. Webster interprets the vectors to mean that, in the case of the first vector, high land with a large area for dispersal of water has soil that is redder and more sandy at the surface than elsewhere; and, in the case of the second vector, that sites with large dispersal areas but close to their local base level have pale, hard, and relatively strongly coalescent subsoil. The relation between soil colour and height is a widely observed phenomenon usually attributed to differences in water regimes. The relation between sandiness and height probably arises from differential winnowing of fines during denudation. Although differences in texture and colour are clear to see on simple catenary sequences, they are not evident in complex landscapes such as at Ginninderra and would have remained "hidden" were it not for canonical analysis.

Webster (1979) also reports an interesting example of use of canonical correlation in a study of the soil and its environment in the catchment of Wungong Brook, a forested area of about 150 km^2 at the edge of the Great Plateau of

Table 6.7. Canonical roots and correlations between soil and variables and physiographical variables at Ginninderra, Australian Capital Territory. (Webster 1979)

Order	Root	Percentage of trace	Cumulative percentage	Canonical correlation
1	0.5945	29.6	29.6	0.77[a]
2	0.5134	25.6	55.2	0.72[a]
3	0.2908	14.5	69.7	0.54
4	0.2080	10.4	80.1	0.46
5	0.1658	8.3	88.3	0.41
6	0.1337	6.7	95.0	0.37
7	0.1008	5.0	100.0	0.32

[a] The first two roots are significant at $p < 0.001$.

Western Australia, some 30 km southeast of Perth. The Wungong Brook and its tributaries have cut into a granite-gneiss plateau with dolerite intrusions. Under the plateau the rock is deeply weathered, the regolith consisting of a ferruginous and bauxitic laterite cap, a mottled zone, a pallid zone, and then more or less fresh rock (Fig. 6.13). These zones in the regolith have been bevelled by erosion. The upper metre or so of soil ranges from loose, yellow-brown sandy material containing lateritic gravel on the highest ground to friable, finer textured, and commonly redder or browner soil on the lower slopes. The catchment was surveyed at 30 m intervals on transects each extending from an interfluve crest to a valley bottom as nearby as possible directly downslope. Some 347 sampling points were examined with a 10 cm diameter anger to a depth of 1 m. Data from the sites used in the analysis were: for the soil – colour on the Munsell scales, clay content and gravel content at 0.5 cm, 15–30 cm, 45–60 cm, and 75–100 cm, the depth at which clay content increases most markedly, and the depth range over which the clay content increase occurs; and for physiography – absolute height, slope gradient and convexity, length of slope, distance to crest, distance to valley floor, height above valley floor, height of slope crest above site, and distance to nearest high-order (fourth order or more) stream. The analysis revealed the first six canonical correlations to be significant ($p < 0.01$), the first being substantially larger than the others (Table 6.8). The first vector is associated with, on the left side (soil variables), colour and clay content in the 70–100 cm layer, and also clay and gravel content at the surface; and, on the right side (physiographic variables), the vertical interval between the site and the slope crest, as well as the length of slope and the height above the valley floor.

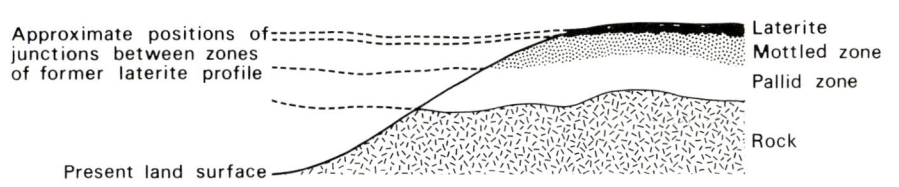

Fig. 6.13. An idealized cross-section through the dissected laterite plateau of Western Australia. (Webster 1979)

Table 6.8. First six canonical roots and correlations for the Wungong Brook catchment. (Webster 1979)

Order	Root	Percentage of trace	Cumulative percentage	Canonical correlation
1	0.7739	41.2	41.2	0.88
2	0.3497	18.6	59.8	0.59
3	0.2615	13.9	73.7	0.51
4	0.1645	8.8	82.5	0.41
5	0.1120	5.9	88.4	0.33
6	0.1046	5.6	94.0	0.32

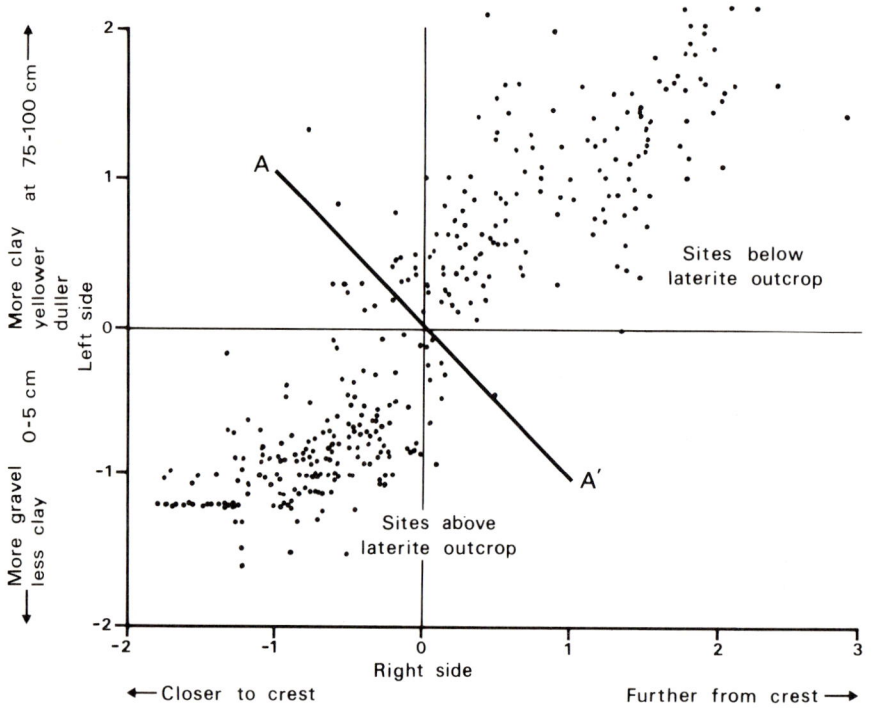

Fig. 6.14. Scatter of sampling sites at Wungong projected into the plane of the first pair of canonical axes. The line AA' separates sites above and below the local laterite outcrop. (Webster 1979)

Particularly revealing is the distribution of sampling points in the plane of the first canonical axes (Fig. 6.14). Two groups of points emerged, neatly separated by the line AA', one tight group consisting of sites above the local level of the laterite outcrop, the other, looser group consisting of sites lying below the laterite outcrop. Thus it seems that, in the Wungong catchment, eroded laterite is transported out of the landscape with not enough debris tarrying on lower slopes to mask the effects of fresh material exposed there. This is an interesting finding because further east on the Great Plateau, eroded laterite does make substantial contributions to soil profiles below its outcrop (Mulcahy et al. 1972).

Table 6.9. Canonical analysis between soil and morphological properties of terracettes. (Vincent and Clarke 1980)

Total data set	Canonical vectors			
	I	II	III	IV
Morphological set:				
Riser angle	−0.28	0.86	0.30	−0.50
Riser width	−0.26	0.05	−0.62	−0.16
Tread angle	−0.58	0.11	−0.49	0.81
Tread width	−0.38	−0.53	1.01	0.27
Soil property set:				
Bulk density	−0.26	0.82	0.22	−0.15
Moisture content	−0.12	−0.24	0.14	−0.88
Cohesion	0.50	0.80	−0.13	−0.77
Angle of internal friction	0.14	−0.18	−0.77	−0.67
Liquid limit	2.59	0.63	−4.30	6.16
Plastic limit	−1.50	0.23	4.31	−6.09
Stoniness	−0.09	−0.22	−0.25	−0.59
Clay content	−1.02	−0.01	−0.05	0.02
Canonical correlation	0.80	0.71	0.63	0.31
$\chi^2_{0.05}$	122.32	69.26	31.76	5.15
Degrees of freedom	32	21	12	5
Probability	0.000	0.000	0.000	0.397

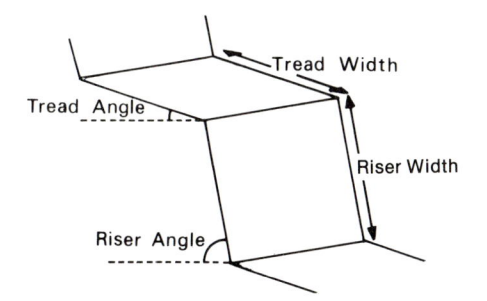

Fig. 6.15. Definitions of variables in the morphological set. (Vincent and Clarke 1980)

Vincent and Clarke (1980) apply canonical correlation analysis to the study of terracettes, step-like features commonly formed on grassy slopes where angles typically range from 25 to 45 degrees. The data, collected at 59 sites in England, Wales, and Norway, were divided into two sets of variables (Table 6.9). The morphological variables are defined in Fig. 6.15. The results of the canonical correlation analysis are given in Table 6.9. The number of canonical vectors corresponds to the number of variables in the smaller, morphological set. The first three canonical vectors are statistically significant; the fourth is not. Together, the first three canonical correlations account for more than 94 per cent of the identifiable relationship between soil properties and terracette morphology. Examination of the loading of the variables on the canonical vectors suggests three main relations. Firstly, soils with high liquid limits, low plastic limits, low clay contents, and low soil moisture contents are associated with terracettes having low tread angles. Secondly, the larger the cohesion and bulk density of soils, the steeper the terracette riser angle. Thirdly, terracettes with large tread widths and small riser widths tend to be associated with soils possessing low liquid limits and high plastic limits.

6.3.5 Problems with Correlation Systems

It will be well at this juncture to remember that the methods just discussed assume that the set of variables comprise a system. The assumption is convenient and once made further investigations fall, more or less, into place. Seldom is there any discussion of whether the variables do form a system. Indeed, Craig (1981) shows that the usual methods for establishing systemhood — correlation analysis, path analysis, and cluster analysis — stand a good chance of inadvertently awarding a nonsystem the status of a system: the identification of systems by examining sets of correlations between observable elements is inherently biased in favour of incorrectly "discovering" a "system". To give credence to his argument, Craig discusses Melton's (1958b) analysis of 15 variables which represented, according to Melton, significant geomorphological, surficial, and climatic elements of a drainage system. Using correlation analysis, Melton found 42 significant relationships but one was discarded on the basis of rational argument leaving 41. Aided by Q-mode cluster analysis, Melton reassigned and simplified the system according to reasonable geomorphological expectation and came up with a description of the drainage system based on just five of the original variables. Craig (1981, p. 271) calculates the probability of discovering a system comparable to that suggested by Melton within a set of variables which do not in fact contain a true system. Given that systems of as few as five elements are acceptable as models of drainage systems, the researcher is virtually certain of identifying some system. The system identified may be reasonable, but structural conclusions based upon analysis of the correlation matrix cannot be taken as support for any conclusion concerning the existence of a true system.

In conclusion, Craig argues that a distinction should be made between the search for systems on the one hand, and the specification of the structure of a known system, insofar as any system can be known, on the other. Because correlational studies and all analogous procedures will almost always suggest the existence of a system, the suggestions they make are useless. However, if it is known that a certain set of elements does comprise a system, then correlation-type studies can be useful in suggesting the precise set of relationships within it. The system hunter, prosetylizes Craig, must fashion other weapons to allow selective capture of the right prey.

Strahler (1980) too is critical of multivariate methods. He points out that process controls by dynamic variables can only be inferred from a multivariate analysis of a set of form variables: the explanation of the physical processes involved will always be conjectural. He concludes that models of process-form systems derived from correlation structures of measured variables may have greater value in emphasizing our ignorance of natural processes than in explaining them (Strahler 1980, p. 25).

CHAPTER 7
Deterministic Models of Water and Solutes

Deterministic models of Earth surface systems normally relate form and process variables in a dynamic manner. They focus on the change in the state of a system from one location to another through time. In other words, they look at system state within some space-time domain under given starting and boundary constraints.

Spatial domains may be modelled as continuous or discontinuous. In a continuous spatial domain, space is regarded as unbroken in all directions. It is commonly defined in the frame of Cartesian coordinates, less commonly in polar and other coordinate systems. For instance, iron concentrations may vary throughout a hillslope mantle covering a drainage basin. The concentration of iron, C, at any point in the three-dimensional spatial domain may be expressed as $C(x, y, z)$ (Fig. 7.1a). As the concentration of iron may vary with time, the full designation becomes $C(x, y, z, t)$. The height of the land surface, h, within the drainage basin may be considered within a two-dimensional spatial domain (Fig. 7.1b). To specify the height at any point and time the notation $h(x, y, t)$ is used. The height of the land surface along any hillslope profile in the basin can be regarded as varying in a horizontal, one-dimensional domain (Fig. 7.1c) and denoted as $h(x, t)$. Similarly, the iron concentration in a soil profile at any point along a

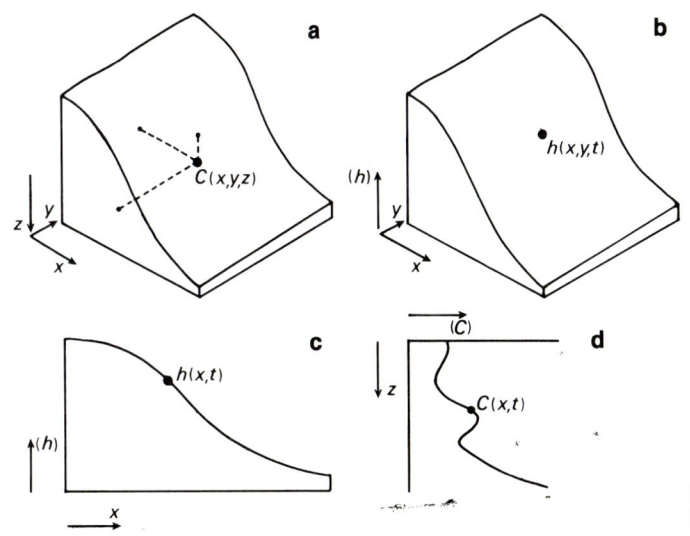

Fig. 7.1. Spatial domains in continuous space

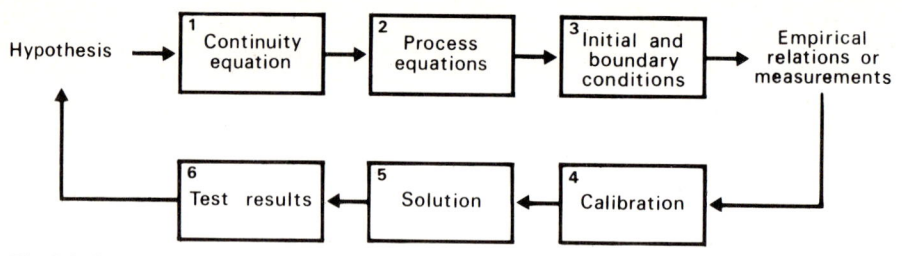

```
Hypothesis → [1 Continuity equation] → [2 Process equations] → [3 Initial and boundary conditions] → Empirical relations or measurements
                ↑                                                                                              ↓
         [6 Test results] ← [5 Solution] ← [4 Calibration] ←
```

Fig. 7.2. Steps in the building and use of a deterministic model

hillslope may be studied in a vertical, one-dimensional domain and denoted by $C(z, t)$ (Fig. 7.1d).

In a discontinuous spatial domain, space is conceived as a set of regular or irregular spatial cells of finite size. The size and shape of the spatial cells (finite elements) is arbitrary and geared to the degree of detail required in the problem in hand. The simplest spatial cell is a cuboid or block, but many other shapes are used.

There are six basic steps involved in building deterministic models (Fig. 7.2). Firstly, the continuity equation is applied to the system of interest. For mass conservation, this will take the general form

$$\frac{\partial M}{\partial t} = -\left(\frac{\partial M}{\partial x} + \frac{\partial M}{\partial y} + \frac{\partial M}{\partial z}\right) \pm Q,$$

where M is mass, t is time, x, y, and z are Cartesian spatial coordinates, and Q is a source or sink of mass within the system. The term on the left-hand side of the continuity equation, $\partial M/\partial t$, is the time rate of change of mass storage. The differential terms on the right-hand side of the continuity equation represent the fluxes of mass in the x, y, and z directions. They are found by subtracting mass outputs from mass inputs in each direction. The second step in building a deterministic model is to define the mass flux rates in terms of a process (transport) equation. One of the standard phenomenological equations will often be suitable (Chap. 2.2.2), though for sediment transport the equations are of a more empirical or heuristic nature. The third step is to define the initial conditions, that is, the distribution of the state variables (mass in the example) at the start of the period of interest when $t = 0$. Also defined in this step are the boundary conditions, that is the values of the state variables at the edges of the spatial domain for all time intervals. The fourth step is to calibrate the model by fitting suitable parameter values as measured in the field or guessed. In the fifth step, the calibrated equation is solved by appropriate analytical or numerical techniques to reveal how the state variables change in the spatial domain over time. Lastly, but very importantly, the predicted results must in some way be tested against empirical data.

7.1 Ice

7.1.1 Glaciers

In the case of a glacier, the continuity equation may be written

$$\frac{\partial Q}{\partial x} + \frac{\partial S}{\partial t} = bW$$

where Q is the mass flux or discharge in the direction of ice flow (the volume of ice passing through a cross-section at x per unit time), x is the distance along the glacier measured from the glacier head, S is the cross-sectional area measured perpendicular to the x-axis, W is the width at the surface, and b is the mass balance (net budget) of the glacier (negative for ablation) averaged across the glacier at x. The quantities Q, S, and W are all a function of distance x and time t. The density of the glacier is assumed constant throughout. The flux, Q, depends on glacier width and ice thickness and ice velocity. The ice velocity, u, depends on the ice thickness, h, and on the ice surface slope, α, in the following way

$$u = A_1 h^{n+1} \sin^n \alpha + A_2 h^m \sin^m \alpha$$

where A_1, A_2, n, and m are constants. The first term in this equation represents differential movement within the ice, whereas the second term represents the slip of the glacier on its bed.

The foregoing formulation of glaciers is the basis of a number of theoretical models built by Weertman (1958, 1961) and Nye (1958, 1959, 1960, 1961, 1963a, 1963b, 1965a, 1965b, 1969). The main applications of Nye's theory, which involves examining the effects of perturbations caused by sudden change of the net budget on a steady-state profile, are outlined by Paterson (1981, pp. 244 – 266). The assumptions behind Nye's formulation are sound but limiting. For instance, the assumption that the flux, Q, is approximated by distance, ice thickness, and surface slope is good only if it is also assumed that the temperature of the glacier is uniform and the effects of changes in the amount of meltwater at the glacier bed are ignored. This limits application of the theory to temperate glaciers. Another assumption of Nye's work is that perturbations are small. The theory is applicable, therefore, to "slow" advances and retreats but is not applicable to ice surges. Models in which the assumption of small perturbations is relaxed, such as those built by Shumskiy (1975), involve exceedingly difficult mathematics.

The chief conclusions of Nye's theoretical studies are that variations in mass balance are propagated down a glacier as kinematic waves of increased or decreased flux, depending on whether the mass balance is increased or decreased. The kinematic waves move at roughly four times the velocity of the ice. A small increase in the mass balance can lead to a thickening of ice near the glacier snout and so to an advance of the glacier. It takes most glaciers several hundred years to adjust to a change in mass balance.

7.1.2 Ice Sheets

In the case of an ice sheet, the continuity equation may be written in one dimension as

$$\frac{\partial Q}{\partial x} + \frac{\partial h}{\partial t} = b(x, t),$$

where Q is, as before, the flux of ice within a flowband, h is ice thickness, and b, is the mass balance. The flow of ice, which is a compressible substance, may be expressed most simply by the power relation

$$\dot{\varepsilon}_{min} = [\sigma/\{D_0 \exp(-F/R\psi)\}]^n,$$

where $\dot{\varepsilon}_{min}$ is the minimum effective strain rate in secondary creep at constant stress, σ is effective stress, n is a property constant known as the viscoplastic parameter and usually lies in the range 2 to 4.5, F is the ice activation energy, ψ is absolute temperature, D_0 is a constant independent of temperature, and R is a gas constant. The term in curly brackets is a "thermally activated ice hardness factor" and usually lies in the range 0.2 to 2.25 kPa yr$^{1/3}$. Using this flow law for ice and the coordinate system shown in Fig. 7.3, Paterson (1972) has derived a set of equations that give the surface profile of an ice mass for cases in which there is no basal sliding. For Antarctic ice sheets, where there is no terrestrial ablation zone, all ice being lost by discharge into the sea, Drewry (1983) has reduced the equations to

$$(h/H)^{2+2/n} + (x/L)^{1+1/n} = 1,$$

where n is the viscoplastic parameter in the flow law.

Assuming that n tends to infinity (that is, the ice is perfectly plastic), the equation for a surface profile simplifies to

$$h = \{(2\sigma_y/\varrho_i g)(L-x)\}^{1/2} \quad \text{and}$$

$$(h/H)^2 = (1 - x/L),$$

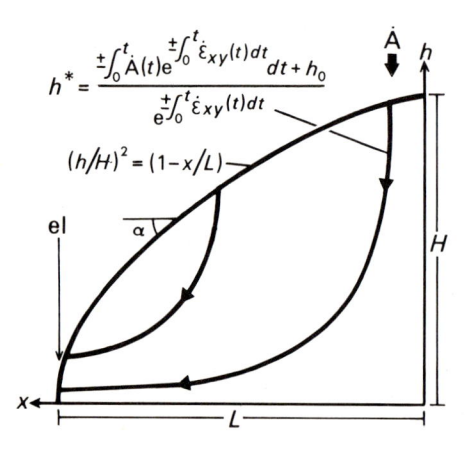

Fig. 7.3. Coordinate system for the ice sheet model. $\dot{A}$ is the surface accumulation rate, el is the equilibrium line. Two internal flowlines and their depth, h^*, can be calculated using the expression derived by Crary et al. (1962), where $\dot{\varepsilon}_{xy}$ is the sum of two principal horizontal strain rates, h_0 is the depth at time $t = 0$; H is the height of the ice sheet at its centre, h is ice thickness, and L is ice sheet half-width. (Drewry 1983)

Ice

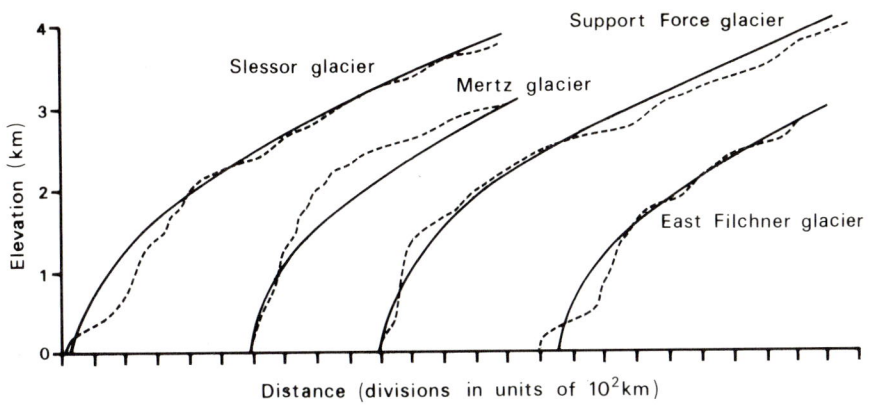

Fig. 7.4. Surface profiles for ice stream flow lines in East Antarctica. Theoretical profiles with $\sigma_y = 50$ kPa are shown by *solid line*. (Drewry 1983)

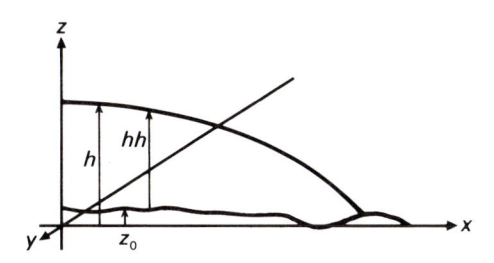

Fig. 7.5. The coordinate system used in Mahaffy's ice sheet model. The z coordinate is parallel to the gravity vector. The height of the ice surface above the x, y plane is h. The value of z at bedrock is h_0. (Mahaffy 1976)

where ϱ_i is ice density, g is gravitational acceleration, and σ_y is the "yield" stress for ice. More generally, Drewry writes

$$h = \gamma(L-x)^{1/2},$$

where γ depends on the value of σ_y and has to be tuned to the properties of a particular ice cap. Figure 7.4 shows theoretical and observed profiles along ice stream flowlines in East Antarctica, the theoretical forms being calculated with $\sigma_y = 50$ kPa. Drewry remarks on the good correspondence between theoretical and observed profiles but warns that detailed examination and discussion of the differences between them is not warranted by the level of sophistication of the original model.

The two-dimensional models of ice flow just described, consider the distribution of mass along a cross-section or a streamline in a glacier or an ice sheet. A three-dimensional, hydrodynamic model was constructed by Campbell and Rasmussen (1969) to study valley glaciers and later extended by them (Rasmussen and Campbell 1973) to allow for a basal sliding power law. The first three-dimensional model of ice sheets was developed by Mahaffy (1976) and other such models have followed (Jenssen 1977; Hughes 1979; Reeh 1982).

Mahaffy's (1976) model can predict the heights of an arbitrary size sheet of ice on a rectangular grid which includes the entire ice cap. The coordinate system used in the model is shown in Fig. 7.5. The mass discharge, q, is a function of x, y, and t. With the specific net budget denoted by $b(x, y, t)$, the continuity equation used in the model is

$$\frac{\partial h}{\partial t} = b(x, y, t) - \left\{\frac{\partial q_x}{\partial x} + \frac{\partial q_y}{\partial y}\right\}.$$

The terms q_x and q_y are the x and y components of the mass discharge and are defined as the integrals of the horizontal velocities, V_x and V_y, over the height of the ice sheet:

$$q_x(x, y, t) = \int_{z_0}^{h} V_x dz,$$

$$q_y(x, y, t) = \int_{z_0}^{h} V_y dz.$$

The mass discharge is calculated from a form of Glen's flow law put forward by Nye (1957). Assuming that the ice sheet moves by shear strain parallel to the geoid, and that the vertical velocity in the ice changes with horizontal distance much more slowly than the horizontal velocities change with depth, Mahaffy (1976) derives the following expressions for mass discharge:

$$q_x(x, y, t) = \frac{-2A^{-n}}{n+2} (\varrho g)^n \frac{\partial h}{\partial x} \alpha^{(n-1)}(h - z_0)^{n+2} + (h - z_0) V_x(x, y, z_0, t)$$

$$q_y(x, y, t) = \frac{-2A^{-n}}{n+2} (\varrho g)^n \frac{\partial h}{\partial y} \alpha^{(n-1)}(h - z_0)^{n+2} + (h - z_0) V_y(x, y, z_0, t),$$

where A is a flow law constant dependent on temperature, n is a flow law exponent, ϱ is the ice density (assumed constant), g is the acceleration of gravity, and α is the absolute value of the slope of the ice surface in the direction of flow. The term α is written as

$$\alpha = \{(\partial h/\partial x)^2 + (\partial h/\partial y)^2\}^{1/2}.$$

With these expressions of mass discharge, with specified starting heights for the ice sheet, with the net budget and basal sliding velocity specified for all times, the continuity equation can be solved numerically.

To test the model, Mahaffy calibrated the parameters and initial states for the Barnes Ice Cap, a sheet of ice occupying 5900 km^2 on Baffin Island in the Northwest Territories of Canada. Bedrock topography and initial ice heights were taken from topographic maps. The bedrock topography beneath the ice was assumed to be a plane with a small general slope connecting the topography on either side of the ice. The specific net budget was defined by a quadratic curve fitted to field data which takes the form (Fig. 7.6):

$$b(i, j) = -314.98 + 54.87 \times 10^{-4} h(i, j) - 2.09 \times 10^{-8} h(i, j)^2$$

where $h(i, j)$ is height above sea level in centimetres and $b(i, j)$ is the specific net budget in centimetres per year. Basal sliding was taken to be zero everywhere, expect at the edge of the ice sheet. The flow law exponent, n, was varied from 1 to 3.5 to see how the ice responded. The value of A was varied for each value of n to study its effects on the ice.

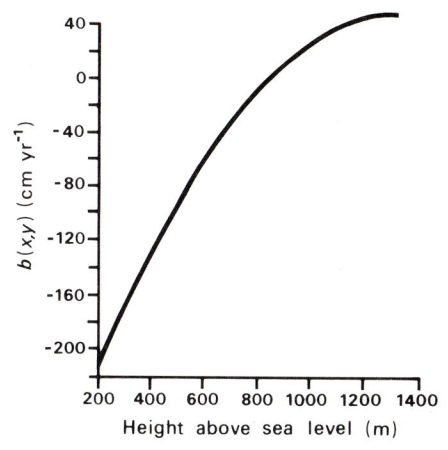

Fig. 7.6. Graph of the specific net budget function with height above sea level. (Mahaffy 1976)

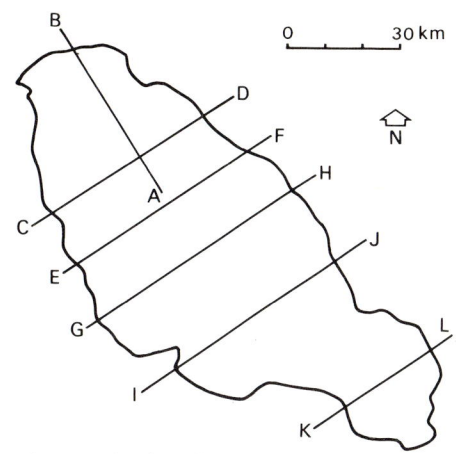

Fig. 7.7. Areal outline of the Barnes Ice Cap. Lines *AB*, *CD*, *EF*, *GH*, *IJ* and *KL* are the lines along which profiles were graphed to compare model results. (Mahaffy 1976)

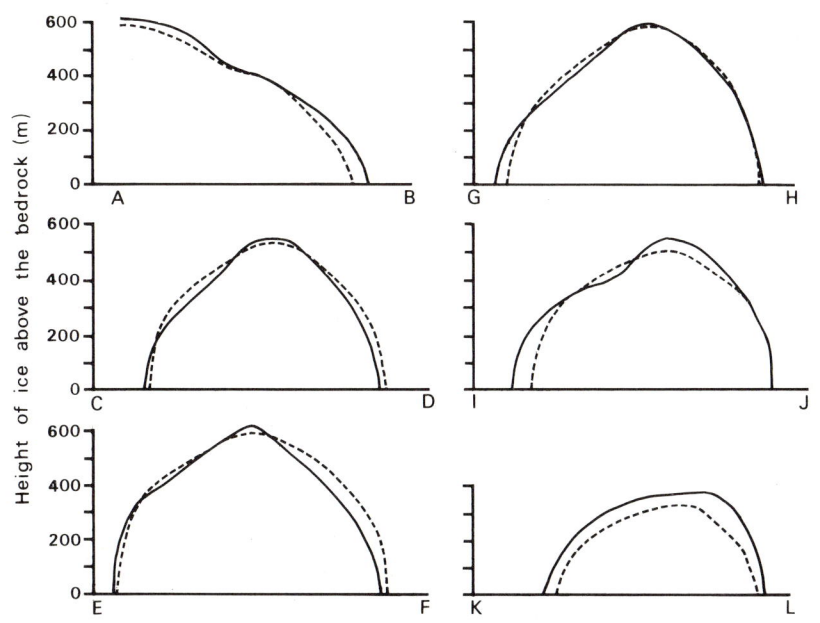

Fig. 7.8. Profiles across the Barnes Ice Cap. Theoretical profiles shown by *broken lines*, actual profiles shown by *solid lines*. (Mahaffy 1976)

To compare the model ice cap with the actual ice cap, Mahaffy took six profiles across the Barnes Ice cap (Fig. 7.7). The large divergence from a parabolic profile in some of the graphs (Fig. 7.8) is due to the profiles not necessarily corresponding to flowlines. In all the graphs, the present ice sheet, which is used as the initial state in the model, is depicted as a continuous line; the level of the ice sheet after 360 years have elapsed is depicted by a dashed line.

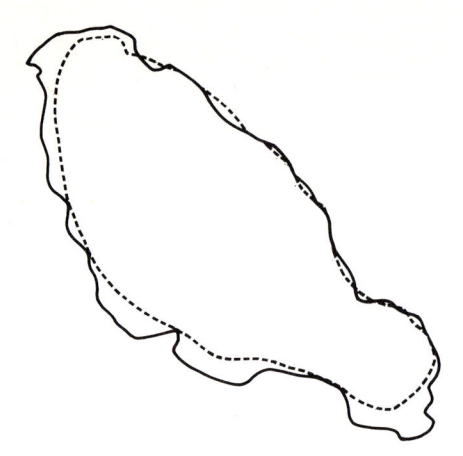

Fig. 7.9. Areal extent of the ice cap predicted by model after 360 years (*broken line*) compared with the present Barnes Ice Cap (*solid line*). (Mahaffy 1976)

The results of the three-dimensional model (Fig. 7.9) are consistent with the field evidence for the late Quaternary glacial history of the Barnes Ice Cap and for its present form. The observed and predicted levels of ice accord within 10 per cent or better for the north dome area for runs with a high flow-law exponent. Mahaffy found that a little surprising given the simple relations assumed for the specific net budget and the lack of basal sliding. The model results for the south dome are consistent with the assumption that the ice has recently surged to the south. The models are unable to replicate the east – west asymmetry of the actual Barnes profile, probably because a significant amount of ice movement beneath the southwest side of the north dome results from basal sliding, whereas ice movement beneath the northeast side of the north dome entails negligible basal sliding.

7.2 Water

7.2.1 Overland Flow

Overland flow is an example of gradually varied, unsteady, nonuniform, free surface flow. It may be modelled by applying continuity and momentum equations to water flowing over a permeable, inclined plane. The one-dimensional forms of these, the de Saint Venant equations may be written per unit width of overland flow as

$$\frac{\partial q}{\partial x} + \frac{\partial y}{\partial t} = r(t) - i(t),$$

in the case of the continuity equation, and

$$\frac{1}{g}\frac{\partial v}{\partial t} + \frac{\partial y}{\partial x} + \frac{v}{g}\frac{\partial v}{\partial x} = S_0 - S_f - \left[\frac{\{r(t) - i(t)\}}{g}\right]\frac{v}{y}$$

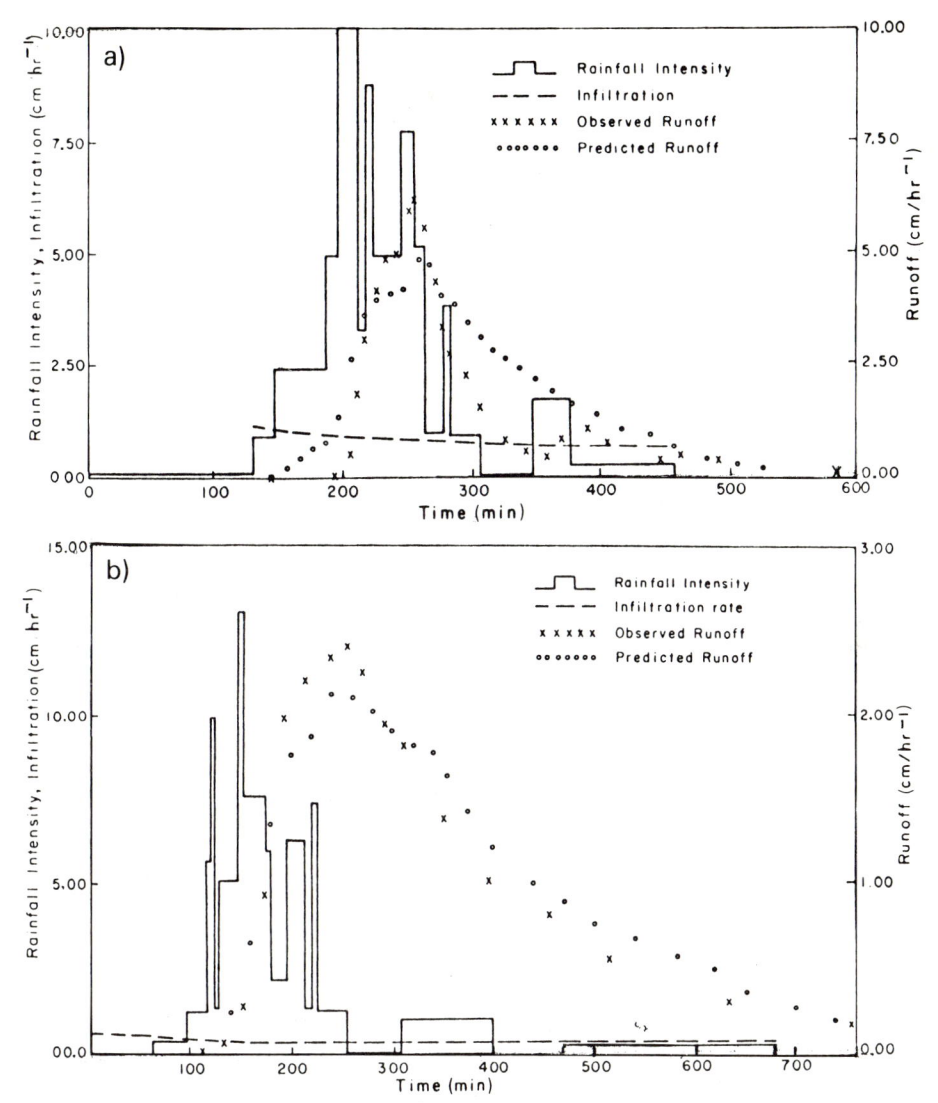

Fig. 7.10a, b. Hydrograph predictions for rainfall event of 29 March 1965 on **a** Watershed SW-17 and **b** Watershed G, Riesel (Waco), Texas. (Singh 1979)

in the case of the momentum equation where y is water depth, t is time, v is water velocity, q is discharge, r is the rainfall rate, i is the infiltration rate, x is distance in the flow direction from the top of the flow plane, g is the acceleration due to gravity, S_0 is the bottom slope of overland flow, and S_f is the friction slope of overland flow. The basic theoretical work on these equations was carried out by Keulegan (1944, 1948). Subsequent studies are too numerous to cite any more than a few cases. A recent example is the work of Singh (1979), who combined the continuity and momentum equations into a single differential equation and then linearized it to have just one independent variable. Singh's linear dynamic model predicts reasonably well the hydrograph produced by rainfall events in two natural agricultural catchments near Riesel (Waco), Texas. As can be seen in Fig. 7.10, the model preserves the shape of the hydrograph satisfactorily and in both cases the error in prediction of the hydrograph peak and its time are less than 30 per cent.

Models of overland flow took a new direction in the early 1970's by the inclusion of the source area concept (for instance, Freeze 1971; Engman and Rogowski 1974). In order to accommodate the source area concept, Ishaq and Huff (1979) revised the continuity equation of overland flow and constructed a model, the results of which are promising and suggest that major portions of runoff are indeed generated by overland flow originating from small parts of a watershed. A successful model of watershed runoff response needs to incorporate a set of distributed parameters defining variable infiltration and soil moisture characteristics in subcatchments (Beven and Kirkby 1979).

7.2.2 Open Channel Flow

The flow of water in open channels is unsteady and nonuniform. The equations of continuity and momentum required to model open channel flow are similar to the equations of continuity and momentum for overland flow. The continuity equation is

$$\frac{\partial Q}{\partial x} + w \frac{\partial y}{\partial t} = q$$

and the momentum equation is

$$S_f = S_0 - \frac{\partial y}{\partial x} - \frac{v}{g}\frac{\partial v}{\partial x} - \frac{1}{g}\frac{\partial v}{\partial t},$$

where Q is the discharge rate, x is the distance in the flow direction, w is surface width, y is depth of flow, q is lateral inflow from stream banks, S_0 is the bottom slope of the channel, S_f is the friction slope of the channel, v is flow velocity, g is the acceleration due to gravity, and t is time (see Chow 1959). It is normally more convenient to express the momentum equation in terms of the Chézy or Manning form of the resistance equation:

$$Q = CA\sqrt{R(S_0 - M)} \quad \text{(Chézy)}$$
$$Q = (1.49/n) R^{2/3} A \sqrt{(S_0 - M)} \quad \text{(Manning)},$$

where C is the Chézy roughness coefficient, R is the hydraulic mean radius, n is the Manning roughness coefficient, and M is the right-hand side of the momentum equation without the S_0 term.

These equations can be used to predict flow at a cross-section or outlet and can therefore be used to solve problems of flood routing (see Raudkivi 1982, pp. 246–253). Alternatively, flow can be predicted from a model which uses existing hydrological theory on the storage characteristics of drainage basins to compute outflow given the inflow. Models of this kind are based on the notion that the stream channel acts like a long thin reservoir: it receives inflow from surface and subsurface flow, stores water in temporary channel storage, and discharges outflow at the outlet (Fleming 1975, p. 78). The continuity relation used in these models takes the form

$$I = O + \frac{ds}{dt}$$

where I is inflow, O is outflow, s is channel and floodplain storage, and t is time. To operationalize these models several techniques are used – unit hydrographs, Muskingham routing, cascades of linear reservoirs, and others (see R. K. Price 1973; Fleming 1975; Beven et al. 1979; Zaghloul 1979; Raudkivi 1982).

7.2.3 Flow in Porous Media

The continuity equation for water moving through rock or soil is written for three dimensions in Cartesian coordinates

$$\frac{\partial \theta}{\partial t} = - \left[\frac{\partial q_x}{\partial x} + \frac{\partial q_y}{\partial y} + \frac{\partial q_z}{\partial z} \right],$$

where θ is the volumetric moisture content and q_x, q_y, and q_z are the fluxes of water in the x, y, and z directions, respectively. The fluxes may be defined by Darcy's Law which can be written either as

$$q_x = -K \partial h / \partial x,$$

where K is hydraulic conductivity and h is total hydraulic head, or as

$$q_x = -K \partial \phi / \partial x,$$

where ϕ is the total water potential and the other terms are as before. In problems concerning saturated flow, units of hydraulic head are normally employed. The basic equation of continuity for groundwater flow, with flux defined by Darcy's Law, is

$$\frac{\partial \theta}{\partial t} = - \left\{ -\frac{\partial}{\partial x} \left(K \frac{\partial h}{\partial x} \right) - \frac{\partial}{\partial y} \left(K \frac{\partial h}{\partial y} \right) - \frac{\partial}{\partial z} \left(K \frac{\partial h}{\partial z} \right) \right\}.$$

Assuming that the hydraulic conductivity is constant in all directions and assuming a steady state, this equation reduces to

$$K \left(\frac{\partial^2 h}{\partial x^2} + \frac{\partial^2 h}{\partial y^2} + \frac{\partial^2 h}{\partial z^2} \right) = 0 \quad \text{or}$$

$$K \nabla^2 h = 0.$$

This is the Laplace equation and ∇^2 is the Laplacian operator. It is of great importance in groundwater hydrology. It can be used to model the configuration of a regional groundwater system assumed to be in a steady state. For example, Toth (1962) has, with a deal of success, applied the model to groundwater systems in central Alberta, Canada.

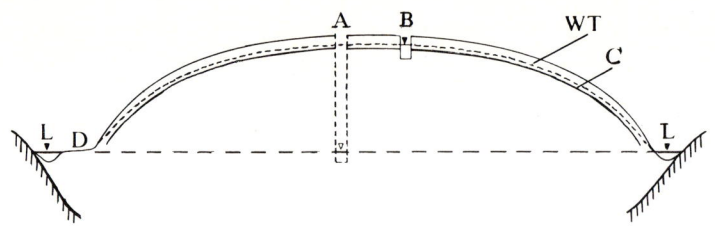

Fig. 7.11. Schematic sectional profile through a raised mire. *A* and *B* are dip wells. *C* is the boundary between the top of the catotelm and the bottom of the acrotelm. *D* is a surface of seepage (a phreatic surface at which water emerges) supporting fen vegetation in the lagg. *L* are lagg streams. *WT* is the water table. (Ingram 1982)

If sources and sinks of water, *R*, are allowed in the groundwater system, then the model is described by the Poisson equation

$$K \nabla^2 h = -R(x, y).$$

Being more versatile than the Laplace equation, the Poisson equation is useful in a wider range of situations. For instance, Jacobs (1943) has used a one-dimensional form of the Poisson equation to simulate groundwater flow beneath Long Island, New York, making allowance for precipitation.

The Laplace equation, Poisson equation, and more complex transient flow models have been applied to a variety of problems involving the management of groundwater resources. These models have a history dating back to the late 1800's. They are described and discussed in a number of recent texts including Freeze and Cherry (1979), Fetter (1980), Mercer and Faust (1981), and Wang and Anderson (1982). The models also have played a role in the understanding of the form and function of certain Earth surface systems such as karst landscapes and raised mires. The models have been applied to karst in a rather general way (Scheidegger 1970, pp. 413–414; Powell 1975; Ford 1980). More specific is Ingram's (1982) application of a groundwater model to raised mires, ecosystems in which waterlogged peat accumulates above the level of surrounding streams to produce surface forms resembling inverted watch glasses. Ingram challenges the suggestion that the domed peat deposit remains waterlogged owing to matric forces and demonstrates that a model involving impeded drainage is better in accord with the structure of the peat and with the basic tenets of soil physics (Fig. 7.11).

The waterlogged zone in raised mires commonly lies within a few centimetres of the surface and most of the peat accordingly experiences a positive hydraulic potential. A water body of this kind is inherently unstable, and without replenishment by precipitation it would in time disperse either by evaporation or by seepage into surrounding lagg streams. It is thus maintained by a dynamic equilibrium between recharge and seepage. Ingram argues that, when developed in homogeneous, isotropic porous media, groundwater mounds have a height, width, and shape which can be predicted from Laplace's equation, Darcy's flow law, and the Dupuit-Forchheimer assumptions (that the flow is horizontal and that the hydraulic gradient is equal to the slope of the water table). In the simplest of cases, as considered by Ingram, an idealized peat deposit, shaped like a raised

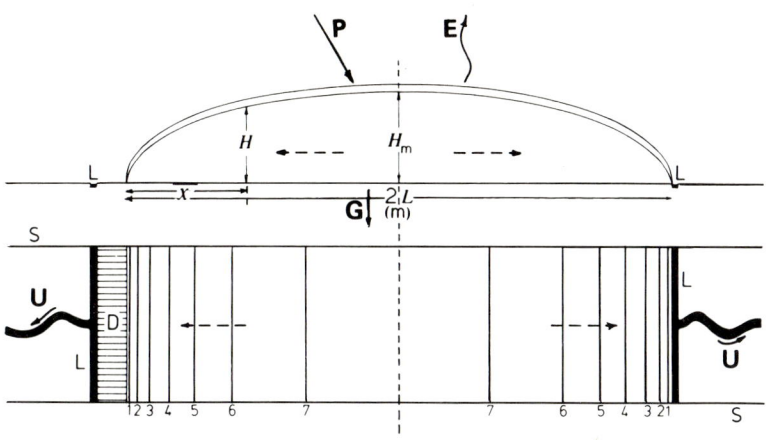

Fig. 7.12. Groundwater mound in a raised mire, in sectional profile (*above*) and in plan (*below*). D and L as in Fig. 7.11; *m* summit and catchment boundary between the two lagg streams; *s* parallel sides of confining valley with level floor; *1 – 7* contours of the water table (equipotentials according to the Dupuit-Forchheimer approximation) spaced at equal intervals of altitude. *Thick arrows* and *bold letters* fluxes of water and flux densities (see text). *Thin arrows* dimensions of groundwater mound (see text). (Ingram 1982)

mire, is confined by parallel-sided valleys (Fig. 7.12). The net recharge is denoted by U and the permeability of the deposit by K. The quotient U/K is related to the dimensions of the groundwater mound by the expression

$$\frac{U}{K} = \frac{H^2}{2Lx - x^2},$$

which takes the form of the equation of an ellipse. By setting $x = L$, the maximum height of the mound, H_m, is defined by

$$\frac{U}{K} = \frac{H_m^2}{L^2},$$

an expression that corresponds to one limit of the exact solution of Laplace's equation.

To put these ideas to the test, Ingram studied Dun Moss, a small raised mire in east Perthshire, Scotland (Fig. 7.13). On the transect across the mire, the highest point, H_m, is 8.2 m above the lagg streams. The length of the transect, $2L$, corrected for 27 m of surfaces of seepage mainly adjacent to Green Burn, is 520 m. So a morphologically based estimate for U/K is 9.95×10^{-4}. Using this estimate gives the ellipse shown as a thick line in Fig. 7.14. Tentative estimates for the conditions prevalent during the driest year on record give U/K as 1.2×10^{-7} and the thin line in Fig. 7.14. Thus the elliptical shape and proportions of the mire surface are in agreement with Ingram's model.

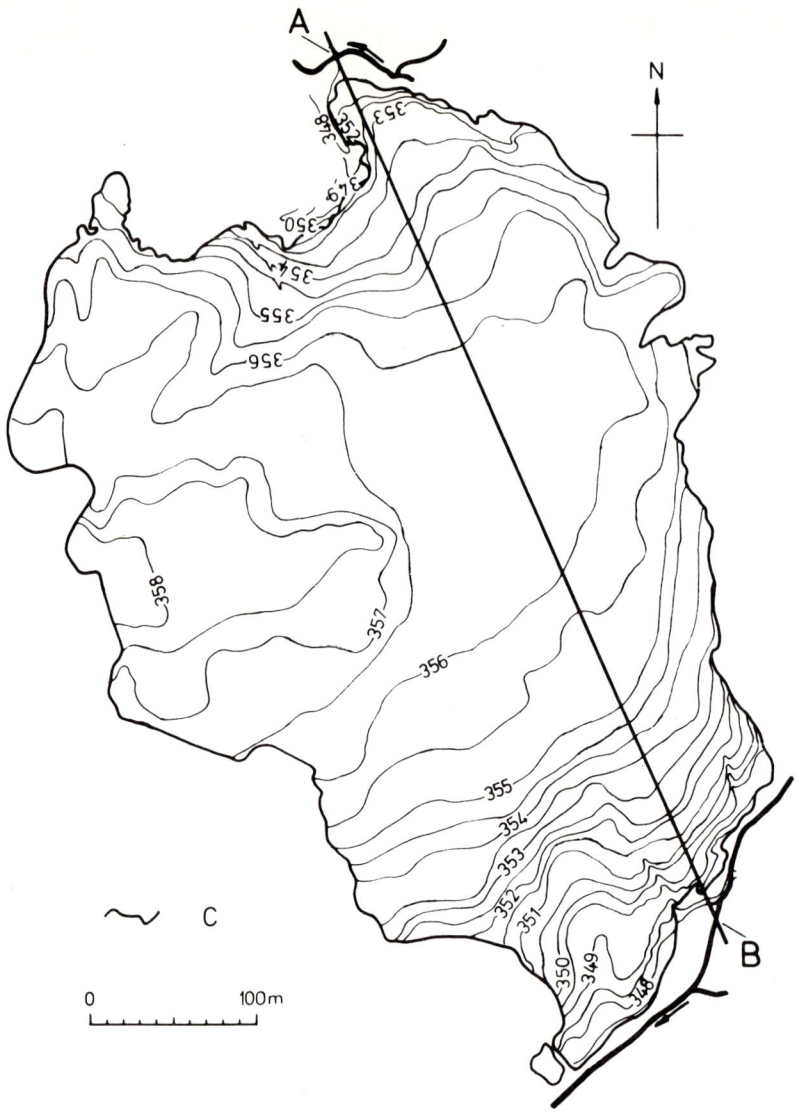

Fig. 7.13. Physiography of the mire expanse at Dun Moss. *AB* line of profile transect connecting the captured headwaters of the Green Burn with SE-NW reach of the Limekiln Burn. *C* boundary of mire expanse. *Arrows* lagg streams and so on. Contours in metres, based on a theodolite survey of the hollows on the uneven surface. (Ingram 1982)

7.2.4 Unsaturated Flow

For the case of unsaturated flow in soils and other porous media, the continuity equation is usually written with Darcy's law written in terms of water potential, ϕ. For the general case in Cartesian coordinates the equation is

$$\frac{\partial \theta}{\partial t} = \frac{\partial}{\partial x}\left(K\frac{\partial \phi}{\partial x}\right) + \frac{\partial}{\partial y}\left(K\frac{\partial \phi}{\partial y}\right) + \frac{\partial}{\partial z}\left(K\frac{\partial \phi}{\partial z}\right).$$

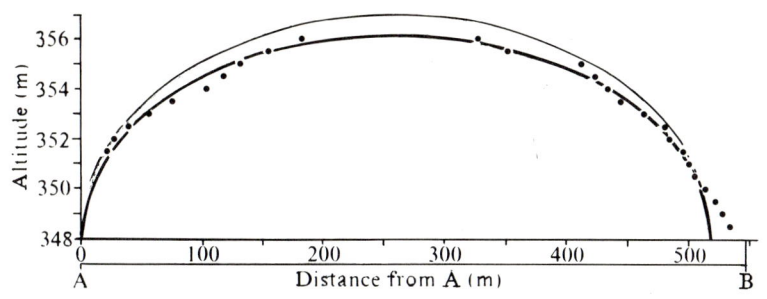

Fig. 7.14. Profile of mean hollow altitudes on the surface of Dun Moss in the vicinity of transect AB (Fig. 7.13). *Plotted points* contour intersections from Fig. 7.13. *Thick line* Dupuit-Forchheimer ellipse for $U/K = 9.95 \times 10^{-4}$, the morphometric estimate, denoting the shape predicted on the groundwater mound hypothesis. *Thin lines* the same for $U/K = 1.2 \times 10^{-3}$, the hydrodynamic estimate for the driest year in a decade. (Vertical scale exaggerated 20-fold). (Ingram 1982)

The total water potential, ϕ, is made up of the sum of at least two separate potentials – the matric or pressure potential, ψ, which is the combined effect of adsorptive and capillary forces, and the gravitational potential, z:

$$\phi = \psi + z.$$

Substituting this expression into the continuity equation and for simplicity dropping the x and y dimensions, gives

$$\frac{\partial \theta}{\partial t} = \frac{\partial}{\partial z}\left\{K\frac{\partial}{\partial z}(\psi+z)\right\}.$$

Since K, the hydraulic conductivity, is a function of matric potential, ψ, the equation is usually written

$$\frac{\partial \theta}{\partial t} = \frac{\partial}{\partial z}\left(K(\psi)\frac{\partial \psi}{\partial z}\right) + \frac{K(\psi)}{\partial z}.$$

This equation describes the vertical one-dimensional isothermal movement of water in a homogeneous porous material. Finite difference approximations to these equations have enabled a number of computer-based solutions to be obtained (for instance, Whistler and Klute 1967). In general, the solutions are complex but show the advance of a sharp wetting front at a declining rate and with some loss of sharpness as the front moves into the soil. A typical solution is shown in Fig. 7.15, which is taken from the study by Watson and Curtis (1975), who in their model allowed for the effect of air compression during infiltration.

The extension of models of soil water movement to two dimensions has led to some interesting findings. An early effort to mathematically model soil moisture movement on a slope was made by Klute et al. (1965). These workers considered the problem of steady-state flow in a uniform, inclined slab of soil wetted by rain. They were able to show that a lateral or downslope component of flow is generated when there is an impermeable layer or a perched water table in the soil, a finding that gave a theoretical framework for Whipkey's (1965) field observa-

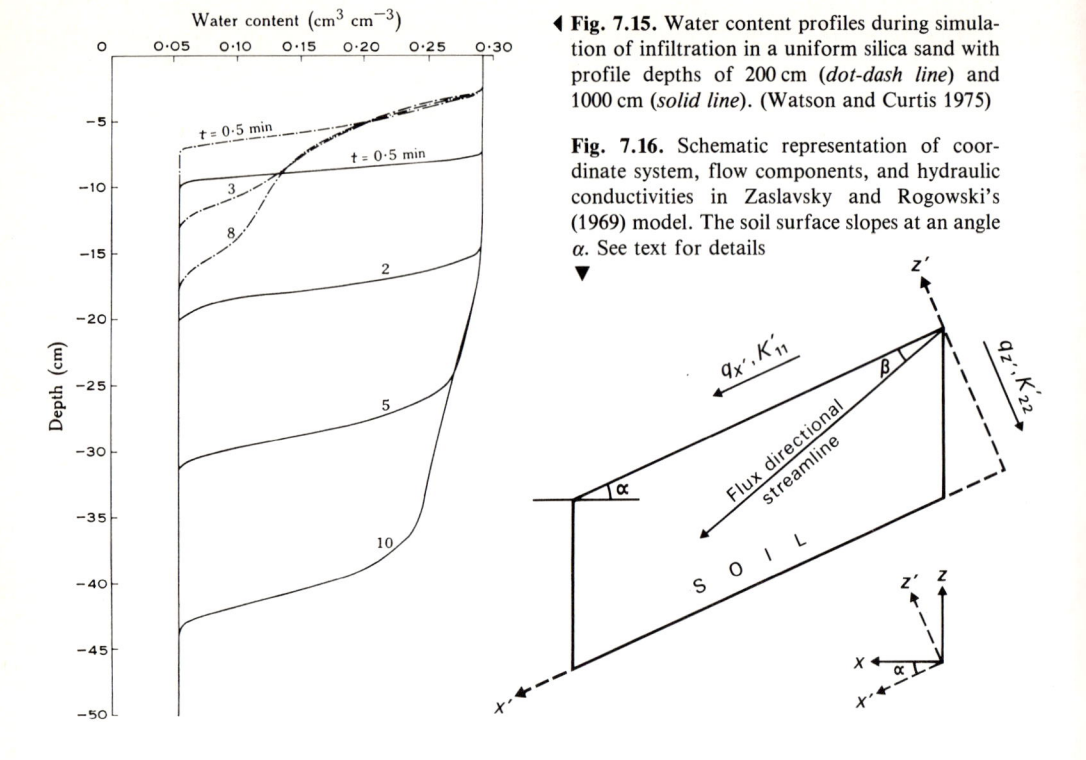

Fig. 7.15. Water content profiles during simulation of infiltration in a uniform silica sand with profile depths of 200 cm (*dot-dash line*) and 1000 cm (*solid line*). (Watson and Curtis 1975)

Fig. 7.16. Schematic representation of coordinate system, flow components, and hydraulic conductivities in Zaslavsky and Rogowski's (1969) model. The soil surface slopes at an angle α. See text for details

tion of throughflow. Zaslavsky and Rogowski (1969) extended the analysis of Klute et al. (1965). They demonstrated, by considering the relative importance of flow normal to soil layers compared with flow parallel to soil layers, that a lateral component of flow will occur, even when the lower layers of the soil are only slightly less permeable than the upper soil layers and during rainstorms which fail to produce saturated flow. Their arguments run so: q'_x is the flow component parallel to the soil layers and q'_z is the flow component normal to the soil layers; K'_{11} and K'_{22} are, respectively, the composite hydraulic conductivities in the x' and z' directions (Fig. 7.16). The ratio between the flow components is given by

$$q'_x/q'_z = \frac{n \tan \alpha}{\left(\dfrac{\partial \psi}{\partial z'} \cos \alpha\right) + 1},$$

where $n = K'_{11}/K'_{22}$ and is the degree of anisotropy, and $\tan \alpha$ is the slope of the hillside. Under steady-state infiltration, or whenever $\partial \psi/\partial z'$ can be neglected, this equation reduces to

$$q'_x/q'_z = n \tan \alpha.$$

In both equations, q'_x does not vanish unless $\tan \alpha$, the angle of slope of the hillside, is zero. Even if $\partial \psi/\partial z'$ is not negligible, and even with the lightest of rainfall, there will be a component of flow parallel to the soil surface. The actual

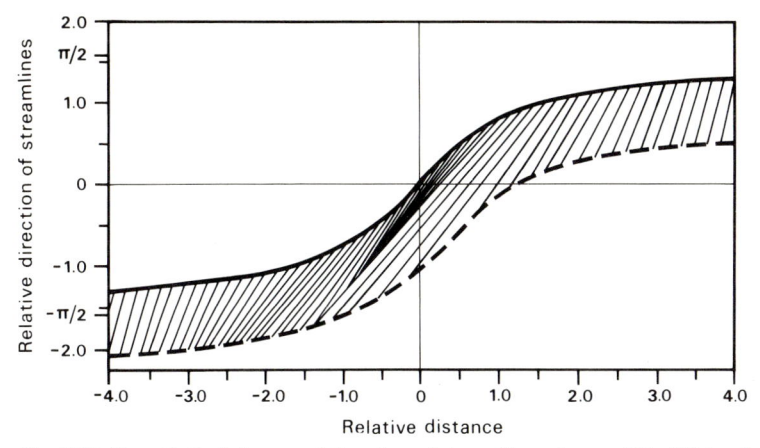

Fig. 7.17. Hypothetical slope, and direction of streamlines, for $n = (K'_{11}/K'_{22}) = 5$. *Dashed line* runs parallel to the soil surface. (Zaslavsky and Rogowski 1969)

path taken by water in the soil is the resultant of the parallel and normal components of flow. The direction of the flux vector is defined by

$$\tan \beta = q'_z/q'_x = 1/n \tan \alpha .$$

In other words, given the assumptions made, β decreases as $\tan \alpha$, the hillside slope, and n, the anisotropy, increase. If $\tan \alpha = 0.1$ and $n = 100$, then $\tan \beta = 0.1$ which means that to reach a depth of 1 m, the water has to move downhill a distance of 10 m. Some results for a hypothetical slope are displayed in Fig. 7.17. Zaslavsky and Rogowski (1969) discussed the implications of their model to flow in three dimensions, where account must be taken of the influence of contour curvature on the flowlines.

Mathematical models of unsaturated flow based on a set of partial differential equations are difficult to apply to a particular field site. An alternative model, devised by Knapp (1974), treats a hillside as a system of linked reservoirs. The storage and discharge of each reservoir are solved in a manner akin to flood-routing techniques. What Knapp's (1974) model lacks in mathematical elegance, it makes up for, so Knapp claims, in its ready adaptability to field conditions.

7.3 Solutes

7.3.1 Seas and Lakes

The continuity equation for the movement of solutes in a river, lake, or ocean is, in Cartesian coordinates,

$$\frac{\partial C}{\partial t} = -\left(\frac{\partial J_x}{\partial x} + \frac{\partial J_y}{\partial y} + \frac{\partial J_z}{\partial z} \right),$$

where C is the concentration of the solute and the J's are the fluxes of solute in the three coordinate directions. The flux of solute may be brought about by two processes. Firstly, currents of water may carry the solute bodily with them. This process may be defined by the product of solute concentration and velocity of flow, that is, by vC. Secondly, the solute may diffuse along a solute concentration gradient as described in Fick's first law of diffusion. This diffusive process may be represented by the expression

$$J_x = D\frac{\partial C}{\partial x},$$

where D is a diffusion coefficient. The total flux is the sum of the two separate fluxes

$$J_x = vC - D\frac{\partial C}{\partial x}.$$

Putting this definition into the continuity equation gives

$$\frac{\partial C}{\partial t} = -\left\{\frac{\partial}{\partial x}\left(vC - D\frac{\partial C}{\partial x}\right) - \frac{\partial}{\partial y}\left(vC - D\frac{\partial C}{\partial y}\right) - \frac{\partial}{\partial z}\left(vC - D\frac{\partial C}{\partial z}\right)\right\}$$

or, expressed more succinctly,

$$\frac{\partial C}{\partial t} = -v\,\text{grad}\,C + D\,\text{div}\,C,$$

which is a diffusivity equation with a mass transport term.

The basic model has a range of practical applications including the simulation of contaminants through groundwater systems (M. P. Anderson 1979) and the simulation of nutrient movement in soil profiles (Boast 1973). A classic application of deterministic modelling to solute movement is Briggs and Pollack's (1967) model of salt concentration and the formation of halite (rock salt) in an inland sea. Specifically, they modelled an evaporite basin which existed in Late Silurian times in the Michigan basin area of the United States (Fig. 7.18). The basin was partially cut off from the open sea by bounding reef banks. Briggs and Pollack assumed that, because of evaporative loss of water in the basin, sea water would have been drawn in to make good the deficit. The sea water would have moved into the basin over reefs and through a limited number of distinct inlets, carrying with it dissolved salt. The saltier water inside the basin (saltier than sea water because of evaporative water loss) would have been prevented by the reefs from returning to the open sea. Continued evaporation would therefore have lead to salt concentrations reaching saturation level and the precipitation of salt.

Briggs and Pollack developed these ideas mathematically by using three deterministic submodels. Firstly, they used the model

$$v = -\text{grad}\,\Phi$$

to describe the steady state, potential flow of water in the basin where v is the "brine velocity vector" and Φ is a velocity potential function. This model is

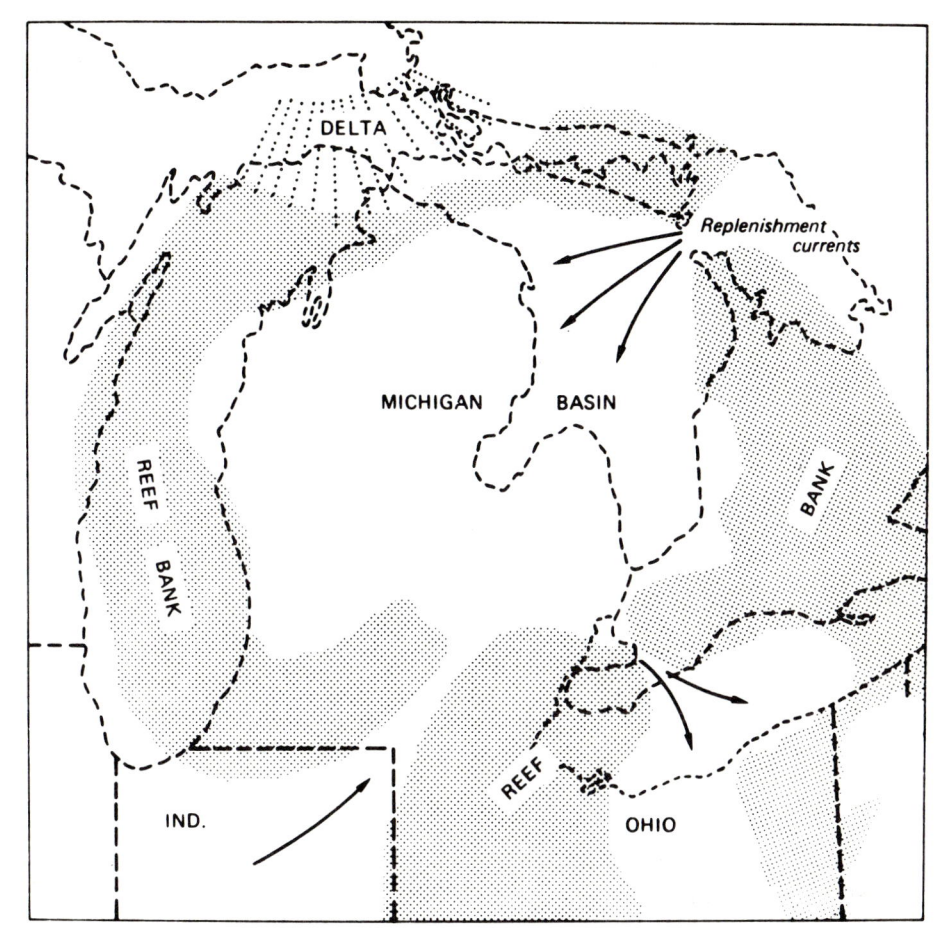

Fig. 7.18. Inferred geography in, and adjacent to, the Michigan Basin during Late Silurian time. Reprinted with permission from (Briggs and Pollack 1967)

essentially a Laplace equation. The loss of water by evaporation from the surface water of the basin was included in this model to give the Poisson equation

$$\text{div grad}\,\Phi = E/h,$$

where E is the evaporation rate and h is the depth of the basin. Secondly, they used the model

$$-\text{div}(-D\,\text{grad}\,C + vC) = \partial C/\partial t$$

to describe the transport of salt in the basin. Thirdly, they allowed for salt precipitation to occur when the concentration reached $311\ \text{g l}^{-1}$.

To simplify the geographical situation, Briggs and Pollack made the initial assumption that just two inlets admitted sea water to the basin and one outlet in the southeast released it; reefs acted as a barrier to the influx of sea water. The equation was solved numerically to produce a map showing the velocity and

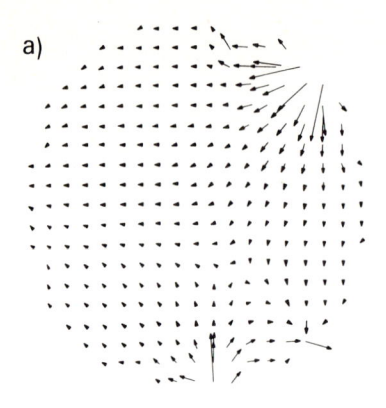

Fig. 7.19. a Flow diagram of sea currents from a simulation run in which two inlets and one outlet to the Michigan basin were assumed. **b** Isosalinity contours produced after a steady state has been attained in a simulation run assuming two inlets and one outlet to the Michigan basin. Values are in g salt l^{-1}. Saturation values of 311 g l^{-1} occur in the western part of the area. **c** Thickness contour map based on borehole data showing distribution of halite in Cayugan series (Upper Silurian) in the Michigan basin. (Briggs and Pollack 1967)

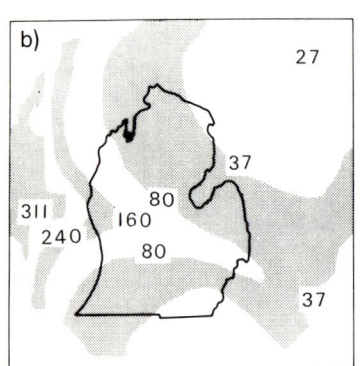

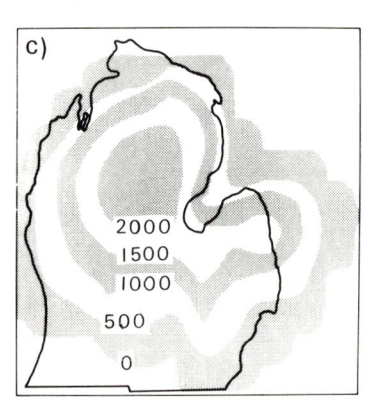

direction of water movement (Fig. 7.19a), the steady-state concentration of salt in the sea water (Fig. 7.19b), and the area of salt precipitation. To test the results, the predicted pattern of halite occurrence was compared to the observed pattern of halite thickness in the region (Fig. 7.19c). The two patterns mismatched, the predicted pattern suggesting that halite would be found in the western part of the basin (the area in Fig. 7.19b with concentrations in excess of 311 g l^{-1}), the actual pattern showing halite deposits in the centre of the basin. No amount of tampering with the relative volumes of sea water through the two inlets and one outlet could shift the site of salt deposition to the central area. However, when, as well as one outlet, a chain of "leaks" through the reefs were built into the model, the salt concentrations reached saturation point in the centre of the basin (Fig. 7.20a) and the predicted salt thickness pattern after 12,000 years of deposition accorded remarkably well with the observed pattern (Fig. 7.20b).

7.3.2 Solutes in Groundwater

A recent development in the deterministic modelling of solutes in Earth surface systems is the building of geochemical reaction models for groundwater. In their simplest form, such models consider a single chemical species. Ford and Drake (1982) and Drake and Ford (1981), for instance, build a model on the basis that

Solutes

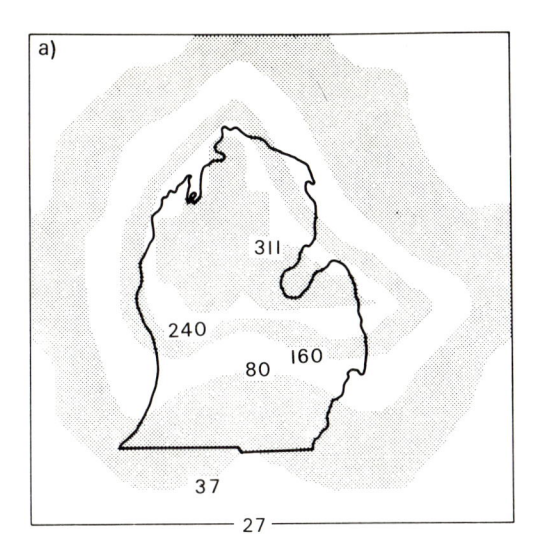

Fig. 7.20. a Isosalinity map showing salt concentrations (g salt l^{-1}) after 4000 years of operation of a model which, as well as two inlets and one outlet, allows for "leaky" reefs. **b** Simulated salt thickness pattern employing "leaky reef" hypothesis. Thickness values are in feet. (Briggs and Pollack 1967)

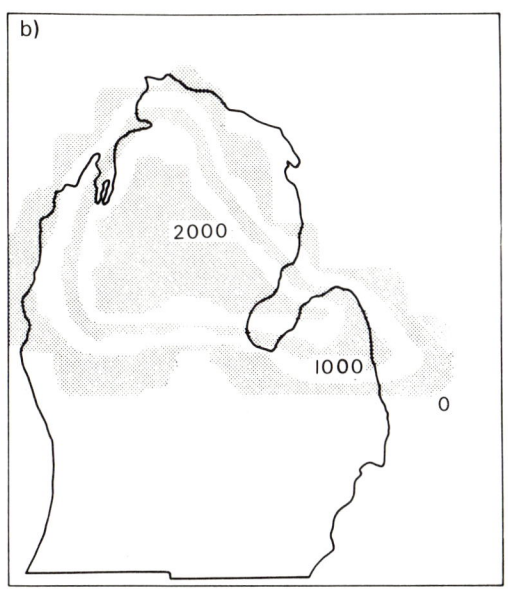

two main factors, temperature and the "evolution system", explain inter-regional differences in limestone groundwater concentrations. Temperature determines the average rate of biogenic CO_2 in the soil and is modelled by

$$\log P_{CO_2^*} = -2 + 0.04\,T$$

$$P_{CO_2} = \{(0.21 - P_{CO_2})/0.21\} P_{CO_2^*},$$

where $P_{CO_2^*}$ is the potential partial pressure of CO_2 (atm), P_{CO_2} is the actual value, T is the mean annual air or groundwater temperature (°C). The first equation expresses the dependence of biological activity in the soil upon temperature;

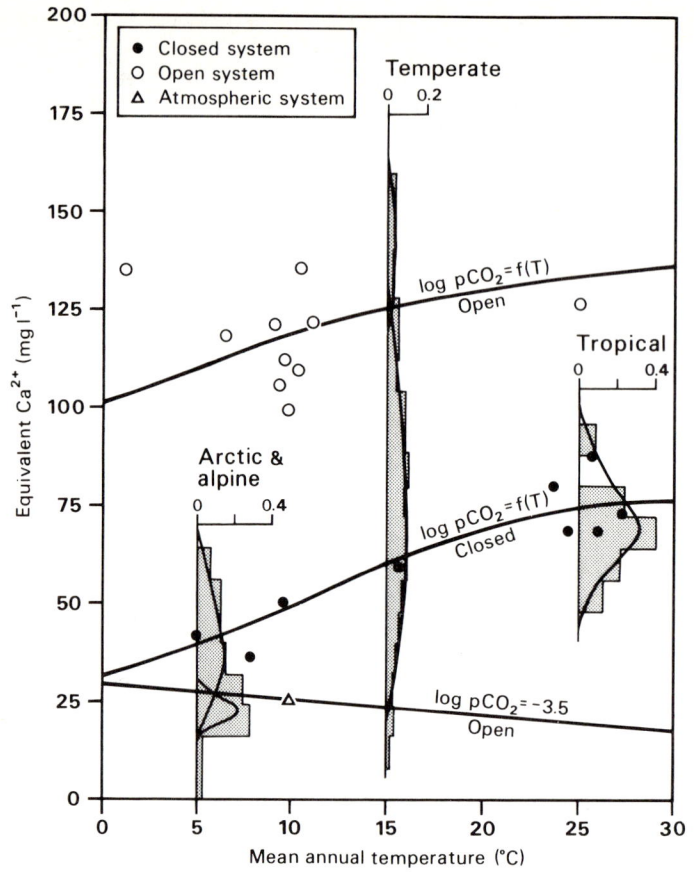

Fig. 7.21. The relation between Ca^{2+} concentrations and temperature in open and closed groundwater systems. *Lines* show the equilibrium Ca^{2+} concentrations for open and closed systems. *Histograms* are based on field data of Smith and Atkinson (1976). *Data points* are mean values selected from the literature by Drake and Ford (1981). (After diagrams in Ford and Drake 1982)

the second equation expresses the inhibition of soil respiration resulting from depressed values of oxygen concentration at high P_{CO_2}. The second factor, the "evolution system", determines, for a given P_{CO_2}, the equilibrium concentration of dissolved limestone.

Ford and Drake distinguish coincident evolution systems from sequential evolution systems. In coincident (open) systems, examples of which include carbonate rich till and a very porous bedrock such as chalk beneath a thin soil, the limestone and water equilibrate with the elevated P_{CO_2} of the atmosphere. In sequential (closed) systems, water first equilibrates with an atmosphere of elevated P_{CO_2} in the absence of limestone (as in a noncarbonate unit above a limestone aquifer) and then equilibrates with limestone in the aquifer in the absence of an atmosphere. Water-logged soils also restrict air supply and lead to a reduction of dissolved annual concentrations in areas of coincident systems with low temperatures and highly seasonal recharge. For this reason, many Arctic-alpine

Solutes

areas mimic sequential systems. The data from published sources fit this scheme well (Fig. 7.21).

Far more complicated geochemical reaction models consider the reaction of several chemical species, using a variety of methods. The philosophy and methology of geochemical reaction models is discussed at length by Plummer et al. (1983). In brief, the method involves using empirical data from selected points along a flow path in a groundwater system to evaluate a number of plausible chemical models which can be applied to the system. To this end, use is made of speciation calculations, reaction-path simulations, and mass balance calculations. Computer programmes are available to aid these calculations (Parkhurst et al. 1980; Parkhurst et al. 1982).

7.3.3 Solutes in Soils

For solutes dissolved in water which is contained in a porous medium and for which there are sources and sinks of solute, the continuity equation, with transport terms included, reads

$$\frac{\partial(\theta C + S)}{\partial t} = -v \operatorname{grad} C + D \operatorname{div} C \pm Q,$$

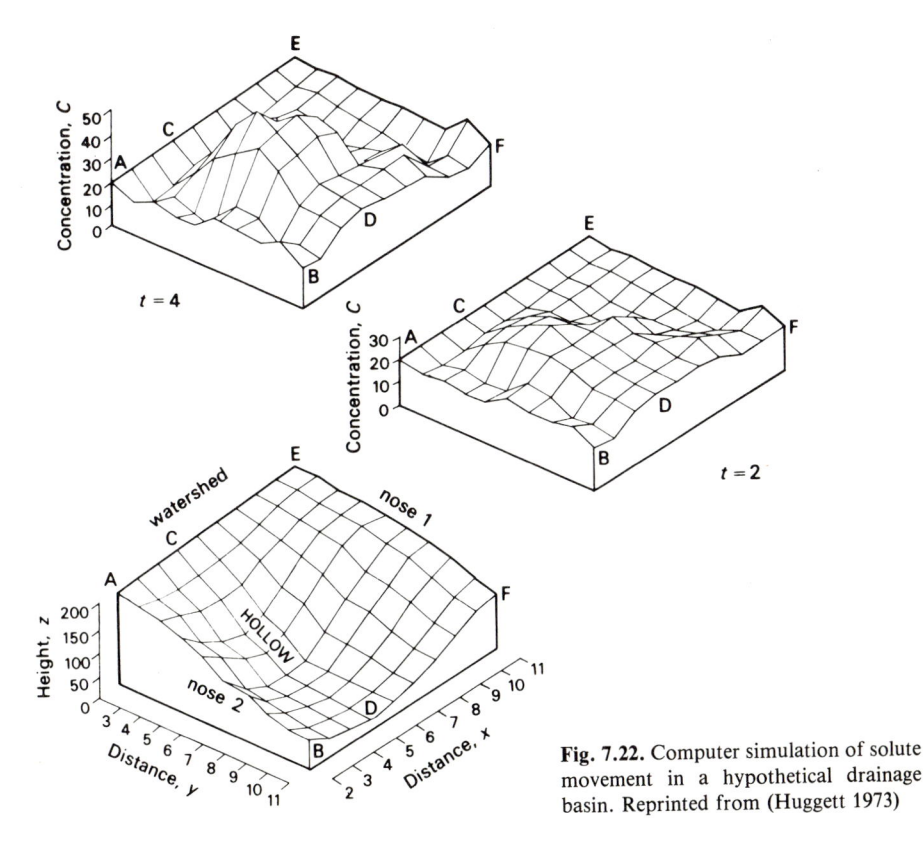

Fig. 7.22. Computer simulation of solute movement in a hypothetical drainage basin. Reprinted from (Huggett 1973)

where θ is the volume of water held in interconnected pores, S is the amount of the solute adsorbed on soil surfaces and located in dead-end pores, Q is a source or sink of the solute, and all other terms are as previously defined.

A simplified form of this equation was used to study solute movement in catenas and soil landscapes by Huggett (1973, 1975, 1982). The simulated pattern of solute redistribution along a catena, starting from a uniform distribution of solute, showed, as would be expected, downslope translocation of material. But is also showed more subtle effects for which some supporting field evidence is available. The downslope movement in the model led to an initial build-up of solutes at the junction of convex and concave portions of the catena, a pattern found in some slope soils by Furley (1971) and Whitfield and Furley (1971). This "peak" of concentration is not a static feature: as time progresses it moves as a wave downslope. The notion of a transient concentration wave along a catena is not unreasonable, since ions do progress through a vertical column of soil in this manner (Yaalon 1965) and seem to move along a soil catena in a similar manner (Yaalon et al. 1974; Rose et al. 1979; Huggett 1976). Huggett also modelled the redistribution of solutes in an erosional drainage basin (Fig. 7.22). Not surprisingly, an overall downslope movement of solutes was modified by convergences and divergences of flow arising from contour curvature. Convergence of movement into the hollow led to a build-up of material there, whereas over the nose divergence of movement led to a removal of material. Superimposed on this general pattern of redistribution in a drainage basin were a number of more subtle effects. Downslope movement along a flowline running into a hollow is relatively faster and shifts more material than downslope movement along a flowline running over the nose. Also, the velocity of throughflow decreases with depth and produces a lag effect: downslope movement is slower in lower horizons than in upper ones. Field evidence for all these predictions is given in Huggett (1976).

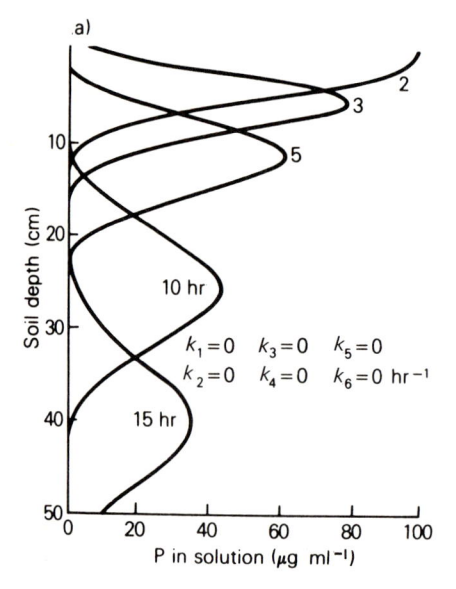

Fig. 7.23. Distribution of solution-phase phosphorus in a soil profile for the case in which all six reaction rate coefficients are zero. The k's are the rate constants: k_1 is from solution phase to sorbed phase; k_2 is from sorbed phase to solution phase; k_3 is from sorbed phase to immobilized phase and k_4 is the reverse process; k_5 is from immobilized phase to precipitated phase and k_6 is the reverse process. (Mansell et al. 1977)

Solutes

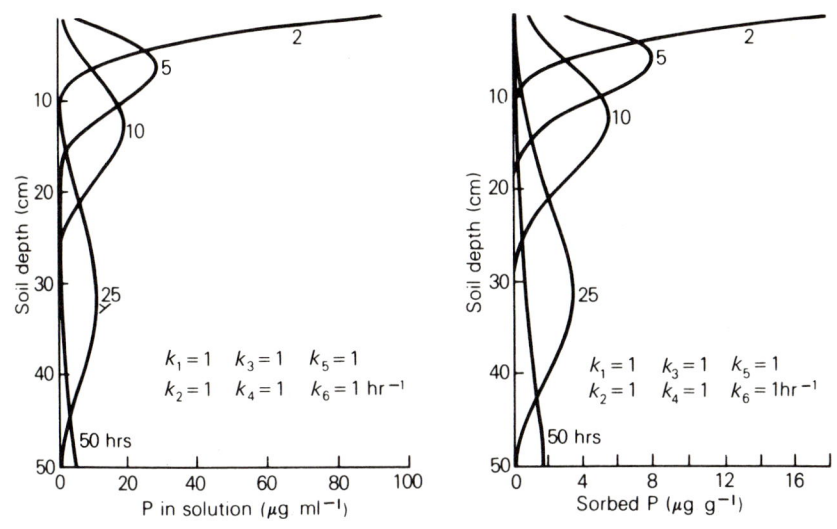

Fig. 7.24. Distribution of solution-phase and sorbed-phase phosphorus in a soil profile for the case in which all six reaction rate coefficients are 1 h^{-1}. (Mansell et al. 1977)

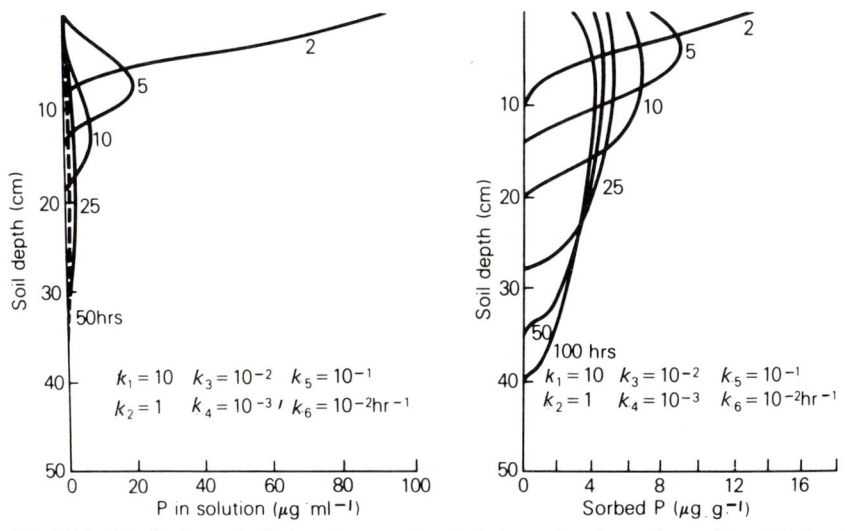

Fig. 7.25. Distribution of solution-phase and sorbed-phase phosphorus in a soil profile for the case in which $k_1 = 10$, $k_2 = 1$, $k_3 = 0.01$, $k_4 = 0.001$, $k_5 = 0.1$, and $k_6 = 0.01$ h^{-1}; this combination of reaction rate coefficients was deemed typical of many soil types and so was referred to as the standard case. (Mansell et al. 1977)

More comprehensive models of soil processes need to incorporate materials in various phases. An example is the model built by Mansell et al. (1977) to describe the storage, transformation, and movement of orthophosphate in a soil profile. Four phases are distinguished in the model: water-soluble phosphorus, physically adsorbed phosphorus, immobilized phosphorus, and precipitated phosphorus.

The phosphorus may be transformed from one phase to another according to processes described by six reversible kinetic equations. The transformations are determined by rate constants and the amount of phosphorus stored in each phase. Plant uptake and leaching are regarded as sinks associated with the solution phase of phosphorus. Rates of change phosphorus storage in each of the four phases are defined for (1) no flow conditions, which apply when the net movement of the soil solution is negligible, and (2) for steady flow conditions in which the movement of solution-phase phosphorus takes place. The model was solved for specific initial and boundary conditions. Initially, the model was applied to a soil column of length 100 cm and devoid of phosphorus. A solution containing 100 µg ml^{-1} of soluble phosphorus was applied (mathematically) to the surface of the column for two hours and then only water was applied to maintain a steady water flux of 1.0 cm h^{-1} for a total simulation run of 100 h. Figures 7.23 to 7.25 show some results. Figure 7.23 is the case where all six rate constants are zero which means that the applied phosphorus does not react chemically with the soil. The rapid transport predicted can be expected from unreactive solutes such as chlorides. Figure 7.24 is the case where all six rate constants are equal to 1.0 h^{-1}. Movement of the solution-phase phosphorus is similar to that in the first case except that zones of maximum concentration move more slowly and magnitudes of maximum concentration decrease more rapidly with time. This is of course because some of the phosphorus joins the adsorbed phases. Case three uses a "standard" calibration of rate constants and shows a rapid attenuation and slow downwards movements of solution-phase phosphorus (Fig. 7.25).

CHAPTER 8
Deterministic Models of Slopes and Sediments

The mathematical study of the development of slopes has been a venerable pastime since the Reverend Osmond Fisher (1866) offered "a slight contribution to the elucidation of questions of denudation, and at the same time an exemplification of the application of mathematics to a geological problem" in the form of a deductive argument as to the form of the profile formed behind the talus which accumulates as a chalk cliff in a quarry disintegrates. From these simple beginnings, a bewildering array of mathematical models of slope development has arisen (Armstrong 1976, p. 20). The models can be divided into three groups. The first group includes all models which, following the lead given by Fisher, focus on the shape of a rocky core beneath a pile of debris and in doing so regard a hillslope as a combination of discrete components. The second group includes all models which focus on a slope profile as a continuous curve which is produced either by an analytical solution to an equation of slope development in which some assumption is made as to the relation between slope processes and slope form, or by an analytical solution to a slope development equation derived from the continuity condition of mass transfer on slopes (cf. Hirano 1976). The third group of models makes similar assumptions to the models in the second group but, instead of deriving analytical solutions, uses computer simulation to predict slope form and slope change, and takes more account of the weathering of bedrock on a slope and the downslope movement of soil. It also includes models of "slopes" formed by wave and wind action.

8.1 Discrete Component Models

In his paper on the disintegration of a chalk cliff, Fisher (1866) showed that the form of the rock slope under a growing pile of talus is a semi-parabola with a vertex at the base of the original cliff (Fig. 8.1). Further efforts to determine the shape of this curve were made by Lehmann (1933), who deduced that where all the debris piles up as scree against the slope, the buried rock slope assumes the shape of a parabola; but that where all the debris is removed from the slope, the rock slope will become rectilinear with an angle equal to the angle of repose of the rock debris. The curvature of the convex slope underneath the debris depends on the original angle of slope (Bakker and Le Heux 1946). All these models concern a slope bounded by a plateau at the top and a plateau at the bottom. Van Dijk and Le Heux (1952) consider the case of a crest at the top of the slope and

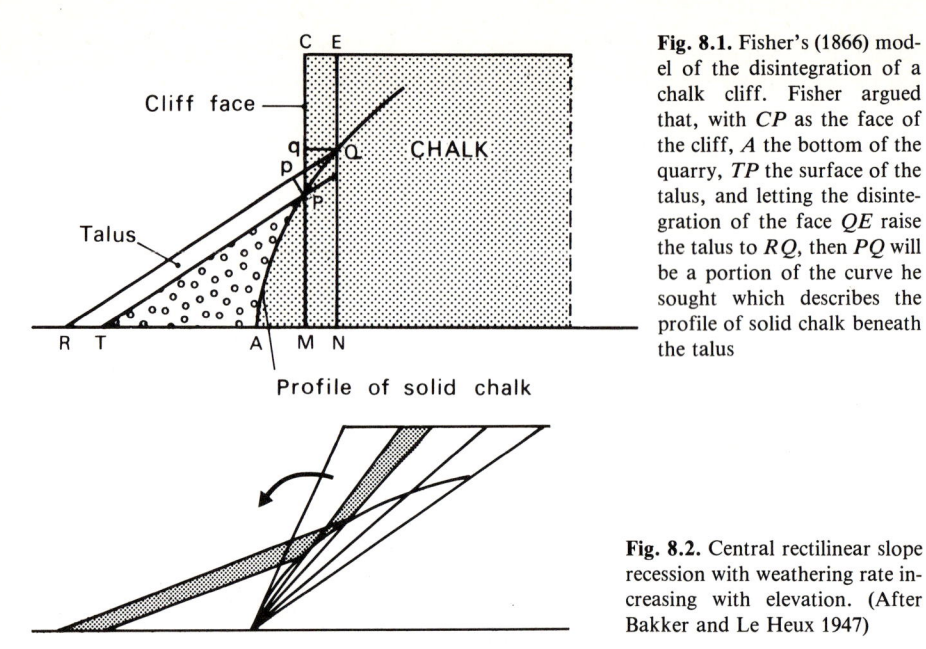

Fig. 8.1. Fisher's (1866) model of the disintegration of a chalk cliff. Fisher argued that, with *CP* as the face of the cliff, *A* the bottom of the quarry, *TP* the surface of the talus, and letting the disintegration of the face *QE* raise the talus to *RQ*, then *PQ* will be a portion of the curve he sought which describes the profile of solid chalk beneath the talus

Fig. 8.2. Central rectilinear slope recession with weathering rate increasing with elevation. (After Bakker and Le Heux 1947)

also the case where a plateau of finite width is weathered to form eventually a crest which, they find, will then recede.

To take into account variations in denudation along the slope, Bakker and Le Heux (1947, 1950) proposed a model of central, as opposed to parallel, rectilinear slope recession. The ideas behind their arguments are seen in Fig. 8.2, where a steep slope is bounded by two plateaux. The material coming off the slope during each time interval (infinitesimal) is bounded by two straight lines which always meet at the original foot-point of the slope. The debris accumulates at the bottom of the slope, forming a pile of screes which protect the slope bottom from further denudation. During denudation, the slope angle is reduced. Eventually, the slopes that develop under the debris assume a curved or straight profile according to the initial slope angle and the proportion of rock removed to the volume of talus: initially steep slopes lead to curved profiles whereas initially shallow slopes, say less than about 25°, lead to straight slopes; slopes on which all debris is removed form curved profiles whereas slopes where the debris accumulates as a veneer of waste of greater volume than the rock tend to be straight and stay at the angle of repose of the rock debris.

8.2 Analytical Models

8.2.1 Heuristic Models

The models of Bakker and his followers deal with a rocky core beneath a pile of debris. The form of visible slopes requires a model of mass transport. One group

Analytical Models

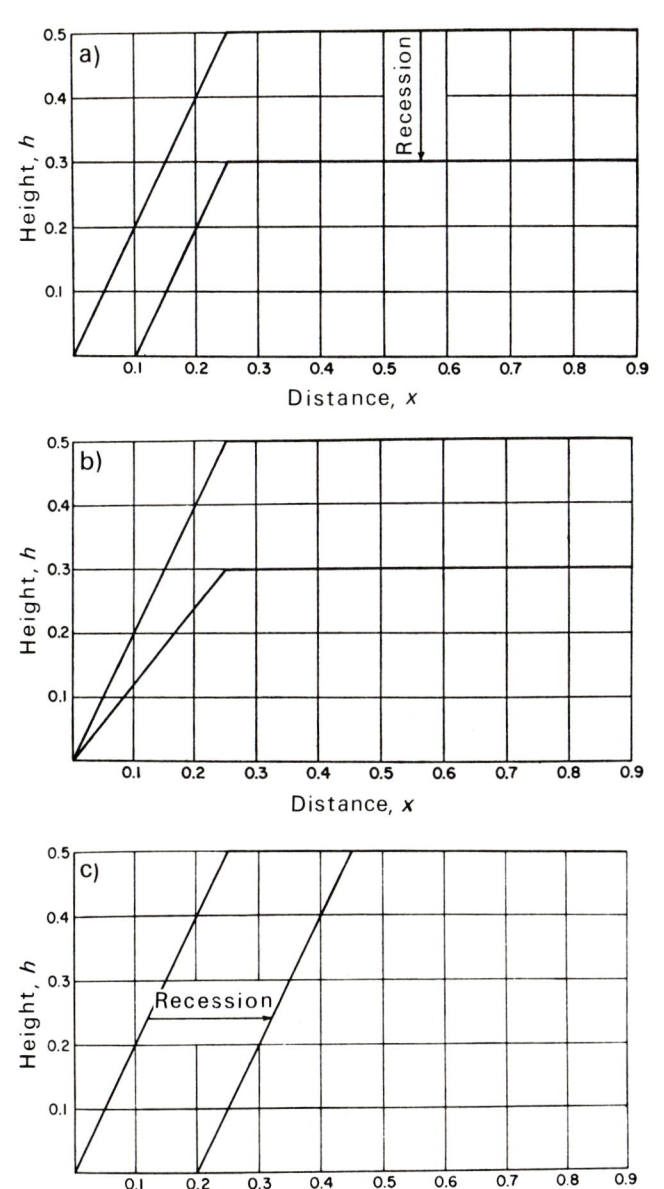

Fig. 8.3a–c. Three cases of slope development according to the linear theory.
a Denudation equal over all exposed parts of slope.
b Denudation is proportional to height above base level.
c Denudation is proportional to the steepness of the slope.
(Scheidegger 1961b)

of attempts at predicting visible slopes assumes certain relations between the rate of erosion and slope form. Early work on these lines was carried out by Scheidegger (1961b) and Young (1963a), but the earliest study is a little-known paper by Harold Jeffreys (1918), who was interested in the development of a land surface by surface wash.

Scheidegger (1961b) takes three possible assumptions. Firstly, that denudation is independent of slope and proceeds at an equal rate along the entire length

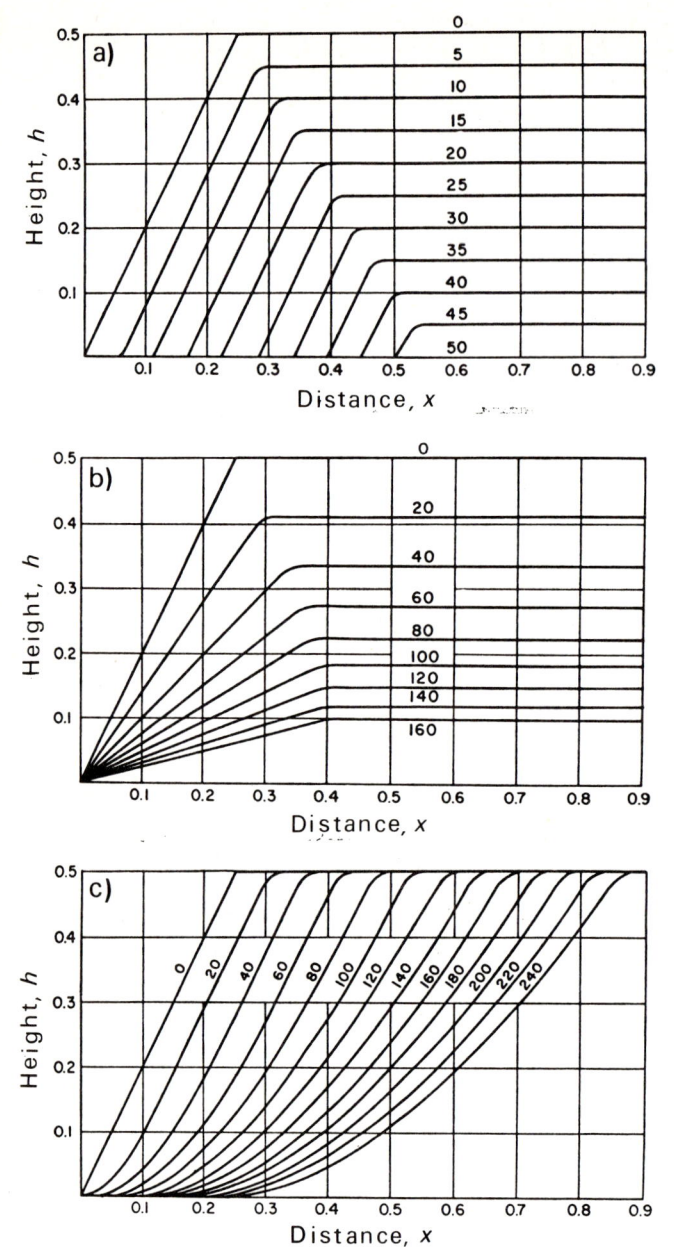

Fig. 8.4a–c. Three cases of slope development according to the nonlinear theory.
a Denudation is equal over all exposed parts of the slope.
b Denudation is proportional to height above base level.
c Denudation is proportional to the steepness of the slope.
(Scheidegger 1961b)

of the slope; secondly, that denudation is proportional to the height of the point along a slope above a base level; and thirdly, that denudation is proportional to slope angle. For an initial rectilinear slope, the results of these cases (Fig. 8.3) are, respectively, parallel slope recession, central slope recession, and parallel slope recession without reduction of the plateau. In all three cases the basic shape of the slope is unaltered. For a change in the form of the slope to take place,

Analytical Models

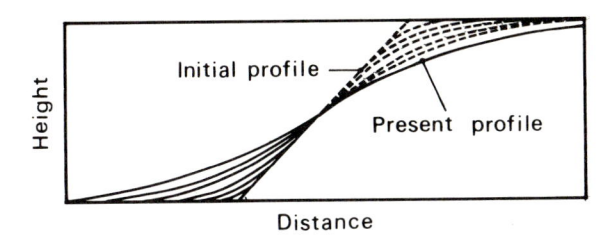

Fig. 8.5. The case of slope development when denudation is proportional to slope curvature

Scheidegger found it necessary to use a nonlinear model which, instead of assuming vertical lowering of slopes, incorporates the fact that weathering acts normal to the slope. Applying this new model to the three cases as before, Scheidegger found that in the first case recession is no longer straight downwards but partly sideways as well; in the second case the slope recession is no longer "central"; and in the third case an originally straight slope cuts back, the toe broadening, the head remaining fairly sharp, and an overall concave form developing, the average inclination of which declines with time (Fig. 8.4). In the first and second cases, the top edge becomes rounded. Scheidegger argued that the third case of the linear theory is especially interesting and therefore subjected the equation for it to a variety of initial and boundary conditions. Specifically, he considered the case of a river eating at the bottom of the slope (Scheidegger 1960), the case in which all the debris coming off the slope is deposited as an apron in front of the slope bank (Scheidegger 1970), both a horizontal apron and a sloping apron, cases in which variations of rock type occur (Scheidegger 1964b), and cases where the effects of uplift are included (Scheidegger 1961b).

Scheidegger does not have a monopoly in heuristic slope models. Other proposals have come from Takeshita (1963) and Young (1963a). These authors use a relation between erosion rate and slope form in which the erosive action is normal to the slope and proportional, not to the slope gradient as Scheidegger (1961b) assumed, but to slope curvature. This assumption leads to the results shown in Fig. 8.5. The same result was obtained by W. E. H. Culling (1963), who treated soil creep as a stochastic process and represented it by a one-dimensional random walk. Hirano (1968, 1975) takes a different tack. He (Hirano 1968) first proposed the model

$$\frac{\partial h}{\partial t} = a \frac{\partial^2 h}{\partial x^2} - b \frac{\partial h}{\partial x} + f(x, t),$$

where a is called a subduing coefficient, b is called a recessional coefficient, and $f(x, t)$ represents the supply of material by, for example, crustal uplift. If uplift takes place instantaneously, as was assumed by Davis (1899), the term $f(x, t)$ may be omitted. Hirano (1975) considers several interesting cases of his equation. In one "experiment", the parameters were set at $a = 0.25$ and $b = 2.0$; the initial slope profile was a horizontal line and might represent a plateau; the boundary condition at one end of the slope represented down-cutting by a river and the boundary condition at the other end represented a watershed where a horizontal slope was maintained. Three rates of river down-cutting were investigated – down-cutting proceeding at an accelerating rate, down-cutting proceeding at a

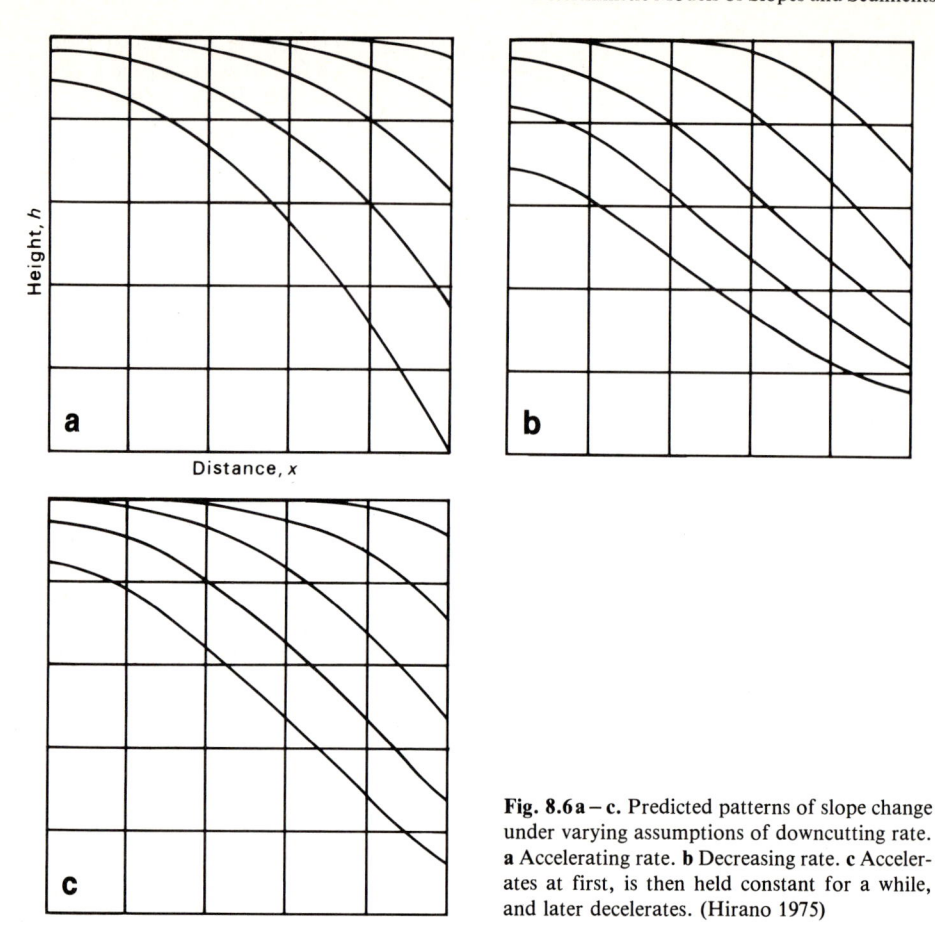

Fig. 8.6a–c. Predicted patterns of slope change under varying assumptions of downcutting rate. **a** Accelerating rate. **b** Decreasing rate. **c** Accelerates at first, is then held constant for a while, and later decelerates. (Hirano 1975)

decelerating rate, and down-cutting accelerating at first, staying constant for a while, and then gradually decelerating. The results of these models are shown in Fig. 8.6. Hirano argues that the first two cases of down-cutting conform to Penck's (1924) *"aufsteigende Entwicklung"* and *"absteigende Entwicklung"* respecticely. In all cases, a constant slope form is never attained. Only if a constant rate of river down-cutting is assumed does a constant or steady-state slope form develop.

The significance of the *a* and *b* parameters in the model is shown in Fig. 8.7. In the two cases shown, the initial form of the slope profile is a straight slope, height at the valley bottom is held constant, and the boundary condition at the watershed is such that $\partial h/\partial x = 0$ at all times. The ratio of a/b is 0.25 in case one and 0.0625 in case two. The effect of this is that in the first case the slope profile tends to wear down, whereas in the second case the slope profile tends to wear back. This is interesting because the parameter *a* determines the relative effect of the curvature term, $\partial^2 h/\partial x^2$, in the model which probably represents soil creep and creep-like processes, whereas the *b* parameter determines the relative effect of the slope angle term, $\partial h/\partial x$, which probably represents surface wash and

Analytical Models

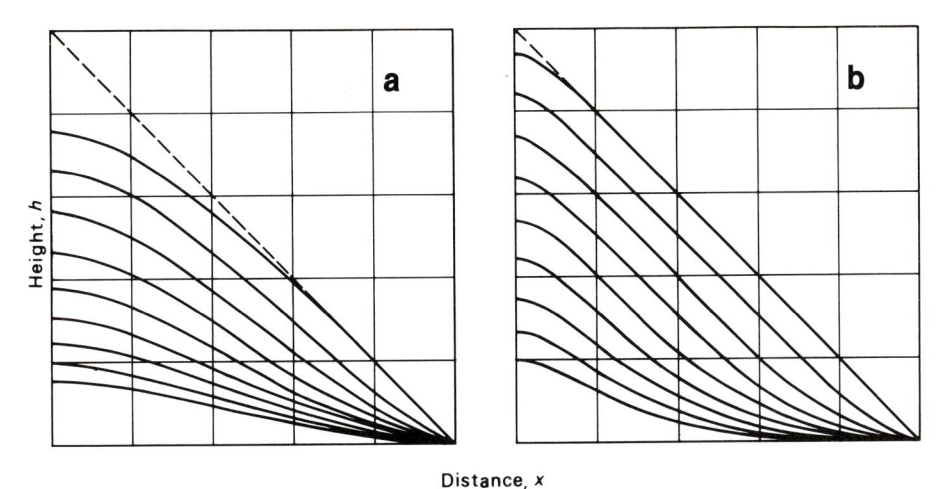

Fig. 8.7a, b. The effect of the parameters **a** and **b** on slope development. (Hirano 1975)

wash-like processes. It would appear that where soil creep is the main process operative on a slope, wearing back takes place. Nash (1981) explains that the model

$$\frac{\partial h}{\partial t} = -b \frac{\partial h}{\partial x}$$

is in effect a modified version of Fisher's (1866) model of parallel hillslope retreat. Nash further modifies Fisher's model, arguing that material loosened from a scarp surface will start to move down slope at a threshold angle, θ_r, and, once in motion, will come to rest where the scarp slope is less than a threshold angle, θ_p. Nash models this situation by the equation

$$\frac{\partial h}{\partial t} = -c_r \left| \frac{\partial h}{\partial x} \right|,$$

where c_r is a constant coefficient of retreat equivalent to Hirano's b coefficient. Figure 8.8a shows a solution of the model for a fault scarp. Debris removed from the scarp accumulates at the scarp base and buries a parabolic core of unmoved material (cf. Fisher 1866). Allowance is made in the model for variations in the packing density of debris and for material transported beyond the debris apron. Figure 8.8b shows the effect of a similar model but with four separate episodes of vertical movement along the fault plane. The fault scarp becomes notched, each notch corresponding to an episode of fault displacement.

Nash (1981) also experimented with a model of "scarp rounding". In this model, the volumetric rate of debris transport, r, per unit length of contour line on a transport-limited slope is assumed to be proportional to the local slope gradient:

$$r = c_d \frac{\partial h}{\partial x},$$

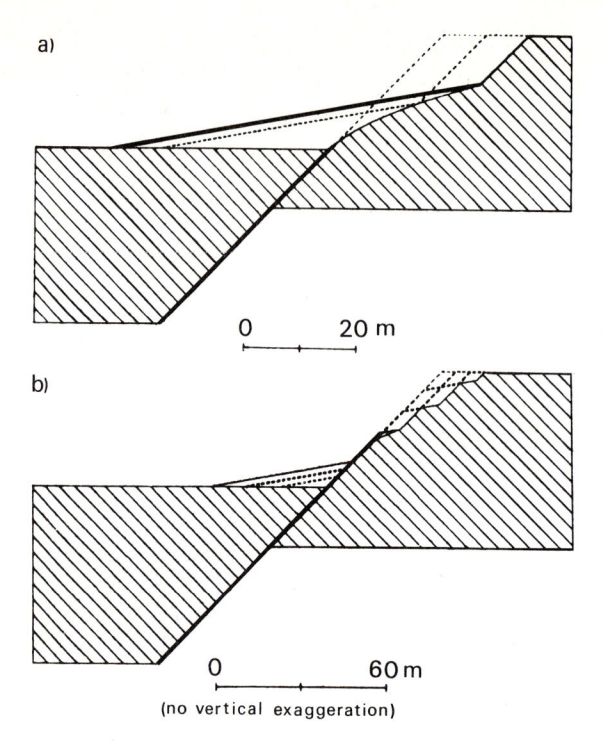

Fig. 8.8. a Parallel retreat of scarp with deposition of debris at base. Scarp is produced by a single movement along the fault. **b** Parallel retreat of scarp with deposition at base. Scarp is the result of four separate episodes of movement along the fault. (Nash 1981)

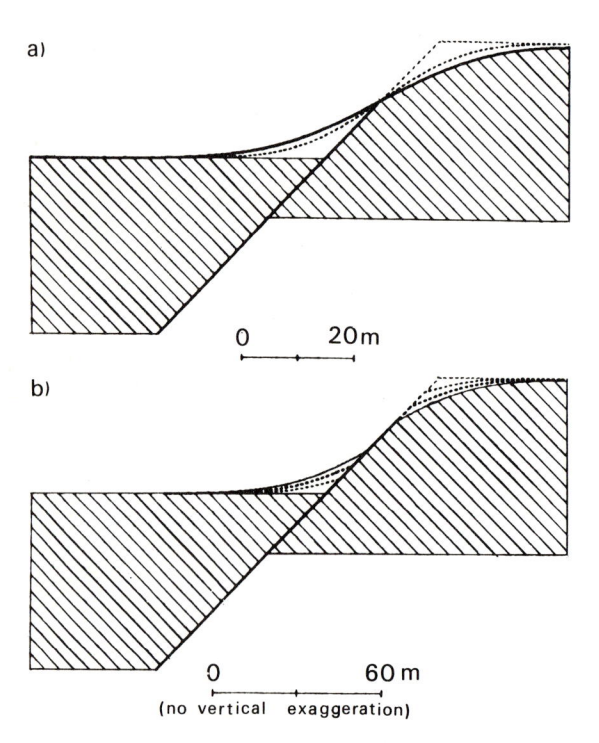

Fig. 8.9. a Rounding of fault scarp that has been produced by one episode of displacement along fault. **b** Rounding of scarp that is produced by four separate episodes of movement along fault. (Nash 1981)

Analytical Models

where c_d is a constant of proportionality termed a constant of debris diffusion by Nash and equivalent to Hirano's a coefficient. The validity of this equation is discussed at length by Nash (1980a, b). The change in elevation at a point on a slope is equal to the downslope divergence of r:

$$\frac{\partial h}{\partial t} = \frac{\partial}{\partial x}(r) = c_d \frac{\partial^2 h}{\partial x^2},$$

an equation which leads to a rounding of a scarp (Fig. 8.9a). Nash (1980a) has shown that this equation is an appropriate model for describing the degradation of wave-cut bluffs along the shores of Lake Michigan. Applied to a fault scarp which suffers four episodes of displacement, the model predicts that each new scarp offsets the centre of the preexisting scarp (Fig. 8.9b).

Pollack (1969) uses a model

$$\frac{\partial h}{\partial t} = \frac{\partial}{\partial x}\left\{K(h) \cdot \frac{\partial h}{\partial x}\right\} + A(x, h)$$

to perform a series of experiments which simulates the development of a hypothetical cross-section of the Grand Canyon, Arizona. The model is essentially a diffusion equation with an elevation-dependent diffusion coefficient $K(h)$ and with an extra term, $A(x, h)$ which determines the rate of downcutting by a stream. $K(h)$ and $A(x, h)$ are proportional to the resistance of the strata in the Grand Canyon. The experiment shown in Fig. 8.10 involves a number of pulses, with rapid deepening occurring when a non-resistant layer is encountered, followed by a period of valley widening as the profile opens out from the site of the stream. Steep slopes are usually, but not always, associated with resistant strata; cliff and bench features come and go as erosion progresses. In a second experiment, Pollack (1969) introduces a second small tributary stream with half the down-cutting ability of the trunk stream (Fig. 8.11). The adjacent sides of the two streams eventually meet as their valleys widen producing mesas which appear and disappear at different stages. Eventually, the trunk stream captures the tributary and continues to erode an asymmetrical valley.

Chappell (1974, 1978) models the development of small valleys in Pleistocene coral terraces in Papua New Guinea. As Chappell explains, the terraces provide a good natural laboratory in which notions of landscape development can be tested because the time of emergence of each terrace is known with a fair degree of precision. The model for the development of a thalweg, in the (x, h) frame of Fig. 8.12c, is

$$\frac{\partial h}{\partial t} = -\left\{A(x+d)e^{-\tau x} + B(x+d)\frac{\partial h}{\partial x} + C\frac{\partial^2 h}{\partial x^2}\right\},$$

and is derived by substituting simple models for solution, A, corrasion, B, and mass movements, C, into Scheidegger's (1970, p. 136) general equation for weathering action normal to a slope and linearizing. The term for solution, $A(x+d)\exp(-\tau x)$, assumes that solution increases downstream with water volume, which itself is assumed to increase linearly with distance and includes overland flow and thoughflow over distance d above the channel head, but

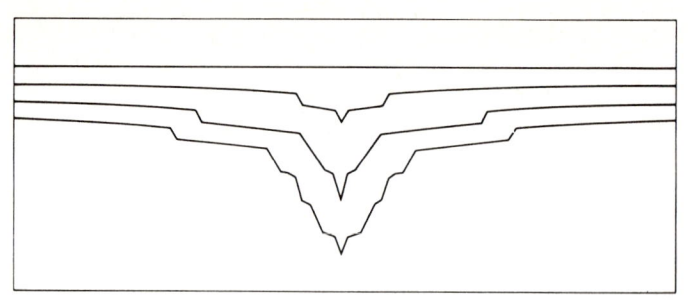

Fig. 8.10. Sample profiles from a model canyon at several stages of development. (Pollack 1969)

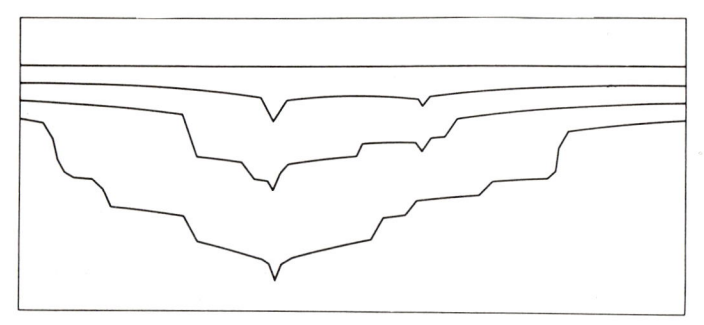

Fig. 8.11. Sample profiles of a model canyon illustrating asymmetrical development. (Pollack 1969)

simultaneously diminishes as travel distance, x, in the channel increases. The term for corrasion, $B(x+d)\partial h/\partial x$, assumes that corrasion is proportional to slope gradient which is a simple Hortonian approximation for erosion in a uniform-section channel. The term of mass movement, $C\partial^2 h/\partial x^2$, assumes that mass movement is proportional to thalweg curvature.

The linearized model was computed by finite difference equations and iterated across the set of eight surveyed valleys. Best-fit values for the coefficients A, B, and C were found to explain 98 per cent of the thalweg profile variance and 94 per cent of the age variance (Fig. 8.13a). Chappell criticizes his model on the grounds that four adjustable parameters, A, B, C, and τ for process functions which have complementary effects are almost certain to provide a good fit and that a different set of plausible process formulations with quite different weightings of A, B, and C would most likely yield an equally good fit. To develop this point, Chappell (1974) develops a more detailed model to describe the process of solution. The resulting pure solution model gives as good a fit as the combined solution-corrasion-mass-movement model (Fig. 8.13b). Applying the principle of parsimony, it is therefore a "better" model.

The fact that the valleys are cut in corralline limestone makes the acceptance of a pure solution model easier. However, weighing up his findings, Chappell (1978) concludes that all the models actually show is that solution is the most likely process forming the valleys and the elementary rules of hydraulics embedded in the models are not contradicted by the "test". He observes that the

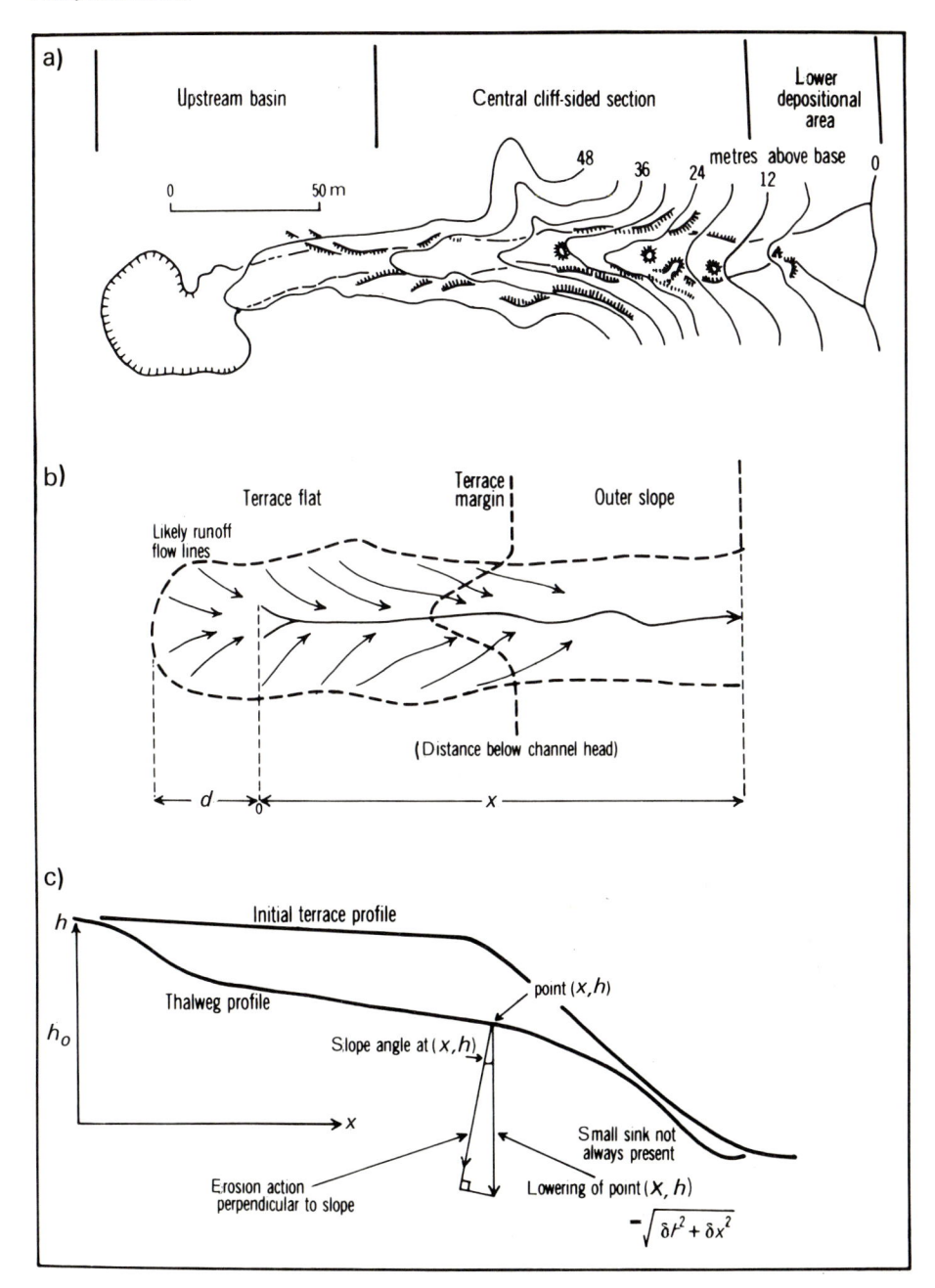

Fig. 8.12a–c. The geometry of small valleys developed on the fronts of coral reef terraces of the Huon Peninsula, Papua New Guinea. **a** Contours of typical valley. **b** Watershed and runoff lines. **c** Geometry of lowering in x, h frame. (Chappell 1974)

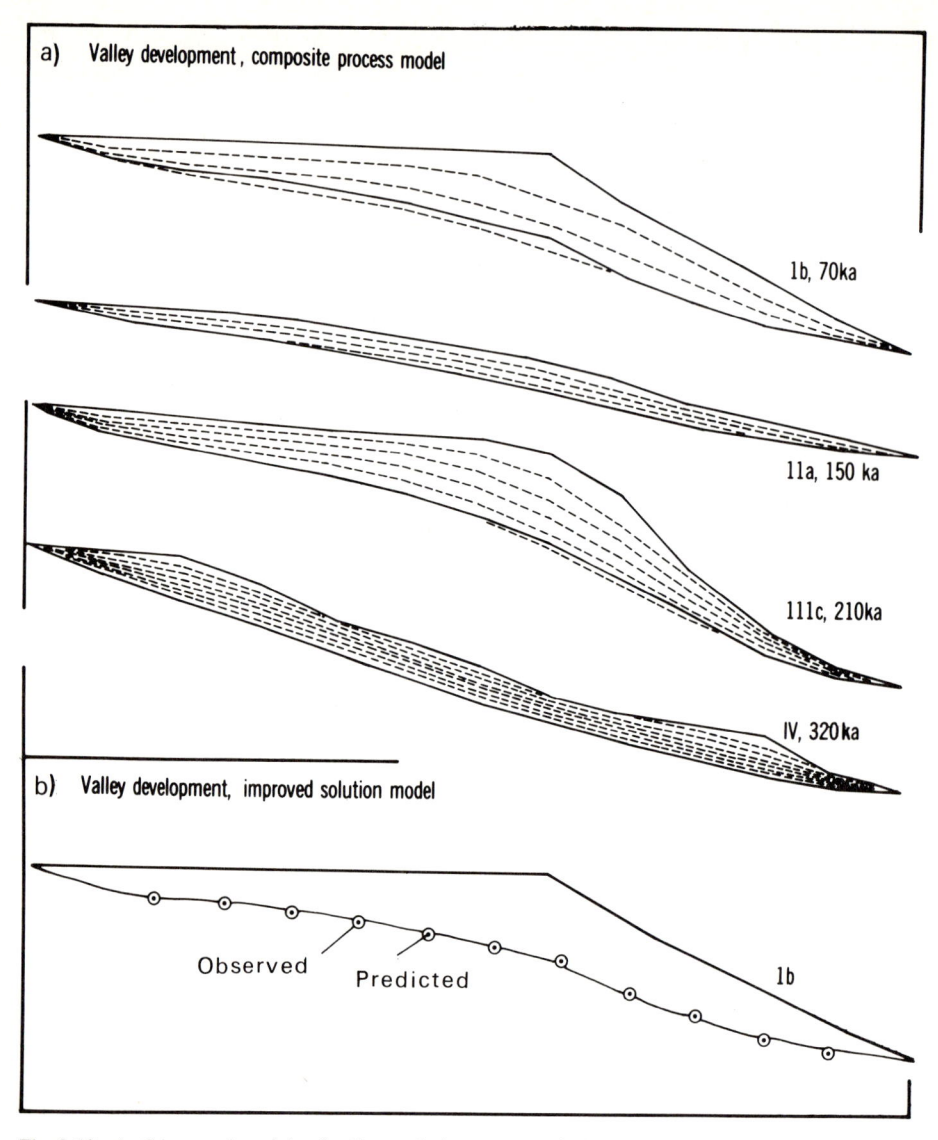

Fig. 8.13a, b. Computed models of valley evolution compared with observation. Initial profile (*upper solid line* in each case) is reconstructed from valley interfluves, which still show the original terrace form; final profile (*lower solid line* in each case) is thalweg. **a** Valley development computed by composite process model is shown by *successive dashed lines*. Model time increment is the same between dashed lines in all cases and the difference between dashed line spacings reflects differences in the actual size of the valley. **b** Valley development computed by solution model for valley 1b according to Chappell's "improved" solution with intermediate time steps not shown. (Chappell 1974)

role played by solution could be more accurately assessed by a small set of field measurements and that if landscape management is the aim of slope-modelling endeavour, then it is better to extend knowledge of fluvial hydraulics by controlled experiment. The extension of these criticisms to other hydraulically reasonable models involving comparison with landscape profiles seems fair. Despite this, Chappell suggests that the desire remains to derive morphological development from process investigations because only by doing so can the present be sufficiently explained and extrapolation to the future (and past) be made.

8.2.2 Models Based on the Continuity Equation

A second group of analytical models of slope development is derived from the continuity equation for debris moving on hillslopes and rivers, first formally introduced by Kirkby (1971), and a debris transport equation. The equation of continuity for debris on a hillside may be written, for one dimension, as

$$\frac{\partial h}{\partial t} = -\frac{\partial S}{\partial x},$$

where h is the height of the land surface and S is the sediment transport rate which needs defining by a transport (process) equation appropriate for the processes being modelled. Generally, a sediment transport equation may be expressed as

$$S = f(x)^m \left(\frac{\partial h}{\partial x}\right)^n,$$

where $f(x)^m$ represents a function proportional to distance from the watershed (roughly the distance of overland flow) and $(\partial h/\partial x)^n$ represents processes in which sediment transport is proportional to slope gradient. Empirical work suggests that $f(x)^m = x^m$ where m varies according to the sediment-moving process operating. The values of the exponents m and n are listed in Table 8.1. The solution of the equation, which takes the general form

$$h = f(x, t)$$

Table 8.1. Exponent values in sediment transport equation. (Kirkby 1971)

Process	Exponent	
	m	n
Soil creep	0	1.0
Rainsplash	0	1.0 – 2.0
Soil wash	1.3 – 1.7	1.3 – 2.0
Rivers	2.0 – 3.0	3.0

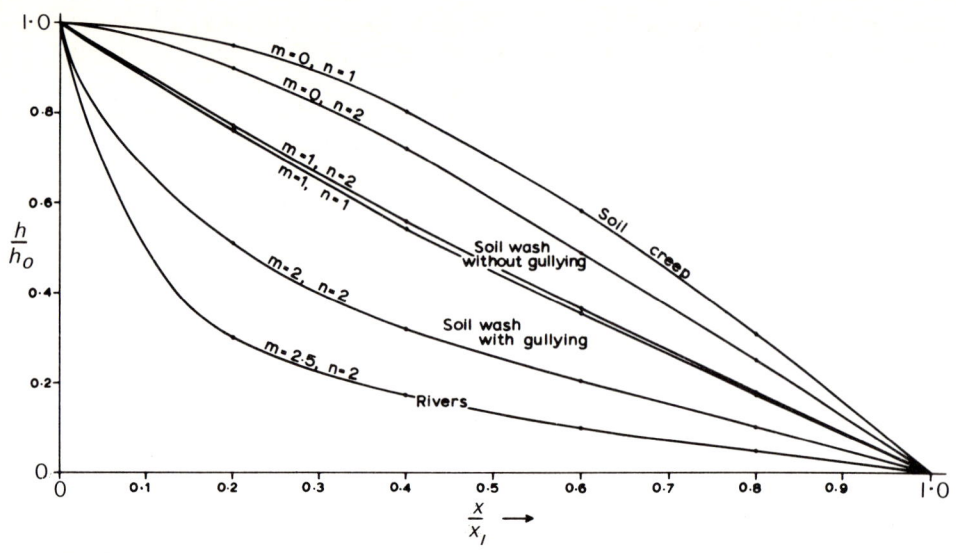

Fig. 8.14. Dimensionless graph showing the approximate characteristic form slope profiles for a range of processes from Table 8.1. (Kirkby 1971)

for a one-dimensional case, describes the development of a slope for an assumed system configuration: specified slope processes, initial state, and boundary conditions (Armstrong 1982).

Hillslopes. Kirkby (1971) finds that in some solutions of the continuity equation for sediment transfer on hillslopes (and rivers) the influence of the initial form of the slope decreases rapidly with time and the slopes tend nearer and nearer to a "characteristic" form in which the height of each point along the profile continues to decline with time but is independent of the initial form (Fig. 8.14). The amount of erosion necessary before the actual slope form becomes indistinguishable from the characteristic form depends partly on the initial slope form. With the continuity equation taken in a form that is a reasonable approximation for ridges, divides, and stream courses, with sediment removal being transport-limited, with no chemical solution, and with the boundary condition of a fixed divide and a fixed base level, Kirkby (1971) obtains approximate solutions for the characteristic form of slope profiles for a variety of slope transport laws.

Kirkby (in Carson and Kirkby 1972) distinguishes two solutions of the continuity equation. The first solution, which is for a constant lower boundary and a fixed slope length, gradually approaches a time-independent characteristic form. The second solution, which is for an exit point which cuts down at a constant rate, shows the development of a constant slope form relative to the moving base. A consideration of the geomorphological conditions under which the two boundary conditions are likely to occur, has led Armstrong (1982) to conclude that a fixed boundary is not unlikely to be found in reality. For a characteristic form to develop with a lower fixed boundary, all the material reaching the slope base must be carried away by an external agency. Armstrong argues that, as both the height of the slope and gradients along the slope decline with time according to Kirkby's solution, so the delivery of material to the slope base will also decline. The development of the characteristic form thus requires that a dwindling supply of material off the slope is exactly matched by a dwindling rate of removal at the slope base so that the height of the slope base remains the same. If removal at the slope base is carried out by a linear process, such as the action of a river, the reduction in removal rate can be achieved by a reduction in gradient. Thus a characteristic form requires at its base a declining gradient at constant height, a condition which can be met only at a single point, not along the entire length of the line process (Armstrong 1982). The constant form, by contrast, maintains its overall height and steepness as it moves through a considerable range of height. All that is required at the base of the slope is a constant rate of removal, a condition which is likely to obtain wherever the rate of change of basal removal is slower than the time taken for the slope profile to adopt the constant form.

At large set of solutions to the continuity equation has been obtained by Gossman (1970, 1975, 1976, 1981). Arguing that too many slope models rely on measurements of present-day processes from vegetated slopes in humid midlatitudes and that too many slope models ignore the effects of variations in lithology, Gossman set out to model slopes in a variety of climatic regions where, owing mainly to a lack of vegetation cover and the predominance of surface runoff, erosional processes are known to be especially effective. Three environments were modelled: the semi-arid and arid environment, the periglacial environment, and tropical wet and dry environments.

The dominant role that surface wash plays in the development of "not-too-steep" slopes in semi-arid regions is well documented (K. Bryan 1923, 1940; L. C. King 1953). Gossman's predicted profiles with a limiting slope angle of 22° and boundary conditions defining a belt of no erosion at the top of the profile accord well with the typical "scarp-behind-pediment" profiles of semi-arid areas (Fig. 8.15). The sequence of development predicted is a straight slope changing to a characteristically concave profile, which under some conditions involves considerable steepening, the gradient of the steepest part remaining constant for a long time. The area of denudation extends onto the lower part of the slope profile and, depending on the precise boundary conditions, the upper part of the profile develops in bedrock. The more dominant wash is relative to other slope processes, the more pronounced is the predicted scarp-pediment form.

In periglacial areas, surface wash, as well as solifluxion, is an important process in the development of slopes (Jahn 1967). Gossman examined the effect of surface wash and solifluxion for three different types of slope: gentle slopes, steep slopes with a rounded upper convexity, and slopes with a marked edge at the upper convexity (Fig. 8.16). The gentle slopes are formed by a combination of creep and wash. Steep slopes with a rounded upper convexity are modelled by reducing the role of solifluxion on steep slopes where, in practice, wash processes are known to predominate (Büdel 1982, p. 88). Slopes with a marked edge at the upper convexity are produced by greatly reducing the role of solifluxion relative to wash processes. The model in this case reduces to the model used to simulate slope development in a semi-arid environment and an interesting question is

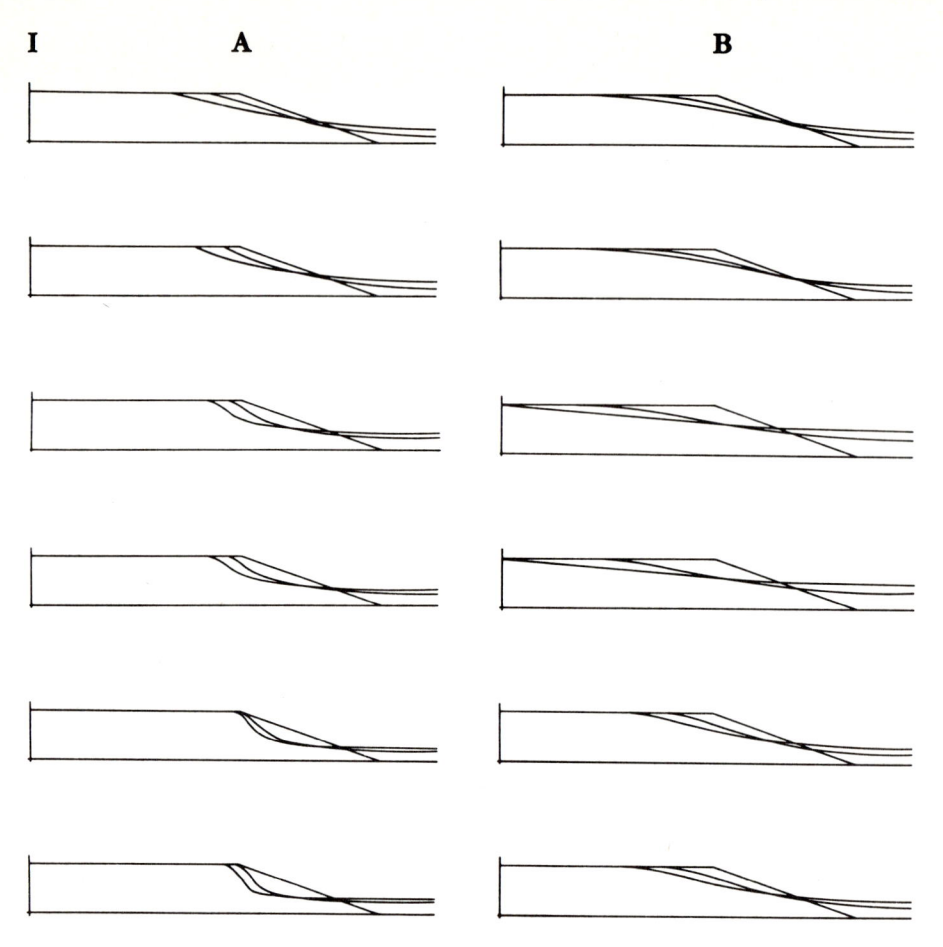

Fig. 8.15 I, II. Two-dimensional slope models based on the continuity equation with different combinations of creep and surface wash and varying boundary conditions. Sediment transport, S, is defined as

$$S = C_1 \cdot d^E \cdot \sin\alpha \cos\alpha + C_2 \cdot \begin{cases} \tan\alpha & \text{if} \quad \alpha \geq \alpha_0 \\ 0 & \text{if} \quad \alpha < \alpha_0 \end{cases}$$

with different sets of values for the parameters C_1, C_2, E, and α. The columns and rows in the diagram have the following meanings. *I* elevation of the base by accumulation ($S_u = 0$ at slope base); *II* lowering of base by erosion ($S_u = S_0 + KS_0$ at slope base, with different values for K). **A** models with a belt of no erosion (no lowering of the watershed by surface wash). **B** models without a belt of no erosion (with lowering of the watershed by surface wash). Parameter values: $a, a', E = 1, C_1 = 20, C_2 = 1$; $b, b', E = 1.5, C_1 = 20, C_2 = 1$; $c, c', E = 2, C_1 = 3, C_2 = 1$; $a, b, c, \alpha_0 = 2°$; $a', b', c', \alpha_0 = 22°$. (Gossman 1976)

Analytical Models

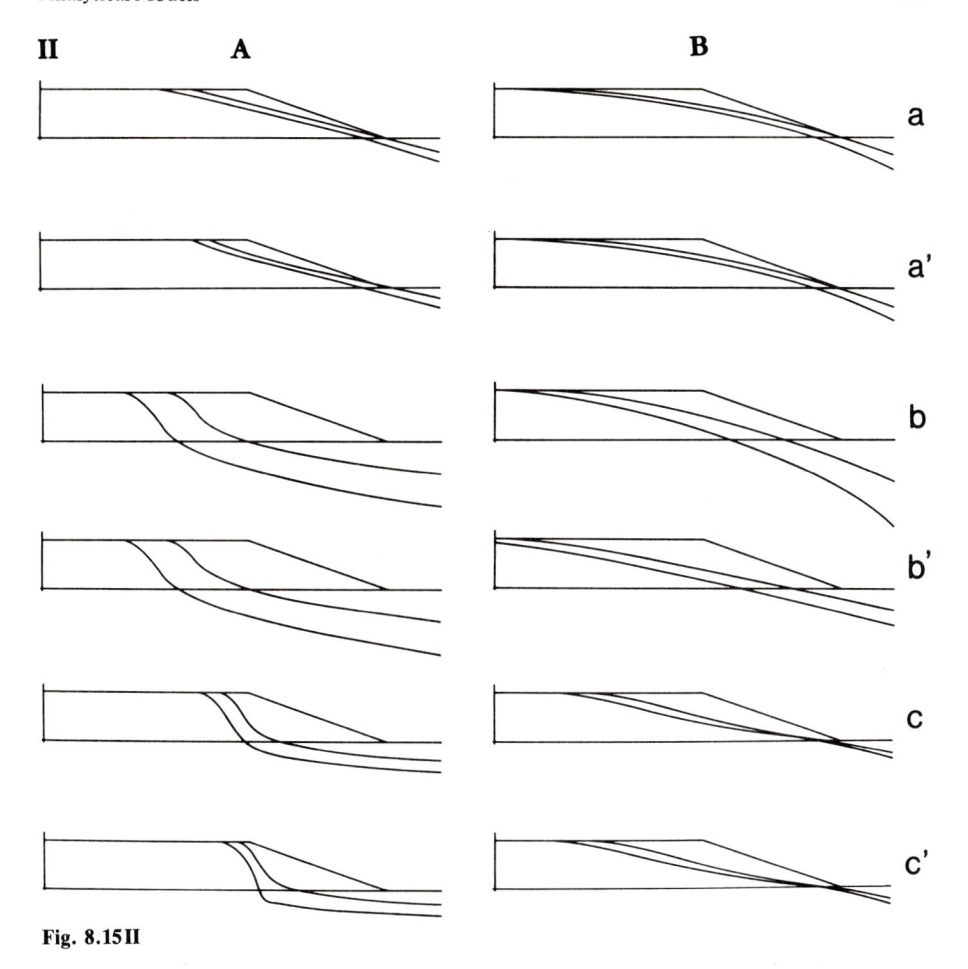

Fig. 8.15 II

raised as to the convergence of form between periglacial and semi-arid slopes where wash processes predominate.

In the semi-humid environment of tropical savanna regions, a contrast is recognized between the physical disintegration of rocks on steeper portions of slope profiles (inselbergs) and intensive chemical decomposition in a thick cover of red loam which forms flat surfaces (peneplains), the chemical breakdown in the peneplain undermining the rocky inselberg (Büdel 1982, p. 133). Gossman attempted to model an inselberg slope profile using a model for surface wash on flat segments (slopes less 5°) and using a weathering-limited transport model for steeper slopes (in excess of 5°). The results show that the slope forms are similar for a wide range of boundary conditions (Fig. 8.17).

Rivers. Three equations are needed to model sediment transport and channel response in rivers (R. J. Bennett 1974). They are the conservation of mass and momentum equations for the flow of water and the conservation of mass equation for the sediment. In one-dimensional form the equations are

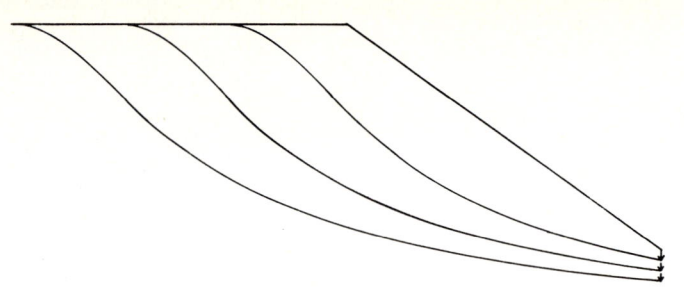

Fig. 8.16. Simplified model for periglacial slopes with solifluxion acting only on flat upper and lower parts, based on the sediment transport equation

$$S = 3d^2 \sin\alpha \cos\alpha + \begin{cases} \sin 3\alpha \cos 3\alpha & \text{if} \quad 0 \leqslant \alpha \leqslant 30° \\ 0 & \text{if} \quad \alpha > 30° \end{cases}$$

Boundary conditions are: lowering of the base; belt of no erosion for surface wash. (Gossman 1976)

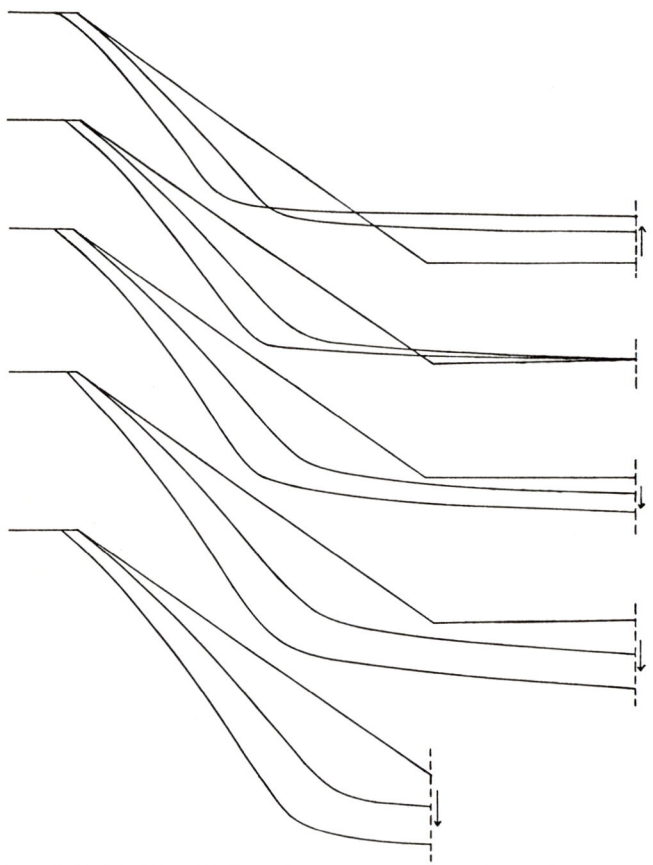

Fig. 8.17. Models with limitation by transport on flat surfaces and limitation by weathering on slopes. Arrangement:

$\alpha < 5°: S = 3d^2 \sin\alpha \cos\alpha$

$\alpha \geqslant 5°: \dfrac{\partial S}{\partial x} = \dfrac{1}{10} 3d^2 \sin\alpha \cos\alpha + \begin{cases} 0 & \text{if} \quad \alpha \leqslant 22° \\ \tan\alpha & \text{if} \quad \alpha > 22° \end{cases}$

(Gossman 1976)

Analytical Models

$$\frac{\partial A}{\partial t} + v\frac{\partial A}{\partial x} + A\frac{\partial v}{\partial x} = q,$$

$$\frac{\partial v}{\partial t} + v\frac{\partial v}{\partial x} + g\frac{\partial y}{\partial x} = g(S_0 - S_f) - \frac{qv}{A},$$

$$\frac{\partial(AC)}{\partial t} + w(1-p)\frac{\partial h}{\partial t} + \frac{\partial(Av_sC)}{\partial x} = \frac{\partial}{\partial x}\left(Ak\frac{\partial C}{\partial x}\right),$$

where A is the cross-sectional area of flow, t is time, v is water flow velocity, x is distance downstream, q is lateral inflow of water per unit width of channel, y is depth of flow, g is the gravitational acceleration, S_0 is bed slope, S_f is the friction slope, w is channel width, h is local bed elevation, p is the porosity of deposited sediment, v_s is sediment velocity, and k is the longitudinal dispersion coefficient for sediments. The flow equations for water are equivalent to those given in Chap. 7.2.2. This set of equations is rather complex to solve. To simplify matters, the conservation of mass equation for sediment is usually written

$$w(1-p)\frac{\partial h}{\partial t} + \frac{\partial Q_s}{\partial x} = i,$$

where Q_s is the sediment transport expressed as volume, and i is the sediment input from bank erosion and slope runoff (commonly assumed to be zero).

The model is applied by defining a water surface slope for an initial bed profile using a gradually varied flow equation starting at a downstream control depth. The flow geometry is then used to predict the spatial variation of bedload transport over a series of incremental reaches by using an appropriate transport equation such as the Meyer-Peter and Müller equation or Bagnold's bedload transport equation. The solution of the continuity equation with bedload transport rate defined then yields bed elevation and cross-section changes along the reach under study.

The basic model has been used to study problems where the continuity of transport is disturbed. Hales et al. (1970) predicted changes in bed elevations following the construction of a dam. Soni et al. (1980) have shown that, following extra load supplied in a stream reach, bed elevations upstream of the point of extra supply are maintained at the same gradient as the original bed elevations but downstream of the point of extra sediment supply the bed elevations steepen in gradient to enable the extra load to be transported. A reduction in bedload leads to degradation. Channel gradient also adjusts to changes of base level. Lowering of base level creates a nickpoint and incision begins. Flow and sediment transport around a nickpoint have been modelled by Pickup (1975, 1977). At a nickpoint, accelerating flow and a steepened slope cause an increase in bedload transport. The result is that erosion occurs at the nickpoint with deposition downstream. The erosion causes the nickpoint to retreat upstream (Fig. 8.18).

The normal sediment transport continuity approach is not applicable to all problems. For instance, the movement of slugs of bedload from a discrete point of input down a river is difficult to model using a conventional finite-difference

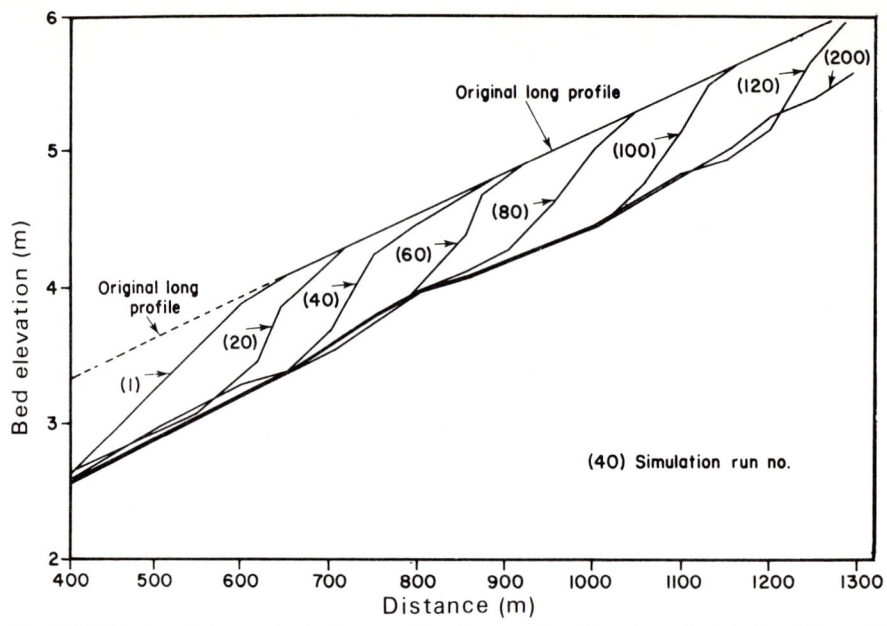

Fig. 8.18. Simulated changes in the long profile of an eroding river channel with time. The profile marked (*1*) is the knickpoint at the start of the simulation run. The reach between 0 and 400 m was affected by boundary conditions, namely, the assumed values of flow depth and energy slope used to start the flow profile computations. It has therefore been excluded. (Pickup 1977)

formulation of the equations of motion. An alternative model has been developed by Pickup et al. (1983), which makes explicit allowance for the mean sediment velocity and, by introducing a dispersion coefficient, the distribution of sediment velocity about the mean. Fluctuations of sediment velocity are important when modelling short-term changes such as the passage of a wave of bed material but they are not allowed for in the conventional continuity of sediment model. Pickup et al. have tested both types of model over a 43-month period using data from seven monitored cross-sections of the Kawerong River on Bougainville Island, Papua New Guinea, a system which receives waste from a large copper mine. They found that their dispersion model performs considerably better than a conventional mass-conservation model (Fig. 8.19).

8.3 Simulation Models

Common to all slope models in the third group is the use of computer simulation techniques to predict slope form and its changes. Some of the models discussed in the previous section make recourse to a computer to find a solution to a slope development equation. However, the models described in this section are, arguably, more versatile as they can take into account a greater range of factors. To some extent, the gains accruing from increased versatility are offset by a loss of generality. But computer simulation models play an important role in the study of Earth surface systems (see Harbaugh and Bonham-Carter 1970) and are likely to become more important as the capabilities and ready availability of easy-to-use simulation packages increases. Examples that will be discussed here include models of three-dimensional landscape development, nearshore bar formation, and sand dune development.

8.3.1 Landscape Simulation

Both Ahnert (1976, 1977) and Armstrong (1976) have developed three-dimensional simulation models of slope development. Armstrong (1976) modelled a small headwater basin in a humid temperate environment of the kind found in upland Britain today. Fieldwork in these areas suggests that the main processes operating are weathering, soil creep, and fluvial erosion. Only these processes were built into Armstrong's model. Soil creep rate, C_s, was modelled as a linear function of the sine of slope angle, θ:

$$C_s = K_s \sin \theta,$$

where K_s is a constant. This transport equation is justified on theoretical grounds by Kirkby (1967) but the field evidence for it is unclear (Young 1972). Fluvial transport rate was modelled using Bagnold's (1966) stream power relation which was derived from a consideration of general physics:

$$C_r = K_r Q S,$$

where K_r is a constant representing the efficiency of the fluvial system in transporting sediment, Q is the river discharge, and S is the slope of the river. This equation is adjusted according to whether weathered material is being moved or whether there is excess river capacity to erode bedrock. Weathering is assumed to be a simple negative exponential function of depth of soil present, D:

$$W = W_{\text{pot}} \exp(-K_d D),$$

where W_{pot} is the potential rate of weathering of a bare rock surface and K_d is a constant. Clearly, this expression is a much simplified representation of weathering but, Armstrong argued, is suitable for the model. Other weathering models are available, the merits and demerits of which are discussed by N. J. Cox (1980).

The landscape to be modelled was divided into a 40 × 40 set of unit cells each representing an area of 200 × 200 m. The mass balance for each cell was computed during 20,000 successive 1-year time steps according to the relations

change in height = inflow − outflow

change in soil depth = inflow − outflow + weathering .

The values of inflow and outflow were evaluated by calculating the volumes of material transported from each point by the erosional processes and then routing these into the lowest adjacent cell. Rivers occupy predefined cells and cannot alter their positions. Some results are shown in Fig. 8.20. Armstrong (1980a)

Fig. 8.19

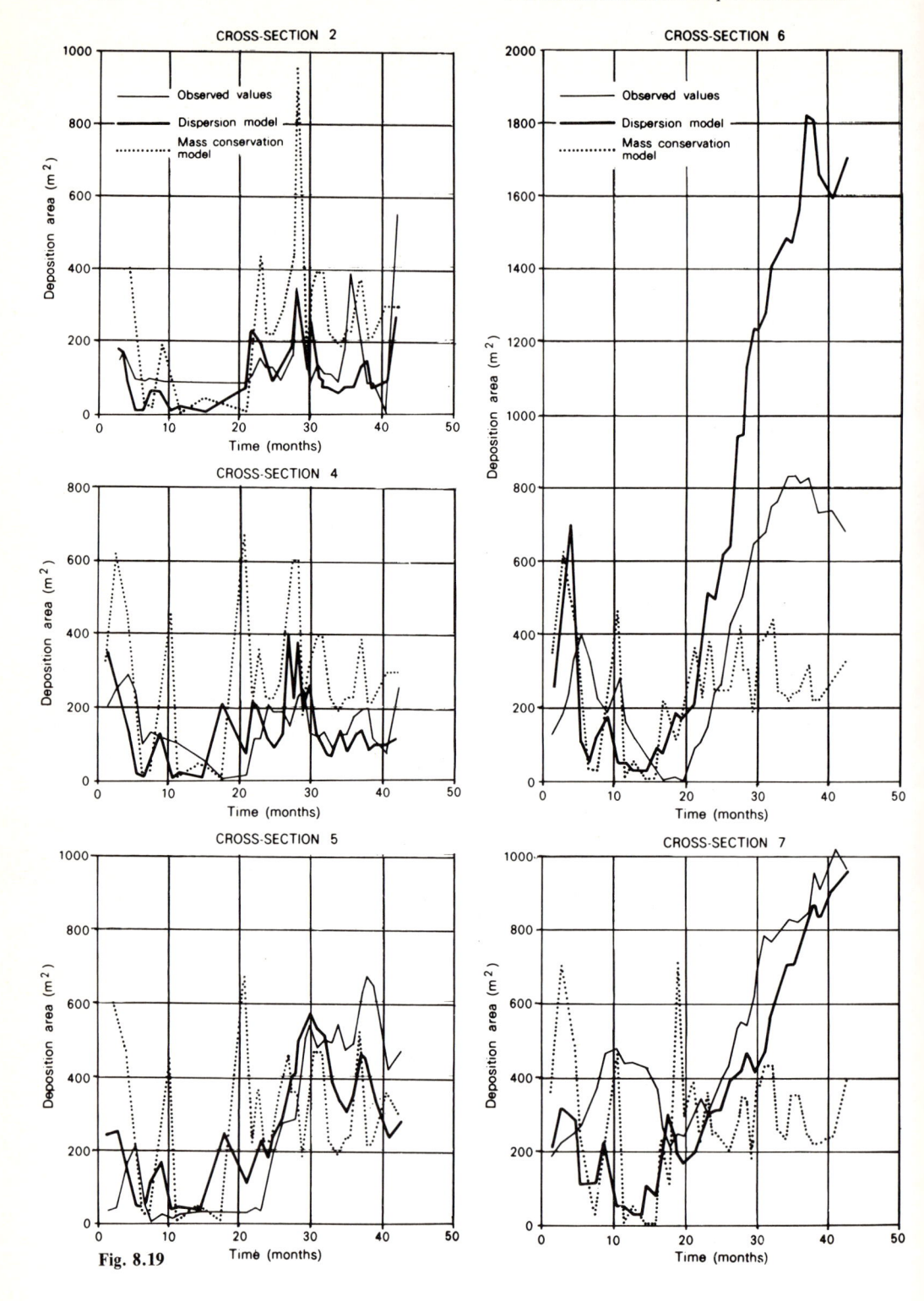

Simulation Models

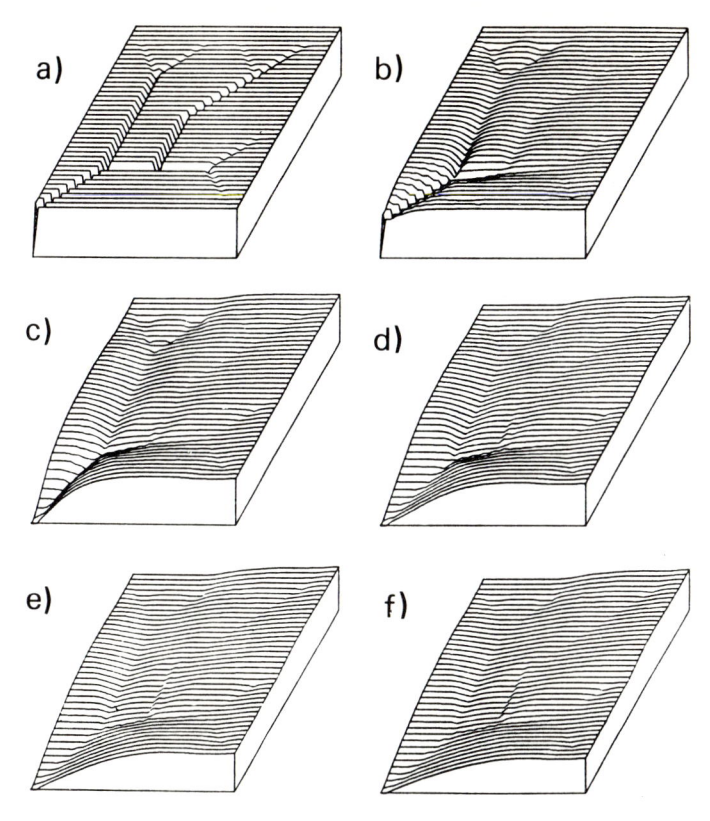

Fig. 8.20 a – f. Computer-generated block drawings of the sequence of landforms. **a** Initial form; **b** 1000 iterations; **c** 5000 iterations; **d** 10,000 iterations; **e** 15,000 iterations; **f** 20,000 iterations. (Armstrong 1976)

later refined the model to consider changes in soil depth during the development of the landscape. The basic steps in the model are the same except that it is calibrated so that one time step is roughly 10 years of real time. Some results are shown in Fig. 8.21.

In another simulation, Armstrong (1980b) set up initial conditions as depicted in Fig. 8.22. The two exit cells of the two drainage basins were incised at a uniform rate for the first 9000 iterations and then kept at a constant rate for a further 3000 iterations. These conditions represent an open landscape in which erosional energy is continuously renewed by incision. The resulting simulation represents the development of a landscape in a state of dynamic equilibrium, a concept first applied to landscapes by Gilbert (1877) and restated and extended by Hack (1960). The general form of the land surface is static but individual land-

Fig. 8.19. Observed and calculated deposition in cross-sections 2 to 7 of the Kawerong River. (Pickup et al. 1983)

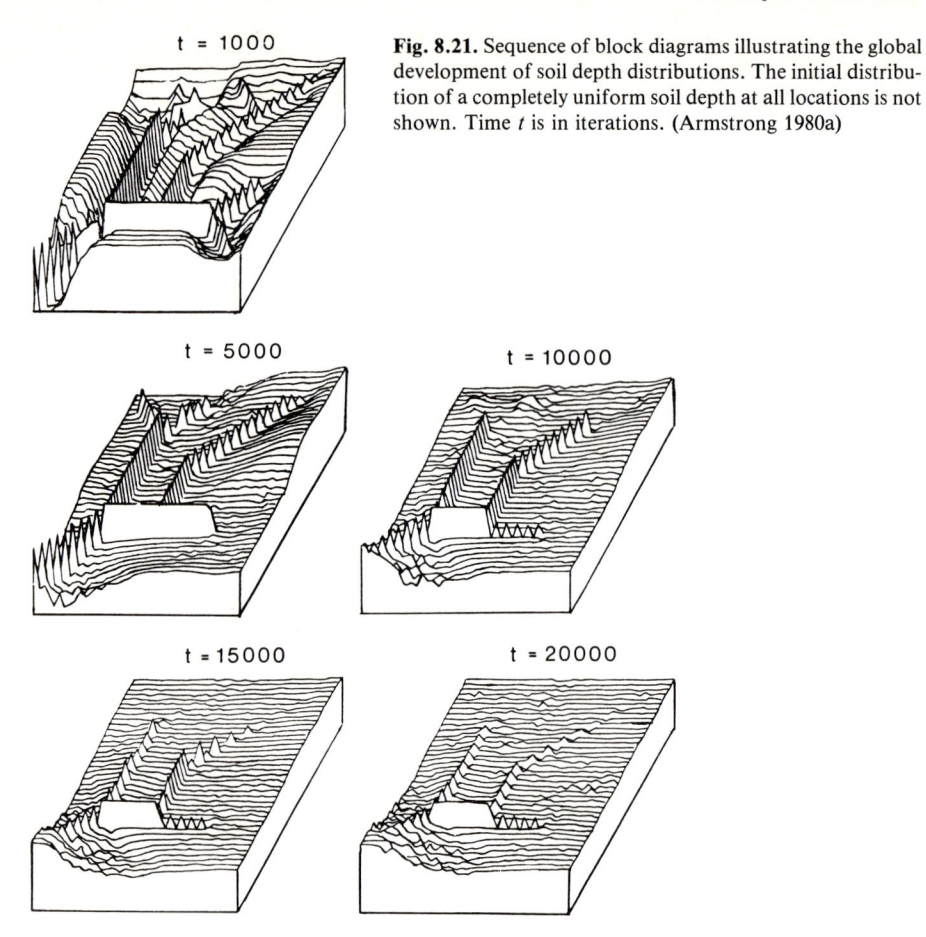

Fig. 8.21. Sequence of block diagrams illustrating the global development of soil depth distributions. The initial distribution of a completely uniform soil depth at all locations is not shown. Time t is in iterations. (Armstrong 1980a)

scape segments slowly adjust to boundary conditions. As incision proceeds, the overall relief of the landscape increases and the slopes become steeper while maintaining the same profile characteristics. The state of dynamic equilibrium is illustrated by the sequence of change along a given profile (Fig. 8.23). From the initial profile, the slopes change to a convex form which resembles the constant down-cutting form predicted by Carson and Kirkby (1972, pp.298–300) for slopes developed under soil creep. Notice that the slope profile does change form during the first 1000 iterations as the initial topography is transformed into the convex form determined by the process law (Fig. 8.24). The transformation involves a mixture of erosion and deposition around a central pivot point as in the manner predicted by the diffusion-decay model (W. E. H. Culling 1963; Scheidegger 1961b; Nash 1981).

An important finding of Armstrong's (1980b) simulation is that downvalley sequences of slope profiles show a constancy of form, the only aspects that vary being the steepness of the individual slopes in response to the continued incision of the exit cells (Fig. 8.23). Individual slopes tend to become steeper with time

Simulation Models

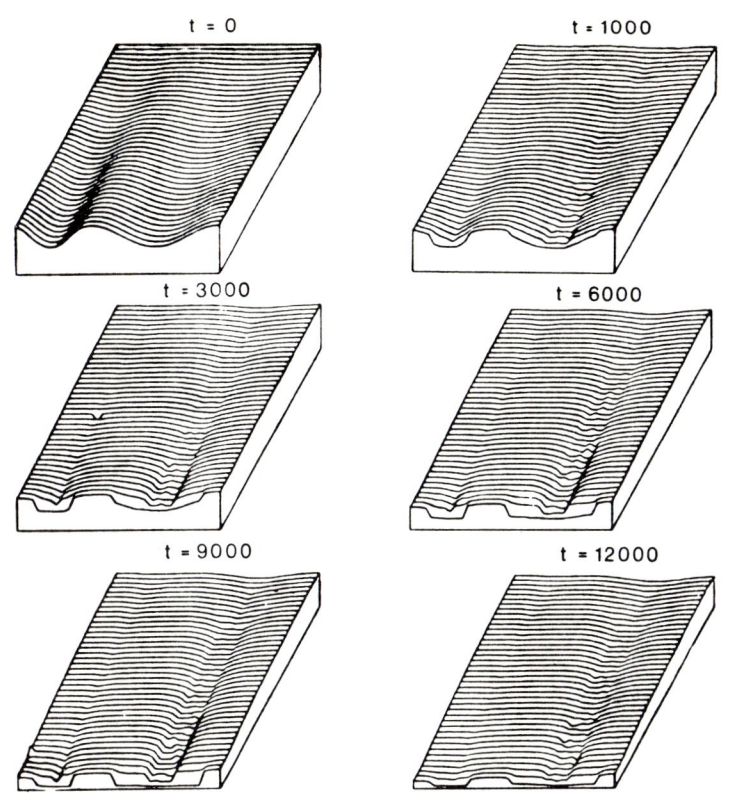

Fig. 8.22. Block diagrams showing the progress of the simulation. Time t is in iterations. (Armstrong 1980b)

and downvalley slope sequences becomes steeper towards the exit cells. However, the downvalley changes are more marked and the two sequences are not interchangeable. The implication of this, assuming that the model is accepted, is that the practice of substituting space for time, of using a downvalley sequence of profiles associated with stream incision to infer a temporal sequence of profiles, as for instance in the studies of Palmer (1956) and Young (1963b), is invalid. As Armstrong reasons, a more appropriate null hypothesis must be that downvalley sequences represent a dynamic equilibrium response to varying boundary conditions.

8.3.2 Drainage Basin Simulation

An early example of a drainage basin simulation model is the work of Moultrie (1970), who predicted the future development of the Poplar Gap drainage basin. A rectangular grid was placed over a topographic map of the basin to provide 106 sample elevations. For each of a series of time steps, totalling 6000 years of actual time, the amount of material removed at each point was calculated on

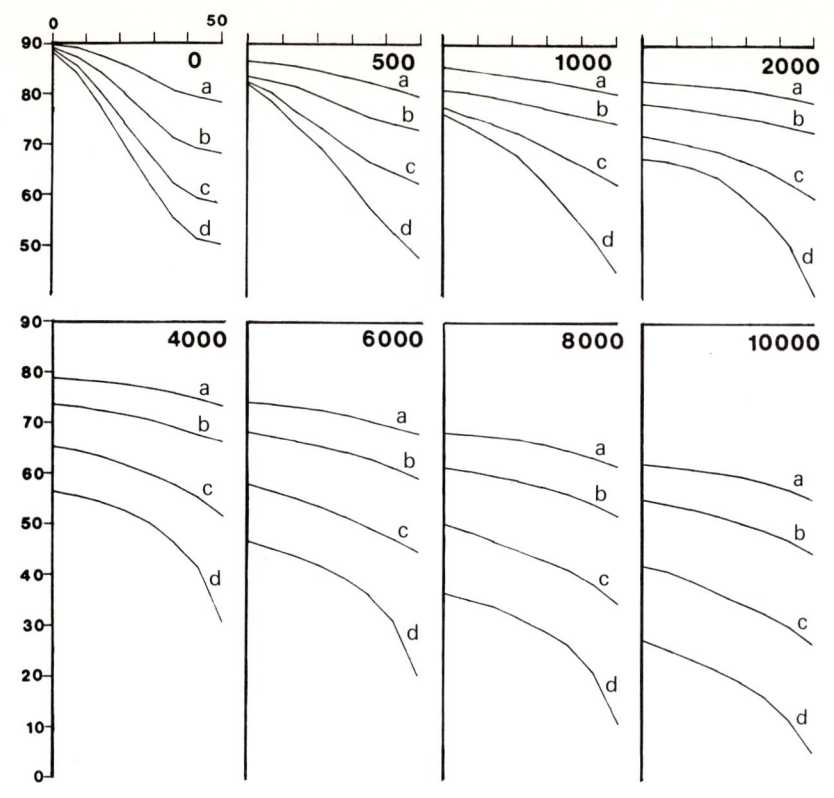

Fig. 8.23. Downvalley sequence of slope profile development. The downvalley sequence of profiles *a* to *d* plotted at iterations 0 to 10,000. (Armstrong 1980b)

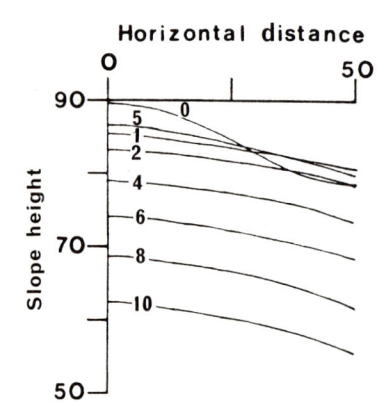

Fig. 8.24. Sequence of slope profiles developed at a single point. Profiles labelled in iterations/1000. (Armstrong 1980b)

Simulation Models

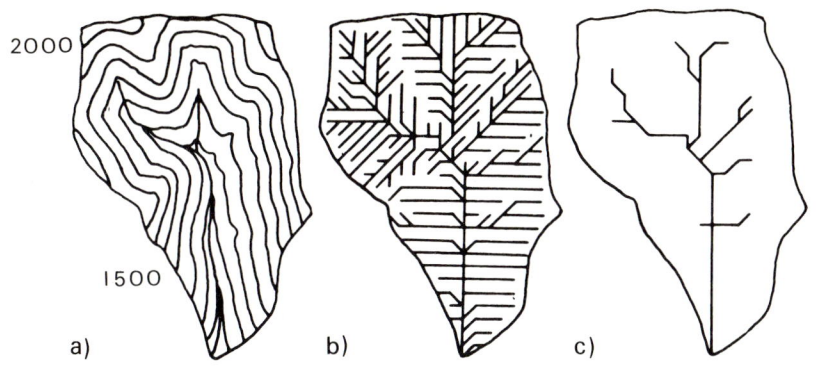

Fig. 8.25 a – c. Automatically derived networks for a basin digitized at points on a 250 ft grid. **a** Source basin defined by 100-ft contours. **b** Computer drawn network showing relationships between grid points. **c** Computer drawn network for all grid points preceded by at least six other grid points on the network. (Sprunt 1972)

the basis of precipitation data and hydraulic flow equations which included terms for slope, depth of overland flow, and shear force. The simulation predicted that slopes will be maintained at roughly their present values.

A more comprehensive and more revealing model was built by Sprunt (1972). In Sprunt's model, the land surface is represented by a 32 × 32 array of elements. Each element is associated with a unit square which may by thought of as a subsystem. The elements are linked by graphs produced by assuming that subsystems (elements) of higher elevation are source areas for elements of lower elevation (Fig. 8.25). Materials moving through the system are routed according to the transport network and according to the "cascade algorithm" of the process being modelled. In one example of the model (Sprunt 1972, pp. 386 – 387), transport was assumed to be proportional to the area drained and a function of slope. The slope function, $f(s)$, was borrowed from Horton (1945, p. 321) and is

$$f(s) = \sin \alpha / (\tan^{0.3} \alpha),$$

where α is the slope angle. Streamlines were arbitrarily defined as possessing flows of (6 + 40/time step) so as to allow stream lengths to increase quickly during the first stages of the simulation and then to attain an asymptotic value. The model considered the erosion of an inclined plane with a base level of 1200 ft and a summit level of 2100 ft. At the start of each time step, 900 units of runoff were distributed randomly among 900 subsystems within the boundaries of the system. The programme was run for 100 time steps. The results (Fig. 8.26) show a system of parallel streams developing on an inclined plane. The amount of material removed per time step reaches a maximum value after just four iterations and then declines rapidly to a small rate of decrease which is maintained from the tenth iteration to the end of the run. Spatial variations in runoff inputs cause local channel deepening. In turn, channel deepening leads to stream capture, at least in cases where an adjacent stream provides a stream with a steeper path than its present course. By the end of the run, 20 first-order basins, six second-order basins, and one third-order basin have formed.

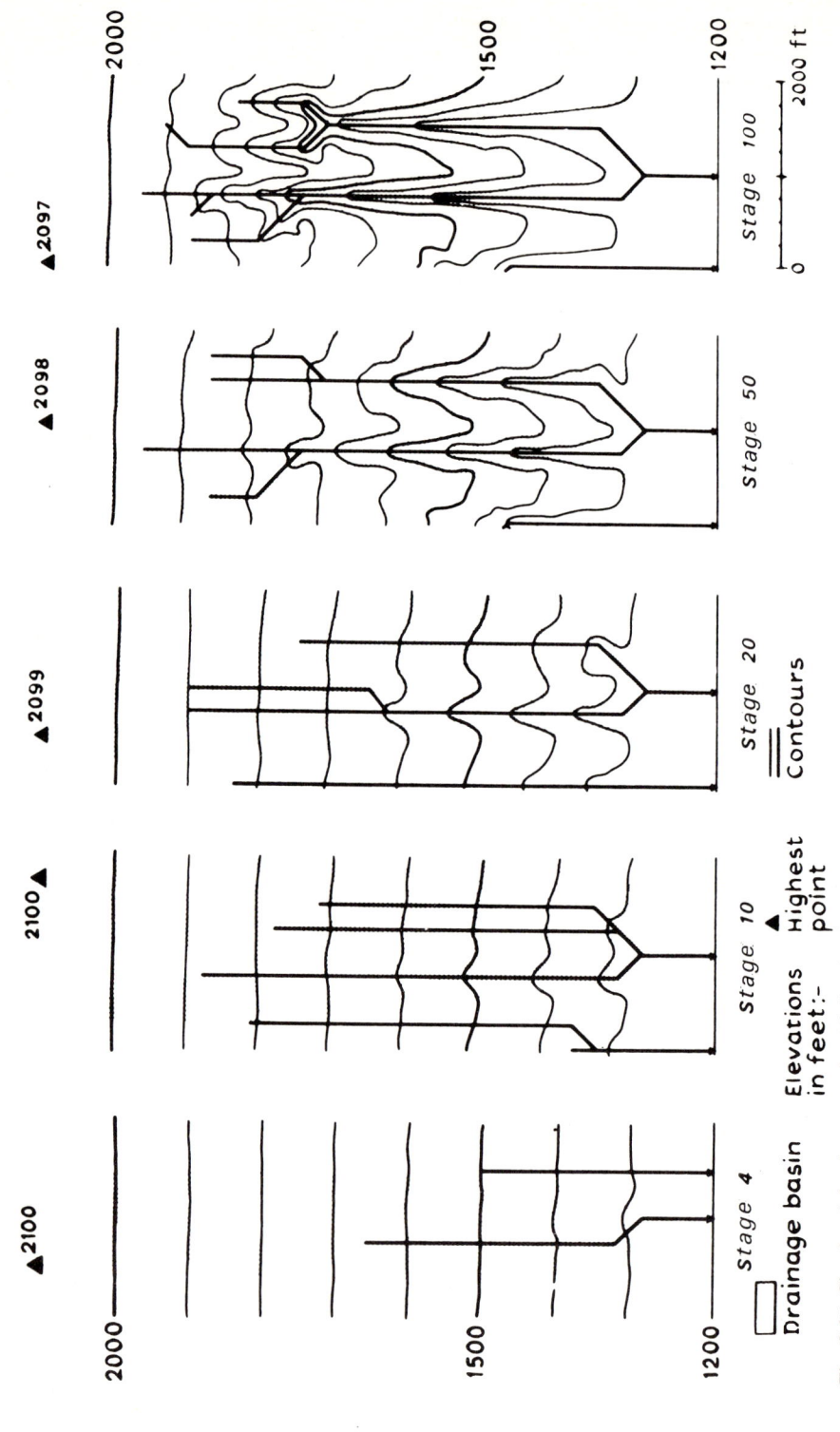

Fig. 8.26. Simulated development of a third-order drainage basin on an initially uniform plane. (Sprunt 1972)

A different approach to drainage basin modelling has been taken by agricultural engineers interested in predicting, with a high degree of accuracy, soil erosion and leaching in specific sites. The ANSWERS (areal nonpoint source watershed environment response simulation) model, developed at the Agricultural Engineering Department at Purdue University, Indiana, enables the movement of water, sediment, and chemicals through a drainage basin to be predicted (Beasley and Huggins 1981). A drainage basin is represented in the model by a grid of square elements, the elements being small enough to ensure that all parameter values within their boundaries are essentially uniform. In practical applications, a size in the range 1 to 4 ha has been found satisfactory. The physical properties used in the model, such as topography, storm intensity, soil type, and vegetation cover, must be specified for each spatial element. The model assumes relationships between the dependent variables (water, chemical, and sediment movement) and the physical properties of the drainage basin. The relationships are essentially empirical equations established from field and laboratory studies. The movement of material between spatial elements is subject to the laws of mass conservation, the output from one element becoming the input to adjacent elements. The hydrological components simulated by the model include spatially varied, unsteady rainfall, interception of rainfall by vegetation cover, surface storage in depressions, tile flow, and groundwater return flow to channel elements. Sediment yield from an area is modelled by equations which predict soil detachment rates due to rainfall impact and overland flow. The detached soil, combined with soil contributed by adjacent elements, becomes available for movement out of an element. An option is available in the model for predicting the particle-size distribution of the eroded sediment.

The model can be used to assess the impacts of alternative management schemes on agricultural land. Dillaha et al. (1982) used it to study problems of runoff and erosion associated with a construction site in Indiana. A detailed study was made of an area of 9.5 ha (Fig. 8.27). The site was divided into spatial cells by superimposing a 15.25 × 15.25 m grid. Input data included soil type, slope steepness and aspect, cover, management conditions, and channel characteristics (if present) for each element in the drainage basin. Simulations were made for 12 drainage basin scenarios, the first three of which represented the agricultural conditions before the onset of construction work. The results of two contrasting scenarios are shown in Fig. 8.28. Scenario 1 (Fig. 8.28a), for tilled corn, represents the average annual conditions of soil cover for agricultural land before construction began. Scenario 4 (Fig. 8.28b), for scalped topsoil and bare roadways, represents conditions when construction is underway. The difference in the pattern of erosion under these two scenarios is evident in the diagrams. The sediment yield in the tilled corn scenario was 13,500 kg ha^{-1} whereas the sediment yield for the scalped topsoil and bare roadway scenario was 16,000 kg ha^{-1}.

CREAMS is another field-scale model designed to evaluate nonpoint pollution of chemicals and sediment (Knisel 1980). The main components of the model are hydrology, erosion and sedimentation, and chemistry, all linked according to the general scheme shown in Fig. 8.29.

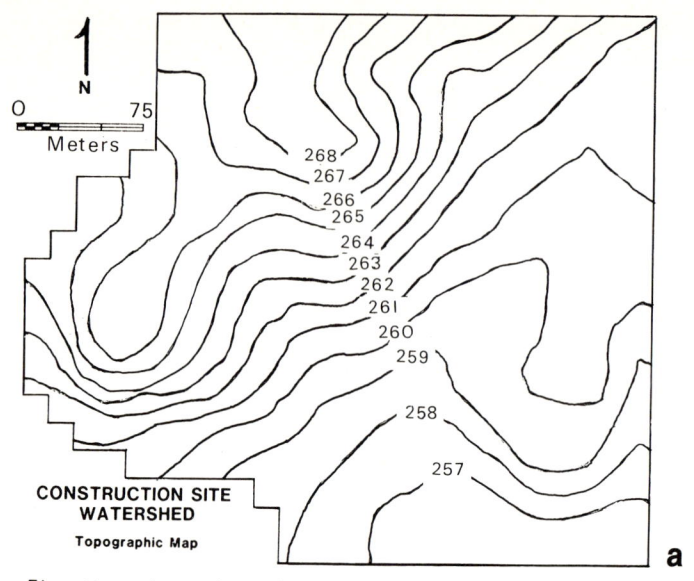

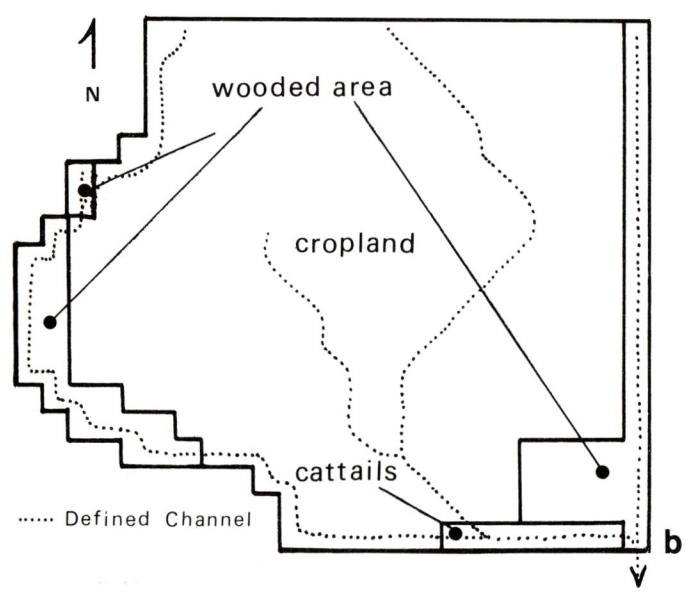

Fig. 8.27. a Topography of the construction site. **b** Land use and drainage patterns prior to construction. (Dillaha et al. 1982)

Simulation Models

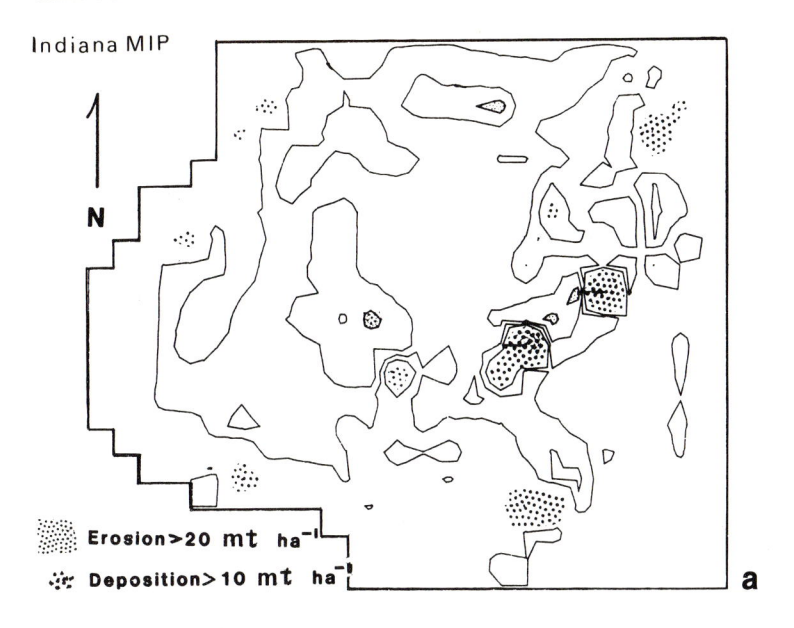

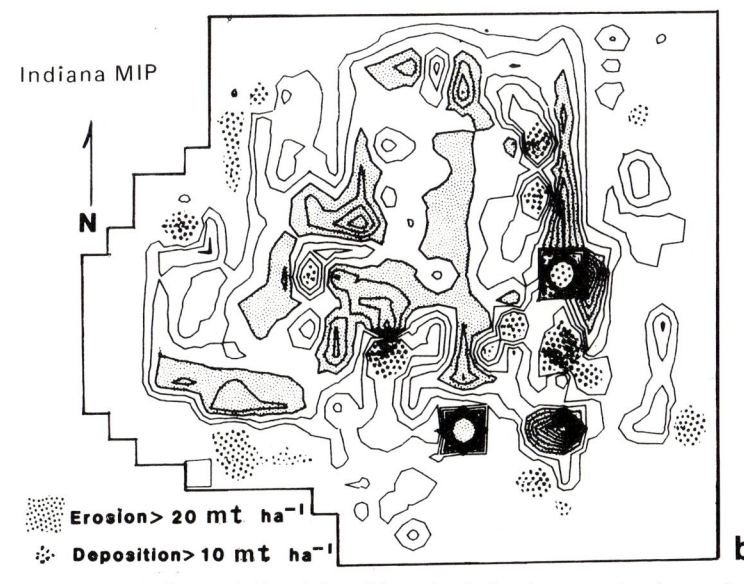

Fig. 8.28. a Sediment yield and deposition, simulation 1, conventional corn. **b** Sediment yield and deposition, simulation 4, scalped topsoil and bare roadways. (Dillaha et al. 1982)

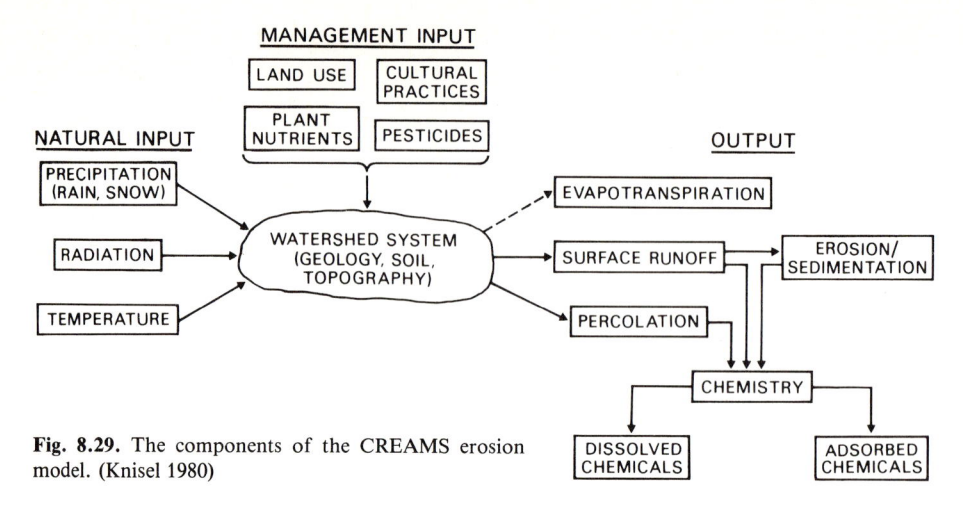

Fig. 8.29. The components of the CREAMS erosion model. (Knisel 1980)

8.3.3 Nearshore Bar Formation

A number of beach features have been modelled using computer simulation techniques. For instance, King and McCullagh (1971) simulated the development of the spit at Hurst Castle, Hampshire; Komar (1973) simulated the growth of a delta; Fox and Davis (1973) simulated the effects of storm cycles on beach erosion on Lake Michigan; Beer (1983) discussed estuarine models; and Davidson-Arnott (1981) simulated the formation of nearshore bars.

Davidson-Arnott's model considers a two-dimensional profile normal to a shoreline. The profile is represented by a series of 150 points spaced at 5-m intervals giving a total profile extending 750 m offshore. Sediment erosion, transport, and deposition are assumed to result from two processes — the net drift velocity associated with wave oscillatory currents, and longshore and rip currents generated by the breaking of waves. In the model, wave shoaling, breaking, reformation, and the generation of longshore currents are simulated and then the magnitude of the net drift velocity and longshore current at each point along the profile are determined. The values for sediment transport can then be determined and the form of the profile adjusted to reflect these values.

The model was initially calibrated with a wave height of 2 m, a wave-period of 6 s, a profile slope of 1°, and no tidal range. The results of the simulation (Fig. 8.30) show an inner bar migrating slowly seawards and disappearing after 1000 iterations but an outer bar showing some sort of equilibrium oscillating around a mean offshore distance of 275 m with about 3 m of water over the bar crest. A state of equilibrium is in fact reached when net seawards movement caused by the higher waves breaking on the seawards side of the bar and generating a rip cell is balanced by a net shorewards motion under unbroken waves.

Further simulations tested the effect of deep-water wave height, wave period, initial nearshore slope, and tidal range on the configuration of the nearshore profile (Fig. 8.31). Increasing wave height (Fig. 8.31a) tends to move bars seawards, increase the depth of water over the crests and troughs, increase the

Simulation Models

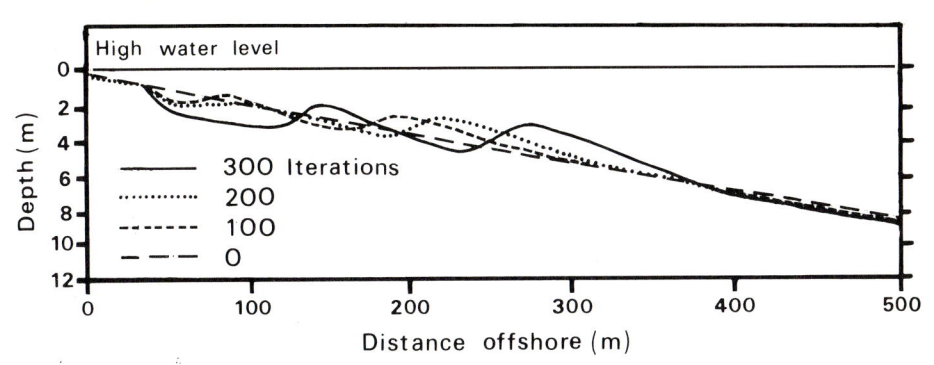

Fig. 8.30. Development of a barred profile from an initial planar profile after 0, 100, 200, and 600 iterations. Input values are wave height, $H_0 = 2.0$ m; wave period, $T = 6.0$ s; initial slope = 1°; and tidal range = 0.0 m. (Davidson-Arnott 1981)

bar height above a trough, and increase the seawards slope. Increasing wave period (Fig. 8.31b) tends to decrease bar height, increase the number of bars formed, and lead to the formation of plane profiles. Increasing initial profile slope (Fig. 8.31c) tends to decrease the number of bars and the distance between them. Increasing tidal range (Fig. 8.31d) tends to increase the number of bars, initiate the generation of further bars offshore, and decrease bar height. All these predictions are, to varying degrees, borne out by field measurements of bar formation (see Davidson-Arnott 1981).

8.3.4 Sand Dune Formation

The formation and migration of sand dunes has been modelled by McCullagh et al. (1972). An assumption made in the model is that the sand-size distribution consists of one class, 0.3 mm, the median size found in many dunes. A consequence of this assumption is that dunes cannot be created from a flat surface. The initial dune form used in the model is therefore a measured dune profile. A second assumption is that the dunes can be represented by 150 cells, each 1 m long, rather than by a continuous surface. A third assumption is that the model dunes are part of a more extensive dune field so that the loss of sand from one end of the dune sequence is added to the other end of the dune sequence forming a closed loop.

The development of the dunes is simulated on a computer through a series of time steps (iterations). Each time step is 1 h long. Calculations during each iteration involve three phases. In the first phase windspeeds are computed. The "regional" windspeed is assumed to follow an exponential distribution through time with a few periods of strong winds and many periods of weak winds. The winds may blow in either direction along the dune field. The variation of windspeed with height over open sand surfaces, $v(z)$, is defined by the logarithmic law of velocity distribution, originally derived to describe water

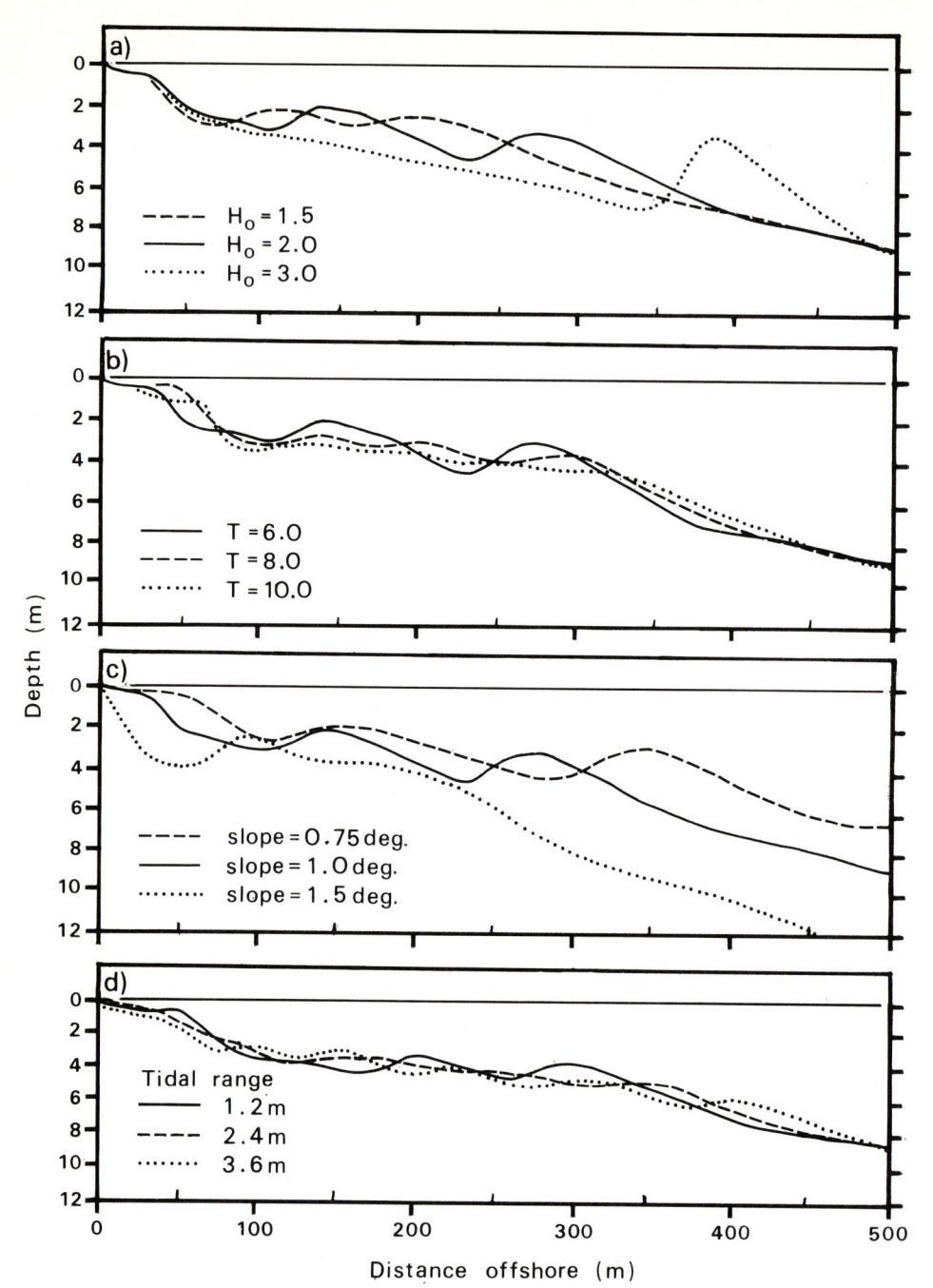

Fig. 8.31a – d. Simulation profiles with **a** wave height varying, **b** wave period varying, **c** initial slope varying, **d** tidal range varying. (Davidson-Arnott 1981)

flowing in open channels but shown by Bagnold (1954) to apply to wind movement,

$$v(z) = 5.75\, v_* \log(z/k)\,,$$

where v_* is the drag velocity, z is height, and k is a constant dependent upon surface roughness and has a value of 0.01 for particles 0.3 mm in size. This equation is not valid in the lee of a dune, where a linear decrease in windspeed from open air velocity to the sand floor of the lee area is assumed.

In the second phase of the calculations, the gradients between each of the 150 adjacent points of the dune field are computed. These gradients will be modified

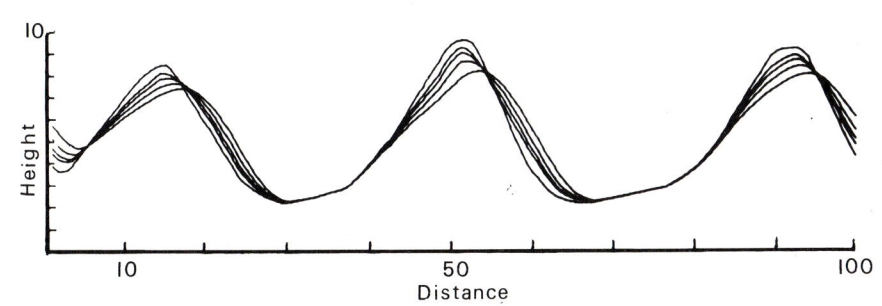

Fig. 8.32. Standard run. Wind direction is from left to right. (McCullagh et al. 1972)

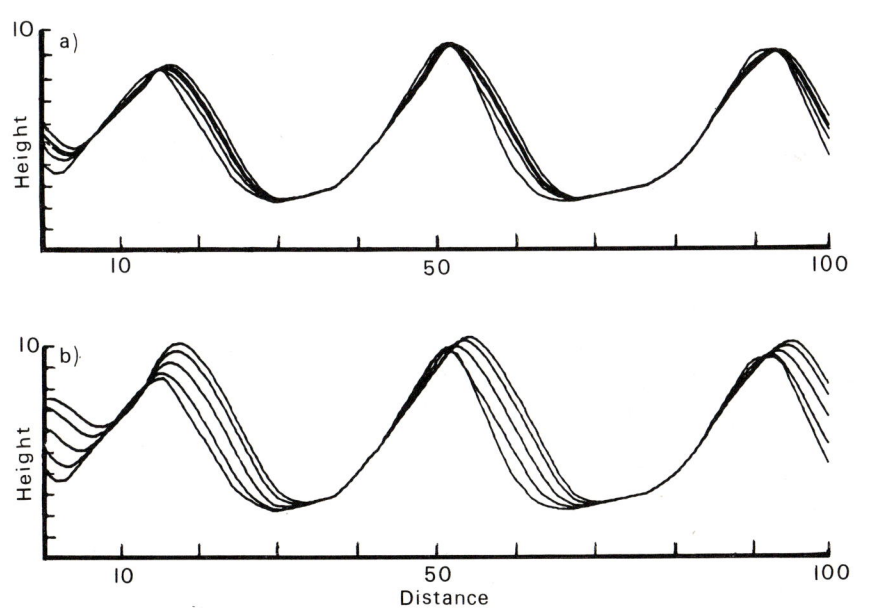

Fig. 8.33. a Slight increase in source material results in constant dune height. **b** Large increase in source material results in increased dune height. (McCullagh et al. 1972)

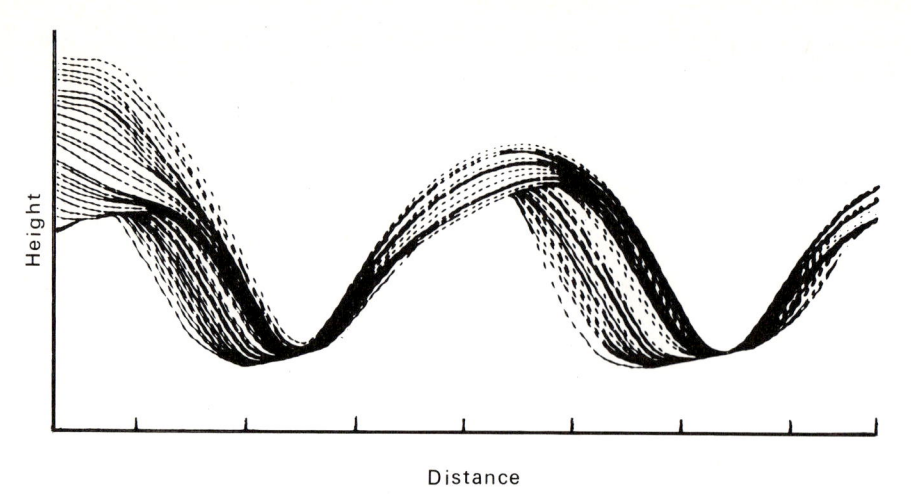

Fig. 8.34. Reduction in effective wind speed with barrier creating an initial lee effect which becomes reduced. (McCullagh et al. 1972)

by the amount of sand deposited if movement takes place. The relationship between gradient and deposition is not linear. McCullagh et al. found, by experimenting with the model, that a logarithmic function worked best giving high rates of deposition on steep, upslope gradients and no deposition on downslope gradients which exceeded the angle of repose (34°). In the final phase of the calculations, the information on wind velocity and gradient obtained in the first two phases is used to predict the quantity of sand movement. The quantity of sand moved, q, varies exponentially with wind velocity according to the relationship found experimentally in wind tunnel experiments by Bagnold (1954, pp. 69 – 70)

$$q = 5.2 \times 10^{-4}(v - Vt)^3$$

where v is wind velocity (m s^{-1}) and Vt is the threshold or critical windspeed above which sand will be in motion. During each time step, the sand in motion is assumed to be distributed downwind by an exponential decay function.

To illustrate the potential of their model, McCullagh et al. presents a series of sand dune profiles produced by holding the initial profile constant in each case and varying the input parameters. In all these runs, it is assumed that the wind has been blowing from left to right so that in the initial profile the right slope of the dunes is steeper than the left slope. Figure 8.32 shows a standard run in which the wind blew at 5 m s^{-1} from left to right. The dune height is reduced and the dunes advance to the right. Variations on this standard run are depicted in Fig. 8.33. A moderate increase in the supply of source material leads to constant dune height (Fig. 8.33a), whereas a large increase in the supply of source material leads to increased dune height (Fig. 8.33b). An increase in windspeed from 5 to 6 m s^{-1} leads to a faster migration and a rapid decrease in dune height. Figure 8.34 shows the effect of longer simulation runs. To avoid the mess produced by superimposing large numbers of dune profiles, cross-sections are plotted in this

Simulation Models

diagram, a procedure which has the disadvantage of losing information about profiles which have subsequently been obliterated by erosion; uneroded bedding planes, however, can be clearly seen. Figure 8.34 shows the result of setting the probability of a forward wind at 70 per cent and having the source direction linked to the probability of a wind blowing backwards (thus confining accretion to times when the wind is blowing from the right). In this case, the middle dune starts to grow and moves to the right, but once the left dune has grown large enough to create a barrier which puts the rest of the dune train in its lee, further growth and migration is to the left.

CHAPTER 9
Dynamical Systems Models

Dynamical systems models are concerned with change and susceptibility to change. They have shown their worth in describing material, and to a lesser extent energy, transfers and transformations in a variety of Earth surface systems. Their applications range from global models of geochemical cycles, through nutrient cycles in different biomes and regions, to models of water and solute dynamics in drainage basins, lakes, and soils. They also provide the theoretical base for the new generation of mathematical models dealing with dissipative structures, multiple equilibria, bifurcations, and catastrophes which hold out the promise of combining short-term and long-term landscape changes within a single framework.

9.1 Model Building

9.1.1 State and State Change

To construct a dynamical systems model, it must be possible adequately to describe the system under study by a set of state variables. In most applications in Earth science, the state variables measure the amount of material in a store (reservoir, pool, or compartment). To express changes through time, an ordinary differential equation is developed for each state variable. The equations equate the time rate of change of the state variables with the sum of inputs minus the sum of outputs. In the general case, for the ith state variable, the equation reads

$$dx_i/dt = \Sigma \text{Inputs} - \Sigma \text{Outputs} .$$

The expression on the right-hand side of the equation, because it determines the change of system state, is called the state transition function.

In the simplest case of a system described by just one state variable, an analytical solution to the state equation normally exists. In more complex cases, recourse normally has to be made to a numerical solution. Take, as an example, the store of litter on a forest floor. Let the amount of litter in store be state variable x (Fig. 9.1). Any change in the state of the system during an interval of time will be brought about by litter fall, z, and litter decay, f. By drawing up litter inputs and outputs, the system state equation is derived:

$$\frac{dx}{dt} = z - f .$$

Model Building

Fig. 9.1

Litterfall, z ↓ [LITTER x] ↓ Litter decay, f

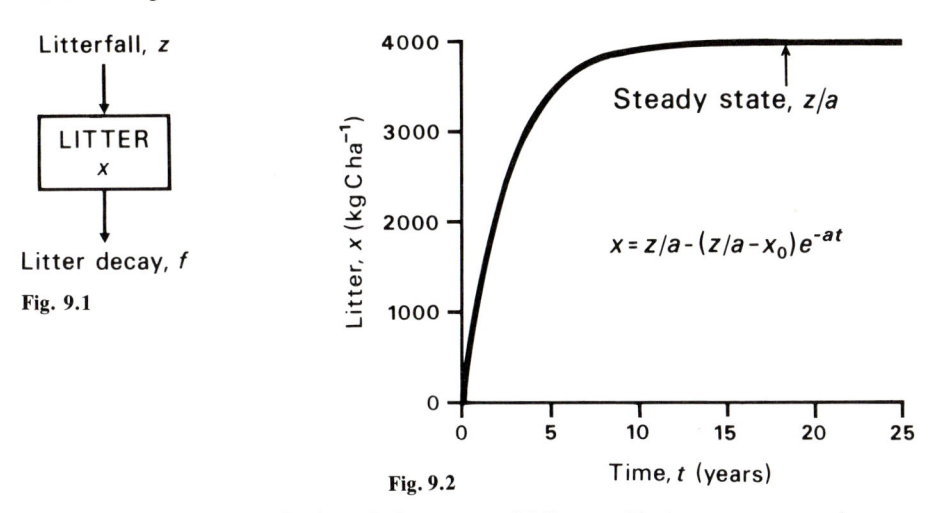

Fig. 9.2

Litter, x (kg C ha^{-1}) vs Time, t (years). Steady state, z/a. $x = z/a - (z/a - x_0)e^{-at}$

Fig. 9.1. Basic components of a dynamical systems model illustrated by inputs, outputs, and storage of litter on a forest floor

Fig. 9.2. Dynamics of the litter system starting from zero storage (which might represent the condition after a forest fire). Parameters: $z = 1600$ kg C ha^{-1} yr^{-1}; $a = 0.4$ yr^{-1}. Notice that the steady state storage $= z/a = 4000$ kg C ha^{-1}

It is reasonable to set the decay of litter, f, as proportional, by a factor a known as the rate constant, to the amount of litter in store, x. The state equation then reads

$$\frac{dx}{dt} = z - ax.$$

Assuming the rate of litter fall, z, is constant, and given an initial state, x_0, the equation can be integrated to provide a solution which expresses x as a function of time, t:

$$x = z/a - (z/a - x_0)e^{-at}.$$

The litter storage therefore changes asymptotically to a certain limit where inputs are balanced by outputs. This limit is the steady or stationary state and is defined by z/a (Fig. 9.2). Simple but effective equations of this kind were being used to study changes in soil fertility near half a century ago (Jenny 1941, p. 256) and have been found useful in ecology (Olson 1963).

An important point in the foregoing exposition of a simple dynamical system model is the idea of the rate constants (see Lerman 1979, pp. 4–14). Rate constants have the dimension of reciprocal time, T^{-1}. They are measured in the field or may be calculated by dividing steady-state storage of the state variable by the input. The reciprocal of a rate constant is called the turnover time and represents the amount of time required to replace the material in store under steady-state conditions. In the physical sciences, the term time constant is used in preference to turnover time, while in the geological literature the term residence time seems to be favoured. All three terms have the same physical significance and express

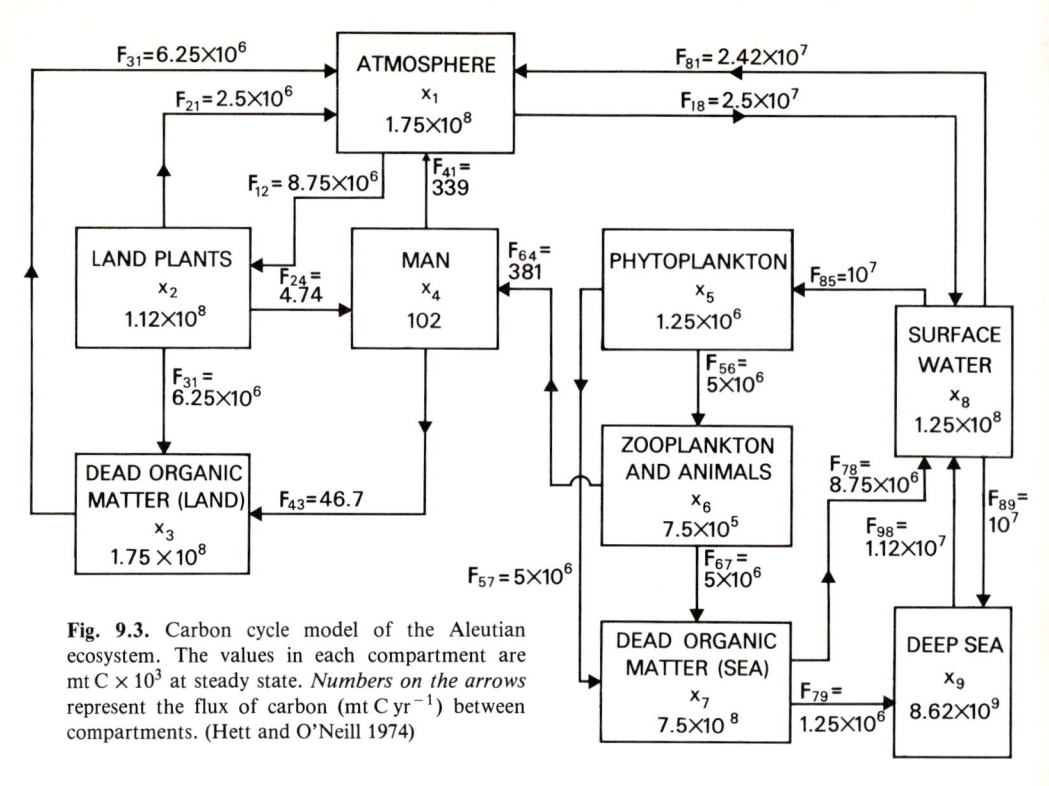

Fig. 9.3. Carbon cycle model of the Aleutian ecosystem. The values in each compartment are mt C × 10³ at steady state. *Numbers on the arrows* represent the flux of carbon (mt C yr⁻¹) between compartments. (Hett and O'Neill 1974)

the ratio of throughput to storage. They thus indicate how much of the input is stored and how much emerges as output during a unit interval of time.

In a more typical case, the system of interest will consist of more than one state variable. Take as an example the carbon cycle in the Aleutian islands as modelled by Hett and O'Neill (1974). The Aleutian ecosystem is represented by nine state variables, each of which is a reservoir of carbon. The reservoirs are the atmosphere, land plants, terrestrial dead organic matter, man, phytoplankton, zooplankton and marine animals, marine dead organic matter, surface water, and deep sea. The interactions between the components of the Aleutian ecosystem take the form of inputs and outputs of carbon (Fig. 9.3). The notation used to identify the fluxes is based on the following logic. For any one flux of carbon, a source (donor) compartment from which the flux emanates, and a terminal (receptor) compartment by which the flux is received, may be defined. The flux between the atmosphere and land plants, because it goes from state variable x_1 to state variable x_2, may be labelled F_{12}. All other fluxes are labelled in like manner. In general, F_{ij} represents the flux of material from compartment i to compartment j. The time rate of change of a state variable depends on the difference between incoming and outgoing fluxes of carbon. In the case of man, state variable x_4, the equation defining the time rate of change is

$$\frac{dx_4}{dt} = F_{24} + F_{64} - F_{41} - F_{43}.$$

Model Building

More generally, the change in reservoir i is written

$$\frac{dx_i}{dt} = \sum_j F_{ji} - \sum_j F_{ij} \quad (i \neq j; i, j = 1, 2, \ldots, n),$$

where n is the total number of reservoirs.

9.1.2 Transfer Equations

The fluxes, F_{ij}, must be defined by suitable transfer or process laws. Many processes in Earth surface systems seem to conform to a linear transport law in which a flux is proportional, by a rate constant, to the state of the donor reservoir:

$$F_{ij} = \lambda_{ij} x_i.$$

Natural processes to which this law applies include the release of water from rock and soil to rivers, the loss of nutrients from biotic reservoirs of ecosystems to abiotic reservoirs, and many weathering reactions. Biotic interactions within ecosystems are better described by nonlinear transport laws of which there are many kinds (Patten 1971; Wilson 1981, p. 97). Applying linear transport equations to the Aleutian ecosystem, and with rate constants as defined in Table 9.1, yields

$$\frac{dx_i}{dt} = \sum_j \lambda_{ji} x_j - \sum_j \lambda_{ij} x_i \quad (i \neq j; i, j = 1, 2, \ldots, n).$$

When calibrated with appropriate parameters and a starting state, the equations can be solved on a computer thus revealing the dynamics of the system (Fig. 9.4).

In the example of the Aleutian ecosystem, the system is closed in that there is no exchange of matter with the environment. In cases where the environment of a system and exchanges with it are modelled, then the general formula is

Table 9.1. Transfer coefficients, λ_{ij}, for carbon model of the Aleutian ecosystem. (Hett and O'Neill 1974)

To	x_1	x_2	x_3	x_4	x_5	x_6	x_7	x_8	x_9
				From state variables					
x_1	–	0.0223	0.0357	3.32				0.194	
x_2	0.05	–							
x_3		0.0558	–	0.458					
x_4		4.21×10^{-8}		–		5.08×10^{-4}			
x_5					–			0.08	
x_6				4.00	–				
x_7				4.00		6.67	–		
x_8	0.143						0.0113	–	1.3×10
x_9							1.67×10^{-3}	0.08	

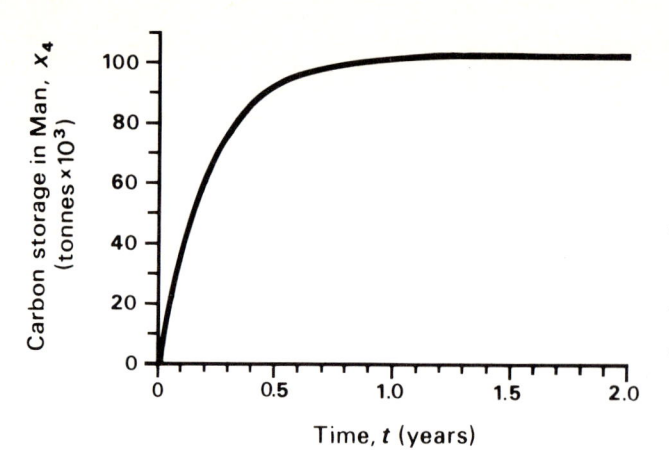

Fig. 9.4. An example of the dynamic behaviour of the Aleutian ecosystem. In this simulation run all state variables were held at their steady-state values, except for carbon storage in Man, x_4, which was assumed to have been instantaneously reduced to zero. The diagram shows the recovery from zero to the steady state

$$\frac{dx_i}{dt} = \sum_j \lambda_{ji} x_j - \sum_j \lambda_{ij} x_i \quad (i \neq j; i,j = 0, 1, 2, \ldots, n),$$

where the environment is labelled as reservoir 0.

9.2 System Stability

The stability of a system is defined in terms of its response to fluctuations. Stability concepts are conveniently examined by tracing changes of system state within the state (or phase) space. A system's state space is the space produced by using the state variables as mutually orthogonal coordinates. In the general case of a system with n state variables, there will be an n-dimensional state space defined by n Cartesian coordinates, one coordinate for each state variable. Any point within the state space uniquely defines a set of values for the state variables and describes the instantaneous condition of the system. If the state of the system should change with time, as it will if taken from equilibrium by a perturbation, a path or trajectory will be traced in the state space, the route taken depending on the nature of the equations governing the dynamics of the system.

9.2.1 State Space

To explain concepts of system stability it is helpful to consider a state space defined by just two state variables, x_1 and x_2. The state space is in this case two-dimensional – a state plane – and can be depicted without difficulty. In the general case, the system-governing equations are

$$\frac{dx_1}{dt} = f_1(x_1, x_2)$$

$$\frac{dx_2}{dt} = f_2(x_1, x_2).$$

System Stability

a) Stable   b) Unstable   c) Metastable   d) Neutral

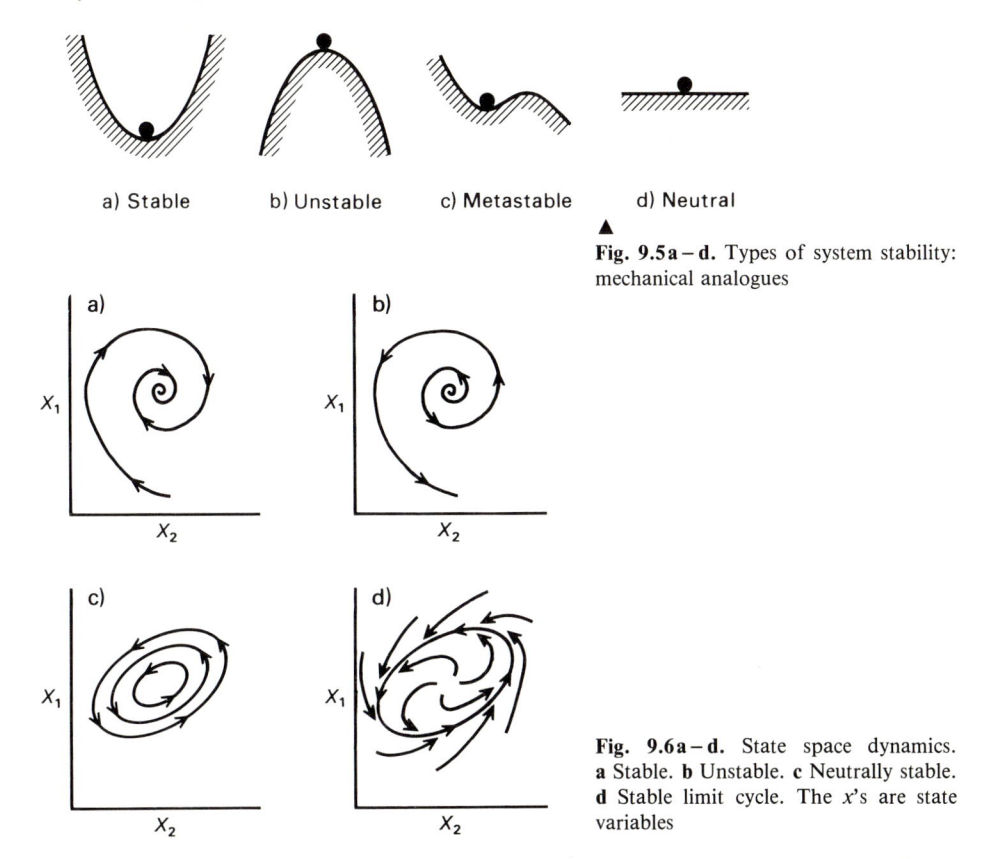

Fig. 9.5 a – d. Types of system stability: mechanical analogues

Fig. 9.6 a – d. State space dynamics. a Stable. b Unstable. c Neutrally stable. d Stable limit cycle. The x's are state variables

Any point in the state plane for which the system-governing equations equal zero is an equilibrium or singular point. Stability concepts involve assessing the nature of stability around these points. Assessment for even simple nonlinear systems can be difficult because the dynamics of the systems is usually complex. However, certain basic types of stability have been identified (Hirsch and Smale 1974). Some of these are readily appreciated if the state plane is visualized as a land surface (Fig. 9.5). A stable point in the landscape is represented by a pit or depression. If a ball lying at the bottom of the pit is forced up the side of the pit and then released, it will return to its original position. This is analogous to the stable steady state in a system in which any perturbation is dampened and the system returns to its undisturbed state (Fig. 9.6). Stable equilibrium conditions of this kind are characteristic of Earth surface systems which tend to maintain mean values. An example is the flow of water in a stream where high and low flows are small deviations about an average state. Examples offered by Thornes (1983) include the ratio of stream width to meander wavelength and the modal spacing of beach ridges. A stable point in the landscape may also lie on a watershed. A ball will remain precariously balanced in this position as long as it is not subjected to any displacements. Once the ball is disturbed it will roll downhill. In actual state planes, the system will tend to move to extreme values of the state variables. Unstable equilibrium conditions of this kind are characteristic of systems which tend to destroy themselves and are therefore uncommon. Thornes (1983) gives as examples gullies, alluvial fans, and caliche. Another example is the system described by Kirkby (1980), in which soil depth influences the rate of weathering such that weathering is at its slowest in either very thin soils or very thick soils. The dynamics of this system determine that neighbouring sites which start with almost the same soil thickness will with time diverge to form exposed bedrock tors and stable soil layers. States in between these two extremes are unstable and so seldom found in the landscape. A metastable point in the landscape is represented by an indentation in a hillside. A ball lying in the indentation will be in stable equilibrium for a limited range of displacements, but beyond a threshold level of displacement it will move to another part of the landscape. This kind of behaviour is characteristic of dissipative structures (Sect. 9.4).

Another basic kind of behaviour found in state planes, but not readily explained in terms of the landsurface analogy, is oscillatory behaviour around a fixed point (Fig. 9.6c). In this case, the state space consists of a set of concentric ellipses, each of which is a possible trajectory of the system. A perturbation will shift the system from one of these trajectories to another. Thornes (1983) believes that meanders display dynamics of this kind. It is also possible for there to be a single, stable limit cycle in the state space (Fig. 9.6d). This cycle will capture all initial states. An example is pool and riffle sequences which appear to oscillate in a regular manner.

State spaces may be very complicated containing equilibrial points of all varieties (stable, unstable, metastable, and so on). In such cases, the system will not be globally stable, that is stable for the entire state space. Local stability prevails and, driven by fluctuations, the system may wander through the state space. Further complications arise if the state space is discontinuous and the system flips from one domain of operation to another. Discontinuous behaviour is tackled in the models of catastrophe theory (Sect. 9.4.1).

9.2.2 Sensitivity Analysis

Having built a dynamical systems model, it may be used to study the kinetics of the system. Commonly more revealing though, is a sensitivity analysis on the system components and interactions. Sensitivity analysis has many facets. Firstly, it enables the relative effect of a change in reservoir x_i induced by a change Δx in reservoir x_j to be compared with the relative effect on reservoir x_i brought about by the same degree of change, Δx, in reservoir x_k. The relative effect is given, assuming indirect flows are ignored, by

$$\frac{\lambda_{ij}\Delta x}{\lambda_{ik}\Delta x} = \frac{\lambda_{ij}}{\lambda_{ik}}.$$

For example, in the Aleutian ecosystem, the relative effect of man, x_4, to a change, Δx, in marine animals, x_6, compared with the same change in land plants, x_2, is

$$\frac{\lambda_{46}}{\lambda_{42}} = \frac{5.08 \times 10^{-4}}{4.21 \times 10^{-8}} = 1.21 \times 10^4.$$

The sensitivity in this case is more than 10,000:1, which is surprising, since the model was constructed with the Aleuts' receiving 95 per cent of their carbon from the marine system and intuition would therefore suggest a value 20:1 (Hett and O'Neill 1974). It does indicate, however, the extent to which Aleuts depend on the sea.

A second facet of sensitivity analysis is examining the relative sensitivity of components to changes in rate constants. This is done by calculating the partial derivatives of the rate constant matrix, Λ, using the following system of equations (Funderlic and Heath 1971)

$$\Lambda(\partial x(\infty)/\partial \lambda_{jk}) = x_k(\infty)(e_k - e_j),$$

$$\sum_{i=1}^{n} \partial x_i / \partial \lambda_{jk} = 0,$$

where e is a vector of zeros except for a 1 in the ith component. Applying this analysis to the Aleutian ecosystem, Hett and O'Neill (1974) were able to show that the ratio $\{\partial x_2(\infty) \partial \lambda_{21}\}/\{\partial x_6(\infty) \partial \lambda_{58}\}$ is 273, which indicated that the terrestrial food supply is some 300 times more sensitive to an uptake of atmospheric carbon than the marine food supply is to the uptake of surface water carbon.

A third facet of sensitivity analysis is the question of rate of system recovery following some perturbation. The rate of recovery is related to the eigenvalues of the rate constant matrix, Λ. These eigenvalues determine the rates of change of reservoir storage. In a stable system they will never be positive. The rate of recovery of a system is determined by the crucial root of the matrix Λ. The crucial root is that eigenvalue, other than zero, which has the smallest absolute value. The crucial root having been found, its sensitivity to changes of rate constants in the system can be determined using the equation

$$\frac{\partial r}{\partial \lambda_{ij}} = \frac{c_{ij} - c_{jj}}{\sum\limits_{k}^{n} c_{kk}} \quad (i, j = 1, 2, \ldots, n),$$

where c_{ij} is the cofactor of $\Lambda - rI$ (Funderlic and Heath 1971). (The eigenvalues are denoted by r because λ has already been used to denote rate constants). Performing this analysis on the rate constant matrix of the Aleutian ecosystem, Hett and O'Neill (1974) showed that the smallest sensitivity coefficients were associated with transfer from the Aleut carbon reservoir. For example, $\partial r/\partial \lambda_{14}$ (the transfer from man to atmosphere in respiration), is -8.34×10^{-10}. The largest sensitivity coefficients are associated with the deep sea and marine dead organic matter; for instance, $\partial r/\partial \lambda_{97} = 0.739$. These findings suggest that the rate of recovery of the Aleutian ecosystem following a disturbance would be virtually unaffected by the human population. These results do not automatically follow from the small size of the Aleut reservoir: the crucial root can be sensitive to small compartments.

9.3 Biogeochemical Cycles

9.3.1 The Global Cycle of Phosphorus

Lerman et al. (1975) model the global biogeochemical cycle of phosphorus using a dynamical systems approach. The model is presented diagrammatically in Fig. 9.7. The storages in the geochemical reservoirs and the fluxes of phosphorus are taken or derived from literature sources (see Lerman et al. for details). The major reservoirs considered are sediments, land (soil), land biota, oceanic biota, surface ocean, deep ocean, and mineable phosphorus deposits. With the exception of the mineable phosphorus reservoir and the fluxes associated with it, the phosphorus cycle is a steady-state system. The actual estimates of some of the reservoir storages and interreservoir fluxes varied by a factor of three or four. Several of the values used in the model, in particular fluxes F_{61}, F_{21}, F_{32}, and F_{46}, had to be assumed to make the system balance. However, to safeguard against unrealistic fluxes entering the model, these four troublesome fluxes were calculated by balancing them against the respective reservoir content when the values of complementary input or output fluxes were obtained by other means.

In the steady state, the residence time of phosphorus in the land biota is 46 years, significantly longer than the residence time of phosphorus in the oceanic biota which is 0.14 years. Of more interest are the transient states of the phosphorus cycle. Lerman et al. explored the effect of a complete cessation of photosynthetic productivity on Earth. This "Doomsday" scenario involves an instantaneous cessation of the uptake of phosphorus by the land and oceanic biota. The outcome of this scenario is shown in Fig. 9.8. The oceanic biota decays in less than a year; the land biota has decayed to a hundredth of its original value in 200 years. The values for land and oceanic biota could find new steady-state values if, say, a nonphotosynthetic biota became established; but this possibility is not considered in the model. During the 200-year period of the simulation, the phosphorus content of the surface ocean increases by a factor of about 2.5; the contents of the land, deep ocean, and sediment reservoirs remain virtually unchanged. So after 200 years, the total phosphorus concentrations in the surface ocean and deep ocean draw closer to one another and, when a new steady state is reached, the phosphorus concentration throughout the ocean is nearly uniform. Thus the role played by terrestrial and oceanic biota in the global phosphorus cycle becomes apparent.

9.3.2 The Global Cycle of Carbon Dioxide and Oxygen

Garrels et al. (1976) present a model of the global cycles of carbon dioxide and oxygen which suggests that a dynamic, steady-state system persists, maintained by effective feedback mechanisms. The model is portrayed in Fig. 9.9. (Details of the assumptions made and the sources of data can be found in the original work.) The model is used by Garrels et al. to study the effect of perturbing the steady-state system. Three perturbations were modelled – a tripling of the erosion rate, a doubling of the rate of photosythesis, and an instantaneous halting of photo-

Biogeochemical Cycles

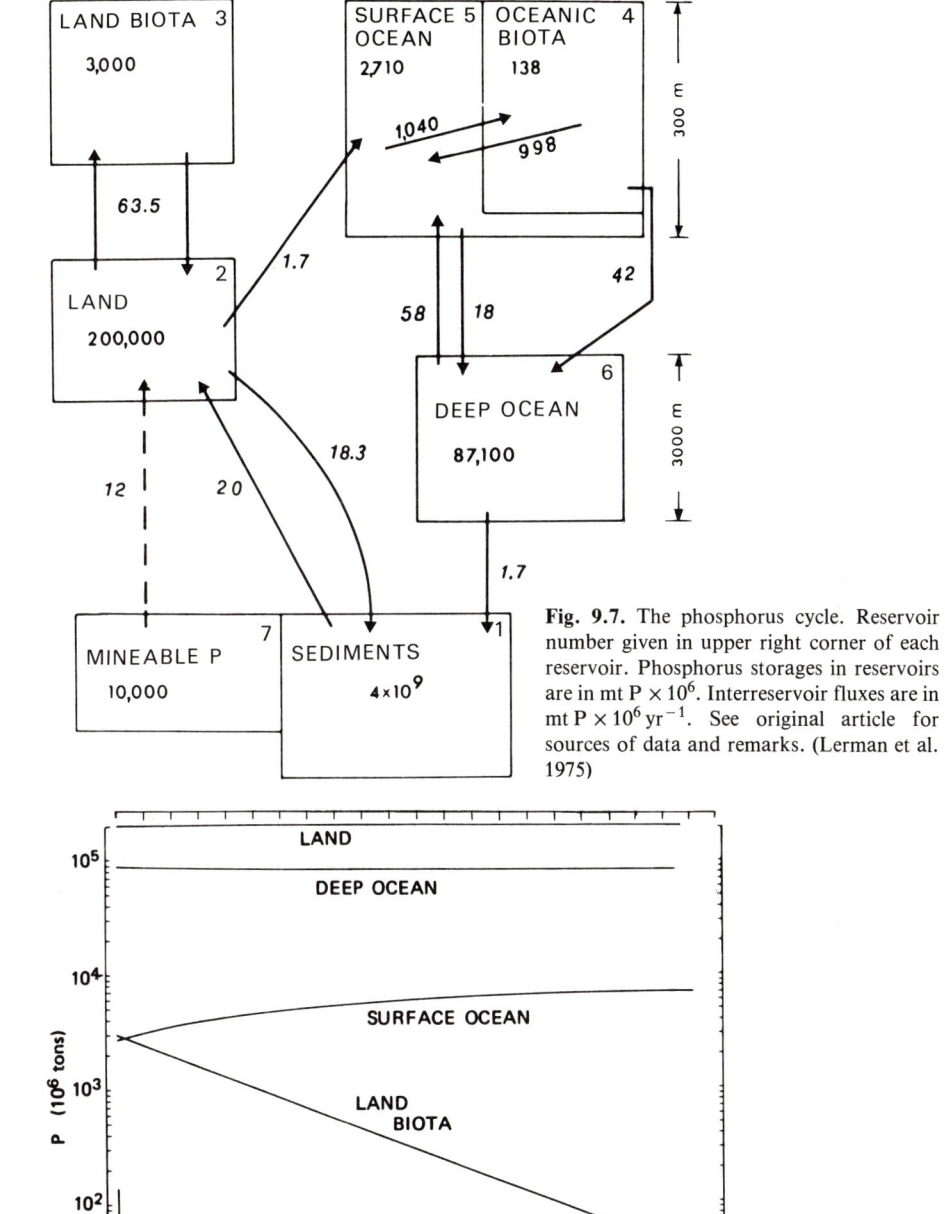

Fig. 9.7. The phosphorus cycle. Reservoir number given in upper right corner of each reservoir. Phosphorus storages in reservoirs are in mt P × 10^6. Interreservoir fluxes are in mt P × 10^6 yr^{-1}. See original article for sources of data and remarks. (Lerman et al. 1975)

Fig. 9.8. Doomsday scenario. Changes in phosphorus content of reservoirs after all photosynthetic productivity on Earth is ceased at time 0. The reservoirs not shown remain virtually unchanged. (Lerman et al. 1975)

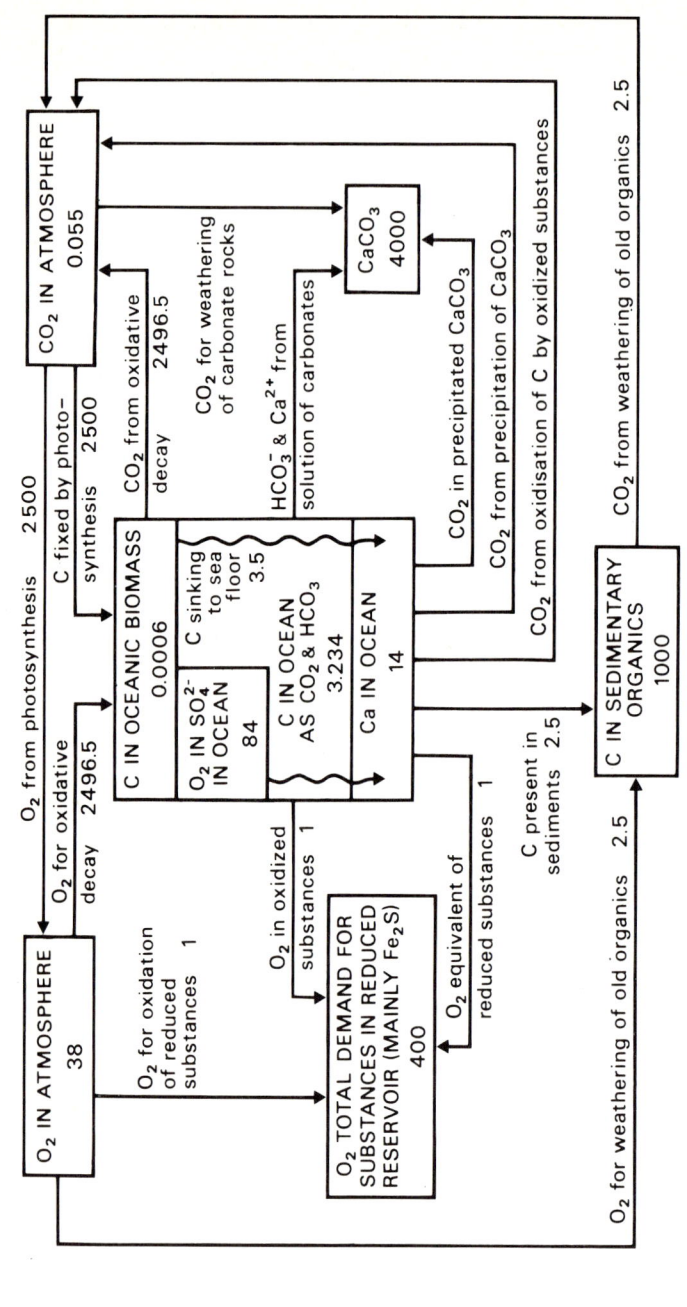

Fig. 9.9. Major components of the exogenic cycle of CO_2 and O_2. The size of the reservoirs is in units of 10^{18} mol; the size of the fluxes is in units 10^{18} mol yr^{-1}. (After Fig. 5 and Tables 1 and 2 in Garrels et al. 1976)

Biogeochemical Cycles

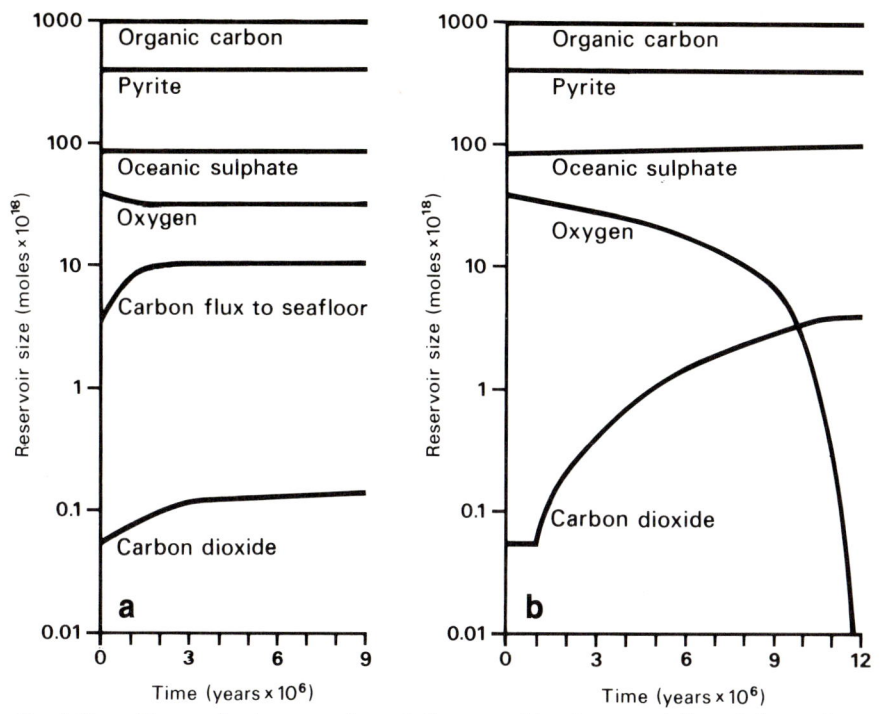

Fig. 9.10. a Changes in the reservoirs and fluxes resulting from a tripling of the erosion rate. **b** Doomsday scenario: changes in reservoirs resulting from cessation of photosynthesis. (Based on data in Tables 4 and 6 in Garrels et al. 1976)

synthesis. The effect of tripling the rate of erosion can be seen in Fig. 9.10a. The atmospheric oxygen reaches a new steady state, some 15 per cent below the present level, within about 2 million years. Atmospheric carbon dioxide rises to about 2.5 times the present level. Although the other reservoirs are not much affected, the flux of organic carbon to the sea floor is tripled. Interestingly then, the effects of accelerated erosion may, on a time scale of millions of years, be greater than the effects of burning fossil fuels in terms of oxygen depletion and carbon dioxide increase. As Garrels et al. explain, "Man's greatest long-term contribution to the exogenic cycle may be his drastic effects on erosion rates".

The effect of doubling the rate of photosynthesis is to increase the size of the organic reservoir and to carbon dioxide's becoming a "limiting nutrient". But the scenario is rather artificial, since it is difficult to envisage a process that would double the rate of photosynthesis. The third scenario considered was a Doomsday scenario: an immediate cessation of photosynthesis. The results, as depicted in Fig. 9.10b, are dramatic. The first event, the disappearance of the oceanic biomass, takes place so fast (within less than a year) that it cannot be shown in the figure. After that oxygen is depleted for 10 million years or so until it finally disappears and the model becomes inoperative.

Berner et al. (1983) constructed a model to study the effects on the carbon dioxide level of the atmosphere, and the Ca, Mg, and HCO_3 levels of the ocean,

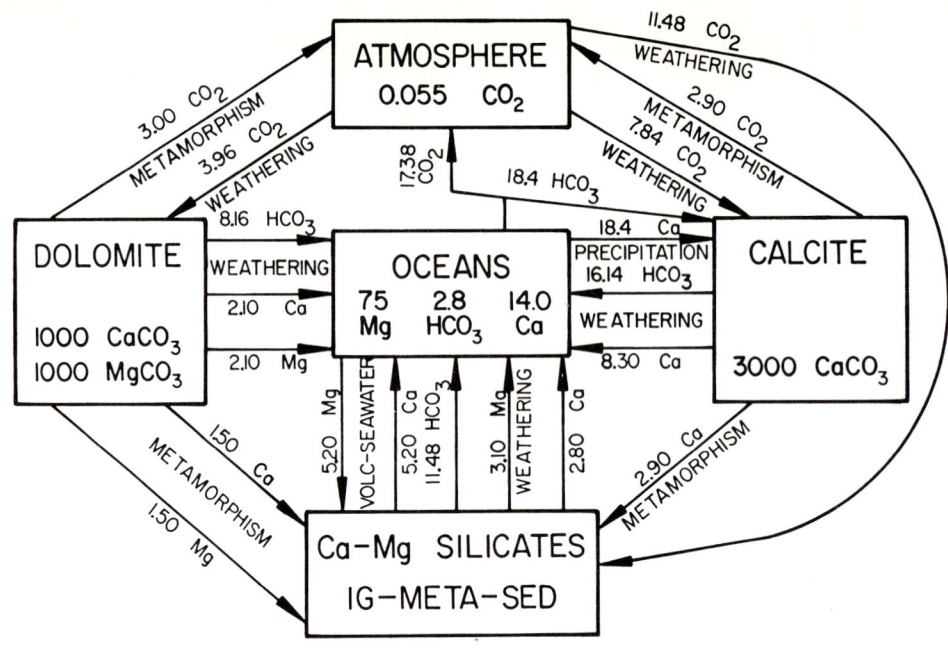

Fig. 9.11. Carbonate-silicate cycle for the present day. Fluxes of CO_2 and HCO_3 are corrected for weathering via H_2SO_4 (from pyrite oxidation) and for sedimentary pyrite formation. Fluxes in 10^{18} mol myr^{-1}; reservoir sizes in 10^{18} mol. (Berner et al. 1983)

of the following processes: continental weathering of calcite, dolomite, and calcium-and-magnesium-containing silicates; biogenic precipitation and removal of $CaCO_3$ from the ocean; removal of magnesium from the ocean via volcanic–sea water reaction; and the metamorphic–magmatic decarbonation of carbonate and dolomite, and the resulting CO_2 degassing, as a consequence of plate subduction. The model is summarized in Fig. 9.11. The present ocean-atmosphere system is assumed to be in a steady state. Rate expressions are derived for all the fluxes to give the following set of nonlinear, mass balance equations:

$$\frac{dD}{dt} = -(k_{W_D} + k_{M_D})D$$

$$\frac{dC}{dt} = -(k_{W_C} + k_{M_C})C + k_{\text{prep}}(M_{Ca}M_{HCO_3}^2 - k_{eq}A_{CO_2})$$

$$\frac{dS_{CaSi}}{dt} = k_{M_C} + k_{M_D}D - k_{W_{CaSi}}S_{CaSi}$$

$$\frac{dS_{MgSi}}{dt} = k_{M_D}D - k_{W_{MgSi}}S_{MgSi} + k_{V-SW}M_{Mg}$$

$$\frac{dM_{Mg}}{dt} = k_{W_D}D + k_{W_{Mg}}S_{MgSi} - k_{V-SW}M_{Mg}$$

$$\frac{dM_{Ca}}{dt} = k_{W_D}D + k_{W_C}C + k_{W_{CaSi}}S_{CaSi} + k_{V-SW}M_{Mg}$$
$$- k_{prep}(M_{Ca}M_{HCO_3}^2 - k_{eq}A_{CO_2})$$

$$\frac{dM_{HCO_3}}{dt} = 4k_{W_D}D + 2k_{W_C}C + 2k_{W_{CaSi}}S_{CaSi} + 2k_{W_{MgSi}}S_{MgSi}$$
$$- 2k_{prep}(M_{Ca}M_{HCO_3}^2 - k_{eq}A_{CO_2})$$

$$\frac{dA_{CO_2}}{dt} = 2k_{M_D}D - 2k_{W_D}D - 2k_{W_{MgSi}}S_{MgSi} - 2k_{W_{CaSi}}S_{CaSi}$$
$$+ k_{M_C}C - k_{W_C}C + k_{prep}(M_{Ca}M_{HCO_3}^2 - k_{eq}A_{CO_2}),$$

where D, C, S_{CaSi}, S_{MgSi}, M_{Mg}, M_{Ca}, M_{HCO_3}, and A_{CO_2} stand for the mass in moles of dolomite, calcite, Ca-silicate, Mg-silicate, ocean magnesium, ocean calcium, ocean bicarbonate, and atmospheric carbon dioxide reservoirs respectively. The equations were integrated through 100 million years using a time step

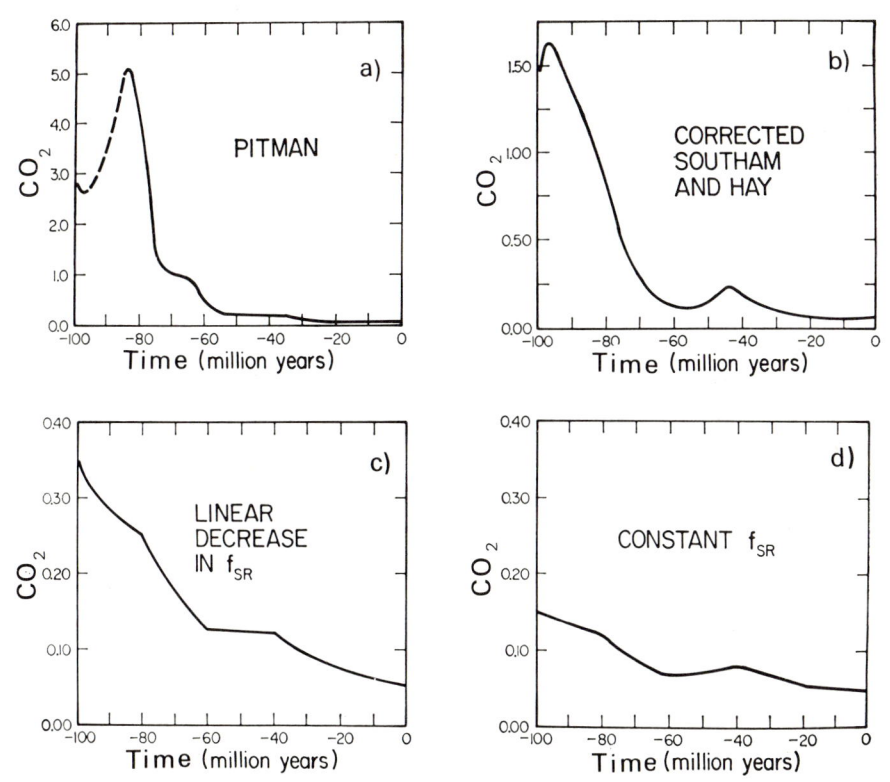

Fig. 9.12a – d. Computer results for the mass of CO_2 (in 10^{18} mol) as a function of time for the past 100 myr using the land area curve of Barron et al. (1980) and various spreading rate, f_{sr}, formulations: **a** Pitman (1978) spreading rate with *dashed line* based on a value of $f_{sr} = 1.70$ at 100 myr *BP* assumed by Berner et al. (1983). **b** Southam and Hay (1977) spreading rate corrected for loss by subduction. **c** Linear decrease in spreading rate corrected for loss by subduction. **d** Constant spreading rate with time equal to the present value. Note that the vertical (CO_2) scales differ. (Berner et al. 1983)

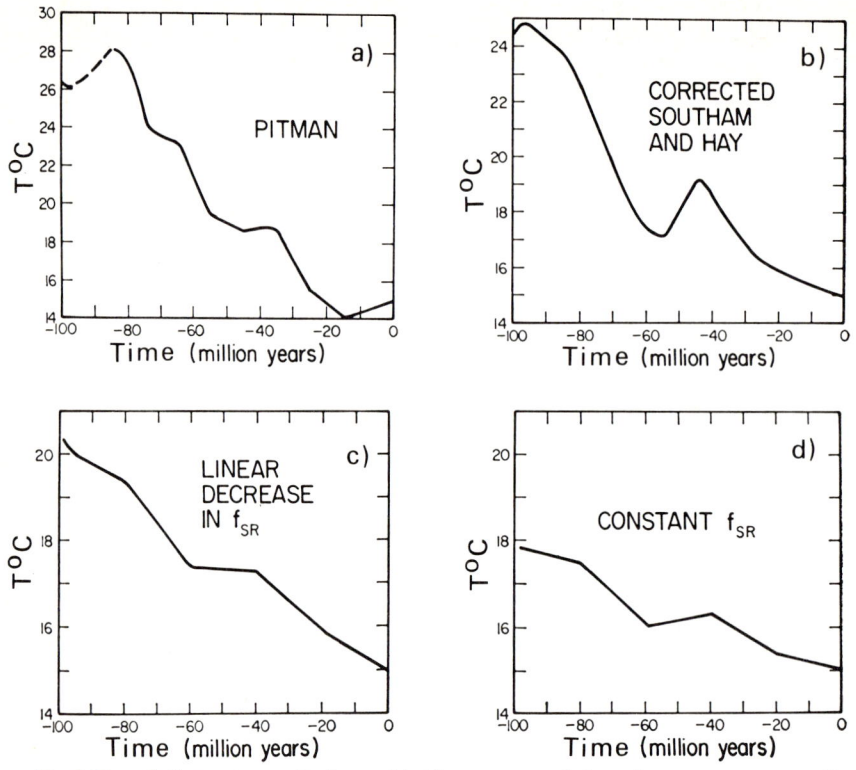

Fig. 9.13a–d. Computer results for worldwide mean annual air surface temperature as a function of time corresponding to the four situations for CO_2 depicted in Fig. 9.12. Note that the vertical (temperature) scales differ. (Berner et al. 1983)

of several hundred years. The integration yields the storage in each reservoir through time. The mean global surface temperature as a function of time can be found by applying the Manabe and Stouffer (1980) CO_2-climate model which uses the relation

$$\frac{A_{CO_2}(t)}{A_{CO_2}(0)} = \exp\{0.347(T - T_0)\},$$

where $A_{CO_2}(t)$ is the mass of atmospheric CO_2 at time t, $A_{CO_2}(0)$ is the mass of atmospheric CO_2 today (0.055×10^{18} moles), T is the worldwide mean annual surface air temperature at time t (°C), and T_0 is the present worldwide mean annual surface air temperature (°C). The reservoir contents of A_{CO_2} and M_{HCO_3} can be used to calculate the pH of the oceans as a function of time using the relation

$$pH = 6.29 - \log_{10}(A_{CO_2}/M_{HCO_3}).$$

The results for predicted atmospheric CO_2 and worldwide mean annual surface temperature are shown in Figs. 9.12 and 9.13 for different assumptions about the

Biogeochemical Cycles

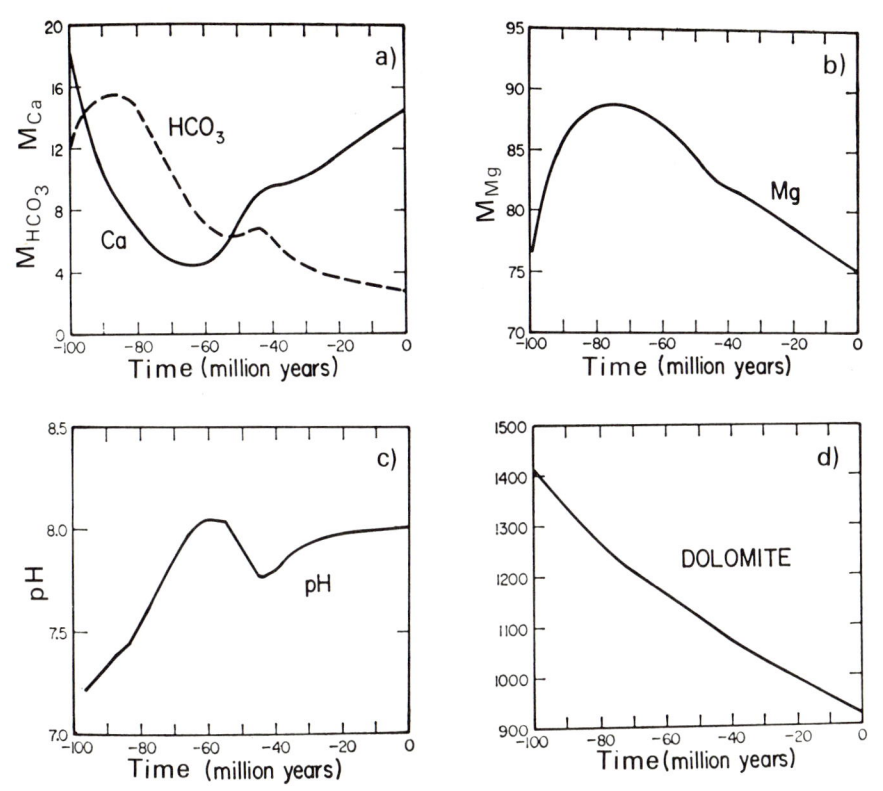

Fig. 9.14 a–d. Computer results for oceanic composition in terms of M_{Ca}, M_{HCO_3}, M_{Mg}, and pH and dolomite mass D, as a function of time, for the corrected Southam and Hay spreading rate formulation and land area versus time data of Barron et al. (1981). Masses in 10^{18} mol. (Berner et al. 1983)

rate of sea-floor spreading. An important point to note is that, whatever the assumption made about sea-floor spreading rate, the levels of CO_2 and mean worldwide surface air temperatures in the Cretaceous period are predicted to have been distinctly higher than they are today. The problem of a warm, equable climate during the Cretaceous period is in fact being actively studied using atmospheric simulation models (Barron et al. 1981; Barron and Washington 1982; Barron 1983).

A salient aspect of the Southam and Hay spreading-rate formula is that, when applied in the model, it predicts a secondary temperature maximum centred around 40 my *BP* during the Eocene epoch, which accords with independent, palaeobotanical evidence. Changes in the Ca, Mg, and HCO_3 content, the pH of the oceans, and the mass of dolomite, for the corrected Southam and Hay spreading-rate formulations and the land-area-versus-time data of Barron et al. (1981) are depicted in Fig. 9.14. The smooth, exponential-like response of the dolomite reservoir is consistent with its large residence time which is in excess of 100 my. The histories of Ca and HCO_3, though roughly mirroring each other, are not correlated in a simple manner. The Mg and Ca reservoirs show changes whereby when one increases the other decreases and vice versa. The most basic finding of the simulations, however, is that tectonic history, through its effects on plate spreading rate and on land area changes, is the most important influence on the carbonate-silicate geochemical cycle.

9.3.3 Strontium and Manganese in a Tropical Rain Forest

Jordan et al. (1973) investigated the long-term dynamics of radioactive isotopes produced by atmospheric testing of nuclear weapons in the late 1950's and early 1960's and now found in ecosystems. The study involved a long-term investigation of the cycling of radioactive and stable isotopes in a Puerto Rican tropical rain forest. Part of the work was the construction of a systems model of the dynamics of the stable isotopes of strontium and manganese. The model was then used to predict the dynamics of radioactive isotopes of strontium, ^{90}Sr, and manganese, ^{54}Mn, in the ecosystem.

The tropical rain forest was represented by four compartments: canopy, litter, wood, and soil (Fig. 9.15). Steady-state fluxes and storages of the stable manganese and strontium were measured in the field and are shown in Fig. 9.15. The model consisted for four equations of the general form

$$\frac{dx_j}{dt} = \sum_{\substack{i=0 \\ i \neq j}}^{4} \lambda_{ij} x_{i,t} - \sum_{\substack{i=0 \\ i \neq j}}^{4} \lambda_{ij} x_{j,t} - \lambda_r x_{j,t} \quad (j = 1, 2, 3, 4),$$

where $j = 1, 2, 3, 4$ are the compartments of the ecosystem (canopy, litter, wood, soil); $i = 0$ is the environment; x_j is the amount of radioactive material per unit area in compartment j; λ_{ij} is a rate constant for transfer of the radioactive material from compartment i to compartment j; and λ_r is the appropriate

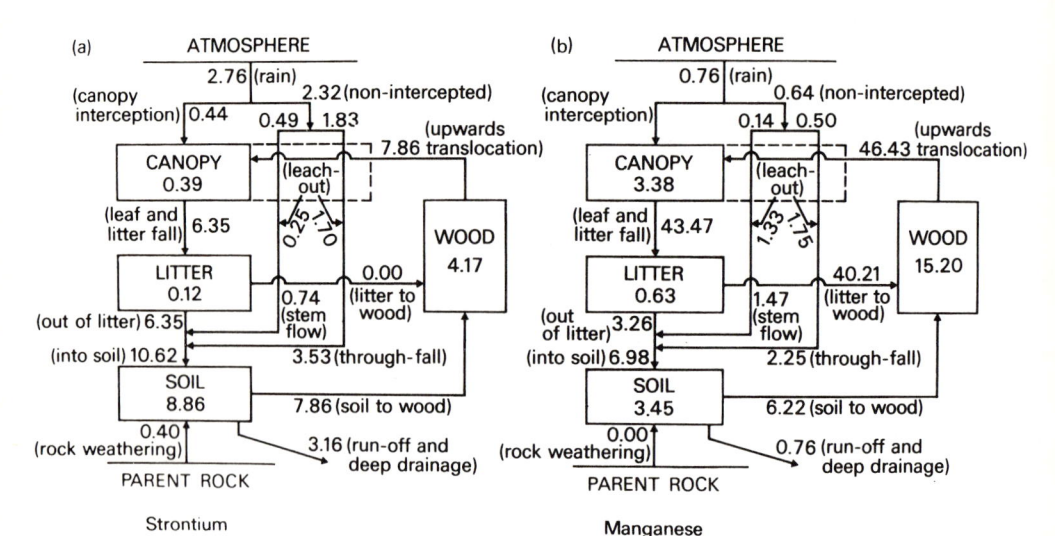

Fig. 9.15a, b. Observed flows and storage of manganese and strontium in a Puerto Rican tropical rain forest. (Jordan et al. 1973)

Biogeochemical Cycles

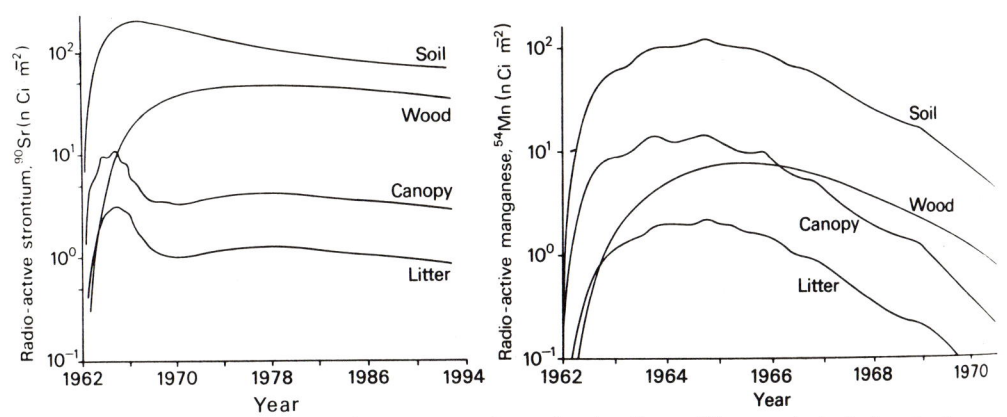

Fig. 9.16. Predicted storages of manganese and strontium in a Puerto Rican tropical rain forest using observed radioactive fall-out patterns as input. (Jordan et al. 1973)

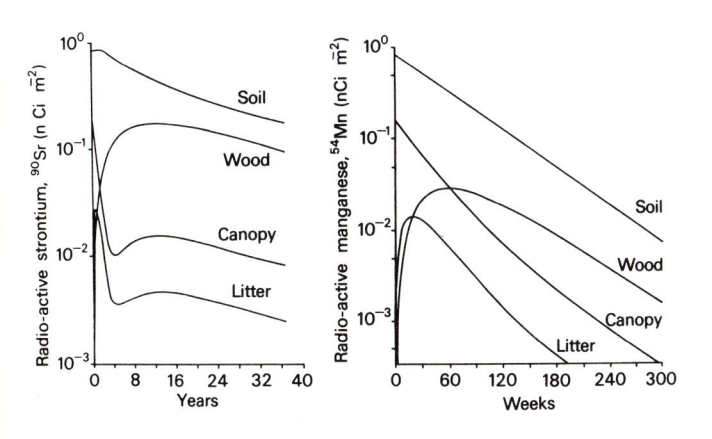

Fig. 9.17. Predicted storages of ^{54}Mn and ^{90}Sr in a Puerto Rican tropical rain forest using an instantaneous injection of 1.0 nCi m^{-2} for each element. (Jordan et al. 1973)

radioactive decay constant. When fully calibrated, the set of simultaneous differential equations were solved on a computer. Two types of simulation were carried out. In the first simulation, the input of the radioactive isotopes was made to resemble the actual radioactive fallout conditions at the Puerto Rican site which resulted from the atmospheric testing of nuclear weapons. In the second simulation, an instantaneous injection of radioactive material was used to represent the possible effect of using thermonuclear devices to excavate harbours or canals in the vicinity. The predictions using fallout patterns as input are shown in Fig. 9.16. In the case of the ^{90}Sr model, the compartments have the following peaks of ^{90}Sr activity: canopy and litter, 1963; soil, 1967; wood, 1976. A secondary peak arises in the canopy and litter, owing to recycling from the soil, in 1977. Although after the initial input the soil stores the largest mass of ^{90}Sr, followed in turn by the wood, canopy, and litter, the concentration of ^{90}Sr (that is, the mass of ^{90}Sr divided by the mass of the compartments), is largest in the canopy, followed in turn by the litter, wood, and soil. In the case of ^{54}Mn, the peaks for the soil, canopy, and litter are roughly the same as in the case of ^{90}Sr; but the wood peaks earlier, in 1964 in fact, owing to the relatively higher rate of physical decay of ^{54}Mn compared with ^{90}Sr. For the same reason, the secondary peaks due to the recycling of ^{90}Sr fail to appear in the case of ^{54}Mn. The model predicts that after 1966, the greatest amount of radioactivity will be found in the soil, followed by wood, canopy, and litter. The concentration of radioactive materials will be greatest in the canopy, followed by litter, soil, and wood. Notice that ^{54}Mn, with its relatively rapid rate of decay, is removed more quickly from the tropical rain forest than is ^{90}Sr.

The second set of predictions, which were from a simulation of an instantaneous input of 1 nCi m^{-2} of ^{90}Sr and ^{54}Mn, are shown in Fig. 9.17. They emphasize the finding that ^{54}Mn is less persistent in the soil and other parts of the tropical rain forest than is ^{90}Sr. This is due to the more rapid radioactive decay of ^{54}Mn.

9.3.4 Water in Soils

Thus far, the reservoirs or compartments in the models have had no particular spatial properties; they have in fact been aggregated. In some cases, spatial disaggregation is of special interest and the compartments are, accordingly, defined to represent spatial blocks or segments. The shape and size of the spatial compartments is somewhat arbitrary, their form varying according to the degree of disaggregation required.

One of the few examples of a spatial dynamical systems model is the model of the movement of water through an old-field ecosystem in Argonne, Illinois built

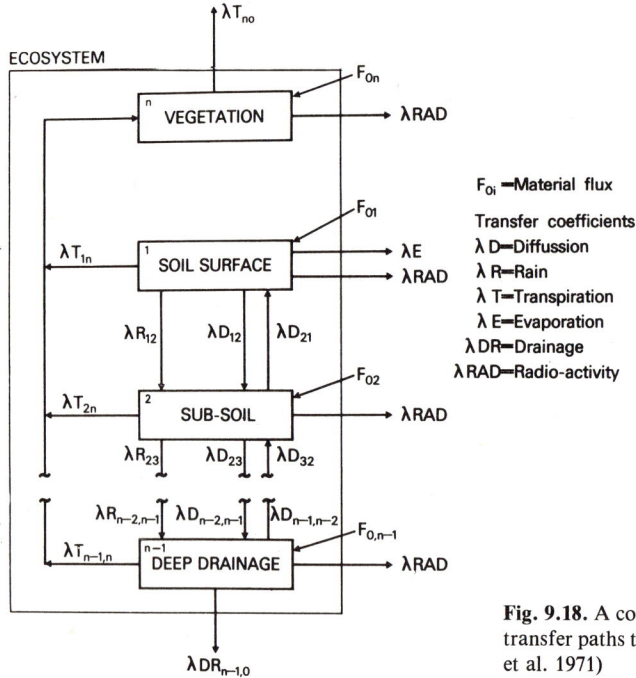

Fig. 9.18. A compartment model of material transfer paths through a soil profile. (Sasscer et al. 1971)

by Sasscer et al. (1971). The system consists of one vegetation compartment and forty-nine soil compartments. Each soil compartment represents a 1-cm-thick soil layer, the full set of soil compartments thus representing a soil profile. The compartments and flows of water between them, and the links of the compartments with the environment, are shown in Fig. 9.18. The dynamics of two kinds of water were modelled: stable water and tritiated water (tritium). The fluxes or flow rates, all of which are considered to pass vertically through the surface of the compartments, are measure in units of ml cm^{-2} h^{-1} for stable water movement, and in units of dpm cm^{-2} h^{-1} (dpm = degenerations per minute) for tritium movement. Flows originating outside the ecosystem are designated F_{0i} and flows between compartments by F_{ij}. The amount of water stored in a compartment per cm^2 of the surface area of the compartment at a given time is x_i. In general, the flux of water between compartments, F_{ij}, is proportional, by a rate constant, λ_{ij}, to the storage of the donor compartment:

$$F_{ij} = \lambda_{ij} x_i.$$

Sasscer et al. assumed that the transfers between compartments are produced by diffusion, mass flow (owing to rain), and evaporation. Each of these processes of transfer is included in the model by specifying a separate rate constant for each: λD_{ij} for diffusion, λR_{ij} for mass flow, and λT_{ij} for evaporation. The overall flux between compartments is thus given by

$$F_{ij} = (\lambda D_{ij} + \lambda R_{ij} + \lambda T_{ij}) x_i.$$

Water may leave the system by evaporation from plants, λT_{n0}, evaporation from the soil surface, λE, and by deep drainage from the bottom-most compartment, $\lambda DR_{n-1,0}$. In the case of tritium, loss from each compartment also occurs owing to radioactive decay, λRAD_{i0}. The continuity equation for compartment j is written

$$\frac{dx_j}{dt} = \sum_{\substack{i=0 \\ i \neq j}}^{n} (\lambda_{ij} x_i - \lambda_{ji} x_j) \quad (j = 1, 2, \ldots, n),$$

or, with all loss and gain terms specified,

$$\frac{dx_j}{dt} = \sum_{\substack{i=0 \\ i \neq j}}^{n} (\lambda D_{ij} + \lambda R_{ij} + \lambda T_{ij}) x_i - \sum_{\substack{i=0 \\ i \neq j}}^{n} (\lambda D_{ji} + \lambda R_{ji} + \lambda T_{ji} + \lambda E_{ji}$$
$$+ \lambda DR_{n-1,0} + \lambda RAD_{j0}) x_j.$$

The time rate of storage change in each of the 50 compartments is thus represented by a first-order, differential equation and the entire system is represented by a set of 50, simultaneous differential equations. The set of equations was solved by computer for given values of rainfall input, initial conditions, and rate constants established by field measurement and experiment. Some results are shown in Fig. 9.19 which displays the tritium concentration in the soil profile as a function of time. The two figures (Figs. 9.19a and b) show the results of separate experiments, the first showing the pattern for 20 mCi of tritiated water placed

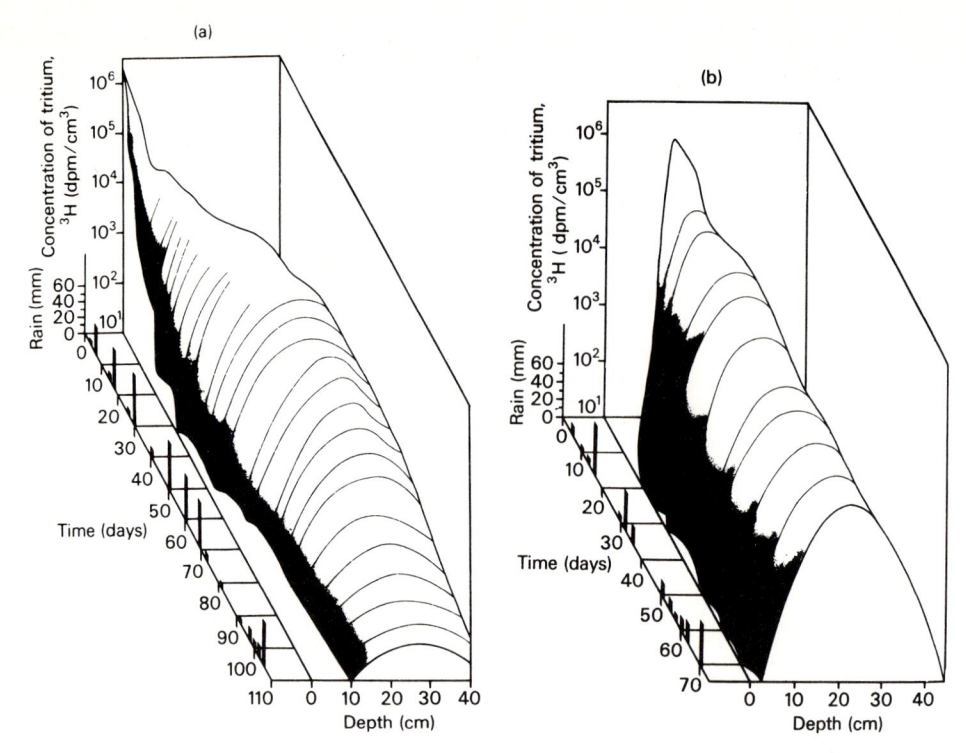

Fig. 9.19a, b. Tritium concentrations in a soil profile as a function of time and depth for **a** deposition of tritium on the soil surface, and **b** deposition of tritium at 15 cm depth. (Sasscer et al. 1971)

on the soil surface during a 1-h period; the second showing the pattern for 17.7 mCi injected at a depth of 15 cm. Experimental data show very good agreement with the predicted results.

9.3.5 Nutrients in Lake Erie

Di Toro et al. (1975) built a spatial dynamical systems model of nutrient movement and storage in the Lake Erie ecosystem, Michigan. The model consists of seven endogenous state variables (Fig. 9.20). The processes which lead to a circulation of nutrients are also indicated in Fig. 9.20. Exogenous variables considered in the model are physical ones – temperature, solar radiation, photoperiod, lake level, inflow from the Detroit river, and water clarity; chemical ones – the chemical quality of the Detroit river and of the Maumee river; and biological ones – the phytomass and zoomass of the Detroit and Maumee rivers. The section of Lake Erie that was modelled is divided into seven spatial segments (Fig. 9.21). The basis of the model is a set of mass balance equations, one for each endogenous state variable, which takes into account flows between endogenous variables, flows between endogenous and exogenous variables, and flows between adjacent spatial segments. With seven state variables and seven spatial

Biogeochemical Cycles

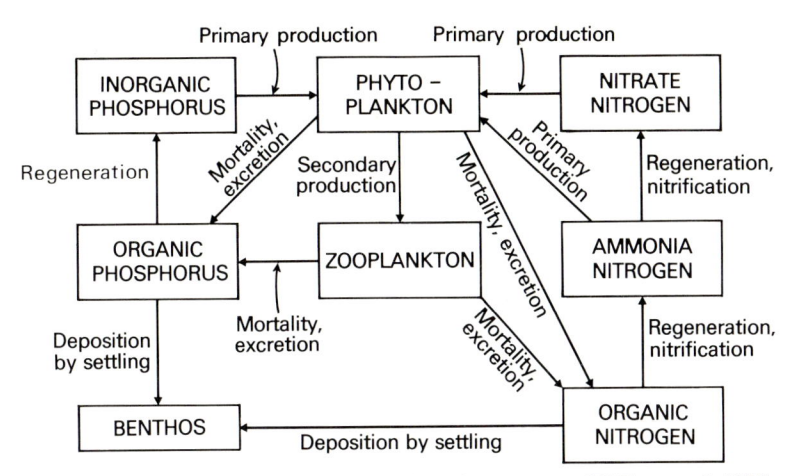

Fig. 9.20. A compartment model of the Lake Erie ecosystem. (Di Toro et al. 1975)

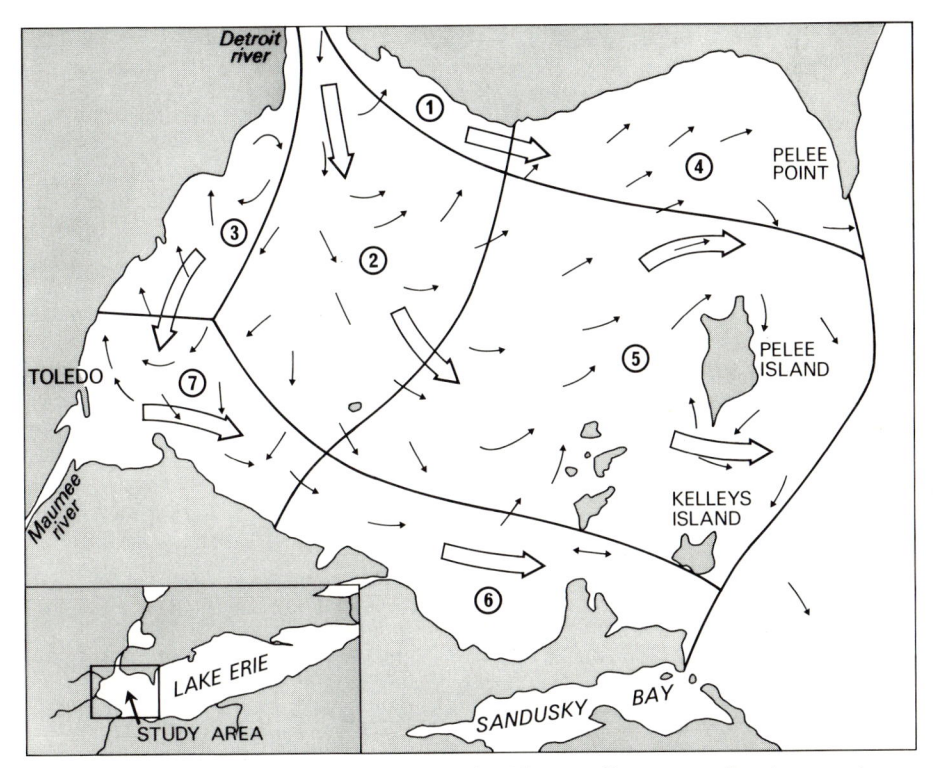

Fig. 9.21. Spatial segments in the Lake Erie model. The prevailing current directions are shown. (Di Toro et al. 1975)

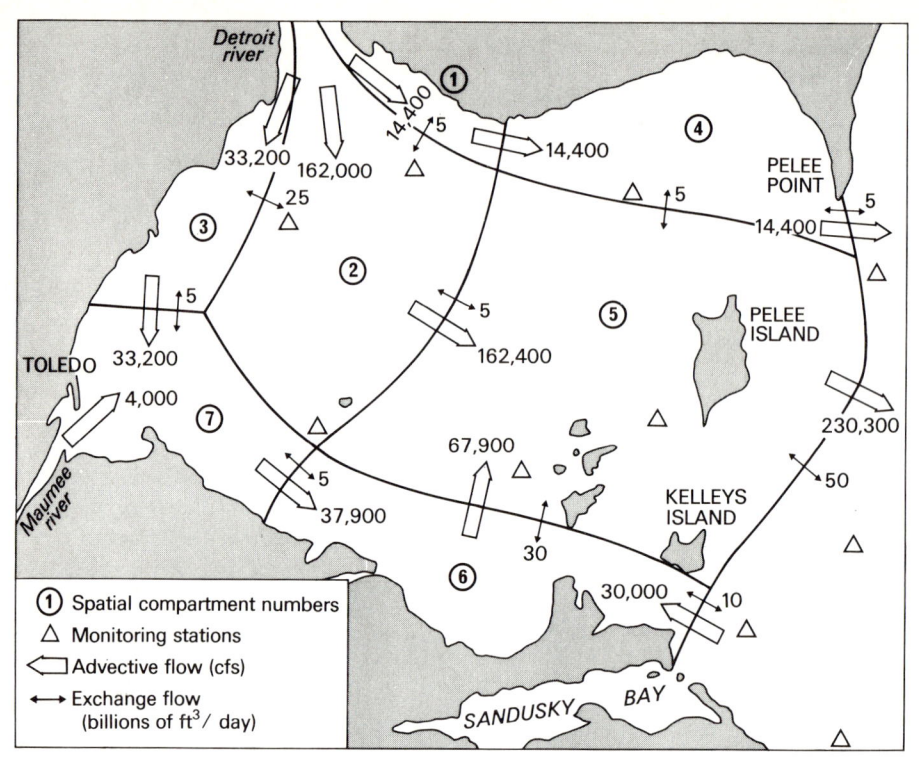

Fig. 9.22. Steady-state water transport for seven-compartment western Lake Erie model. (Di Toro et al. 1975)

segments, the model consists of a set of 49 mass balance equations. Flow between spatial segments is assumed to be the result of advection (the bulk movement of water by lake currents) and dispersion (the diffusion spread caused by small-scale circulations of water which lead to mixing). Advective and dispersive term are shown in Fig. 9.22.

The mass balance equations take the general form

$$V_j \frac{dc_{ij}}{dt} = \sum_k Q_{kj} c_{ik} + \sum E'_{kj}(c_{ik} - c_{ij}) + \sum_k S_{ijk},$$

where V_j is the volume of spatial segment j; S_{ijk} is the kth source $(+)$ or sink $(-)$ of substance i in segment j; E'_{kj} is the bulk rate of transport of c_{ik} into and c_{ij} out of segments j for all segments, k, adjacent to segment j; and Q_{kj} is the net advective flow rate between segments k and j. The equation for phytoplankton biomass, measured as chlorophyll-a concentrations, which equates rate of phytomass change to transport of phytomass by advection from adjacent segments, $\sum_k Q_{kj} P_k$, and by dispersion into, $E'_{kj} P_k$, and out of, $-E'_{kj} P_j$, segment j, as well as the growth rate, $G_{Pj} P_j V_j$, and the death rate, $D_{Pj} P_j V_j$, in segment j:

$$V_j \frac{dc_{ij}}{dt} = \sum_k Q_{kj} P_k + \sum E'_{kj} P_k - \sum E'_{kj} P_j + G_{Pj} P_j V_j - D_{Pj} V_j.$$

Biogeochemical Cycles

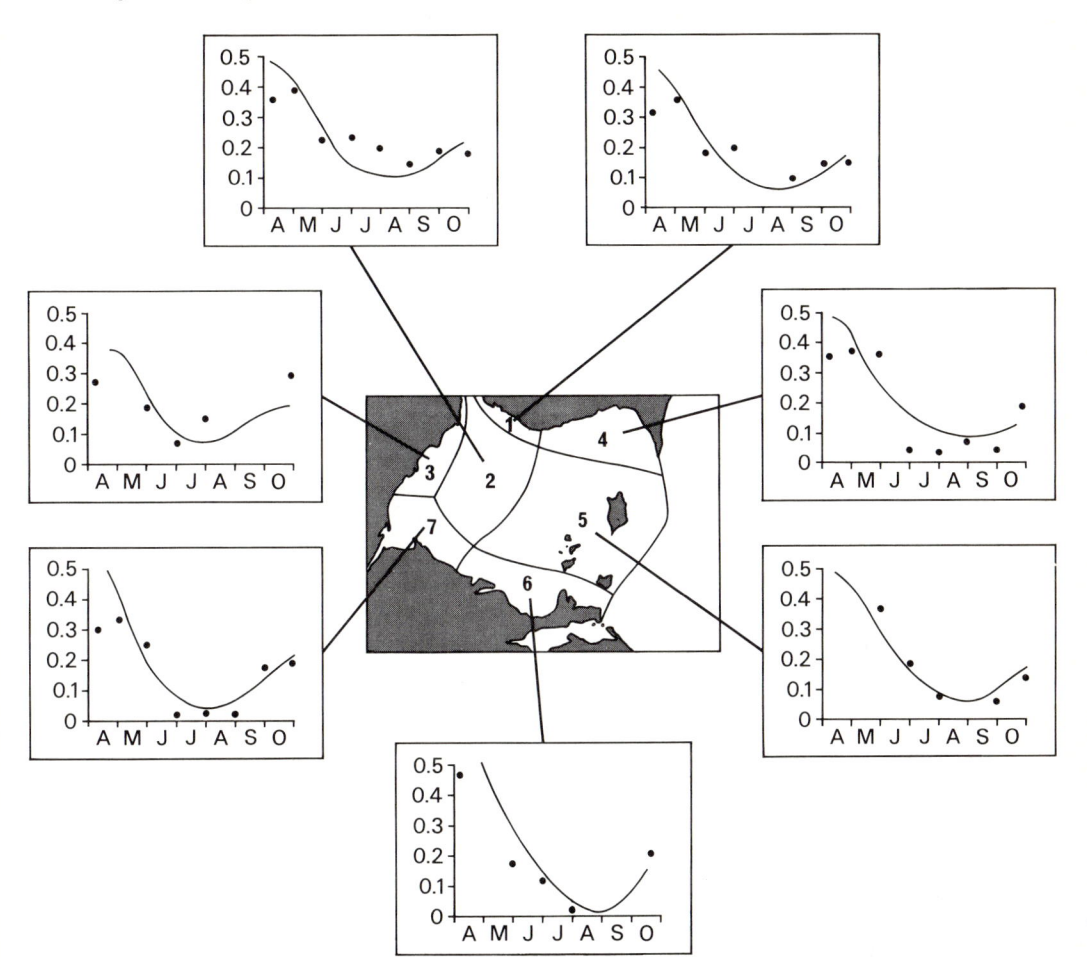

Fig. 9.23. Nitrate nitrogen in seven segments of western Lake Erie: observed and predicted data. Units are mg N l^{-1}. (Di Toro et al. 1975)

A similar equation applies to the inputs and outputs of zoomass in each segment. Boundary conditions were specified for the tributary rivers and the open water boundaries of spatial segments 4, 5, and 6. The parameters of the full set of equations were established from laboratory and experimental data.

The results consisted of predictions of nutrient concentrations in all state variables in each segment for the period April to November. The predictions were tested by comparing them with observations of nutrient concentrations made in Lake Erie in 1968 and 1970. Agreement between observed and predicted values was generally good, though some deviations were found. Figure 9.23 shows the nitrate nitrogen predictions and the observed values.

9.4 Dissipative Structures

Many Earth surface systems seem to be inherently unstable. This is especially noticeable in phenomena related to the flow of fluids and processes of sediment erosion, transport, and deposition. Systems of this kind are driven by imposed forces to states removed from equilibrium until a threshold is reached and the system enters a new regime and adopts a new configuration. Hydrodynamic and alluvial bedform systems are thus pictured as being driven through an orderly series of configurations and regimes, the thresholds between each member of the series commonly being expressed by dimensionless ratios such as the Froude number and the Reynolds number. The processes involved here are the same as in the convective cell mentioned earlier in the book (Chap. 2.3.2). To recapitulate, water at rest between two parallel planes will exhibit fluctuations about a mean state. If the temperature gradient between the two planes is maintained below a critical or threshold value, then the fluctuations are dampened and disappear. If the temperature gradient should exceed the threshold limit, then the fluctuations become amplified and a convective cell develops − the system enters a new regime and adopts a new configuration. In terms of system dynamics, the water is in a steady state below the threshold temperature gradient − fluctuations which perturb the system are dampened. The force imposed above the threshold temperature gradient drives the system to a new steady state − the fluctuation triggers an instability and the system reorganizes itself to accommodate the instability. The kind of dynamics conceived here, in which self-organizing processes lead to an orderly series of configurations, is what Prigogine (1980) calls order through fluctuations. The series of configurations are called dissipative structures.

These arguments show that studies of systems far removed from equilibrium must consider instability and, if possible, predict the threshold beyond which an instability-driven transformation can be expected to arise. Important though such studies are in a number of fields of science (see Prigogine 1980; Haken 1982) their application to Earth surface systems has only just begun.

9.4.1 Bifurcations and Catastrophes

A system with constraints imposed upon it is driven away from equilibrium towards a new steady state. If the constraints are weak then the system will respond in a linear manner. If the constraints are strong then the system may change smoothly along a thermodynamic branch (Fig. 9.24) into nonequilibrium states in which the theorem of minimum entropy production still applies. At a certain distance from equilibrium, called the thermodynamic threshold, nonlinear relationships emerge and the steady states along the branch are not of necessity stable. Beyond the threshold the solutions of the equations governing the dynamics of the system may no longer be unique: the system may enter one of several new regimes. The threshold is therefore a bifurcation point. The path followed by the thermodynamic branch beyond the threshold may involve further thresholds and hence bifurcations. In passing through a bifurcation point

Dissipative Structures

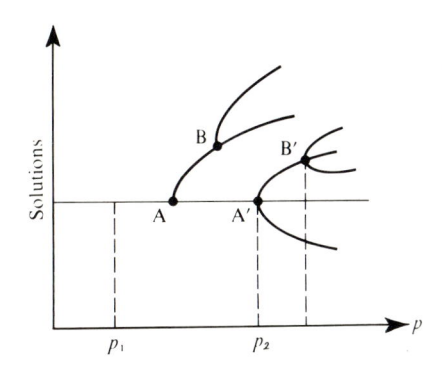

Fig. 9.24. Successive bifurcations. A and A' represent primary bifurcation points from the thermodynamic branch. B and B' represent secondary bifurcation points. (Prigogine 1980)

the system loses its structural stability and undergoes a sudden or catastrophic change to a new form.

The implications of bifurcation theory are profound. Systems which possess bifurcations can be described by deterministic reaction-diffusion equations, but the presence of bifurcations implies that the dynamics of the system involves a chance element. When driven to some critical value the system can move in more than one way. The route taken is probably, in essence, a matter of chance fluctuation. The theory of bifurcations permits the same system to pass through a different series of states and, in a sense, introduces an historical dimension to system development (cf. Prigogine 1980, p. 106).

Applications of bifurcation theory to Earth surface systems are at present rather crude. They are based on a special branch of bifurcation theory developed by Thom (1975) and called catastrophe theory. Thom's elegant treatment of bifurcations and structural stability applies only to potential systems with a finite number of degrees of freedom and described by ordinary, rather than partial, differential equations. So its application to Earth surface systems, where partial differential equations describing changes in space and time are of paramount importance, seems unpromising (Karcz 1980). Undeterred, geomorphologists have tried to use Thom's ideas (for some reason his cusp catastrophe has proved a favourite) to explain certain processes at the Earth's surface.

W. L. Graf (1979, 1982) has applied a cusp catastrophe model to the conditions at stream junctions in the northern Henry Mountains, Utah. The catastrophe surface (Fig. 9.25) is plotted as a smoothed version of data for the refill of sediment in 88 main channels (the response variable) versus the force available in the tributary to deliver sediment and the free available energy in the main channel to carry the sediment farther downstream (the control variables). The diagram demonstrates the differential movement of sediments in the example area. It can be seen that, at any given junction, when the tributary force is low, little sediment is delivered to the main channel. Under these conditions, little sedimentation occurs, regardless of the force available in the main channel. When the tributary force is high, the delivery of sediment to the main channel increases. The fate of the sediment entering the main channel under these conditions depends on the force available in the main channel. If the main channel force is low, much sediment remains; if the main channel force is high, much sediment is carried downstream. However, between these two extremes, two conditions are possible:

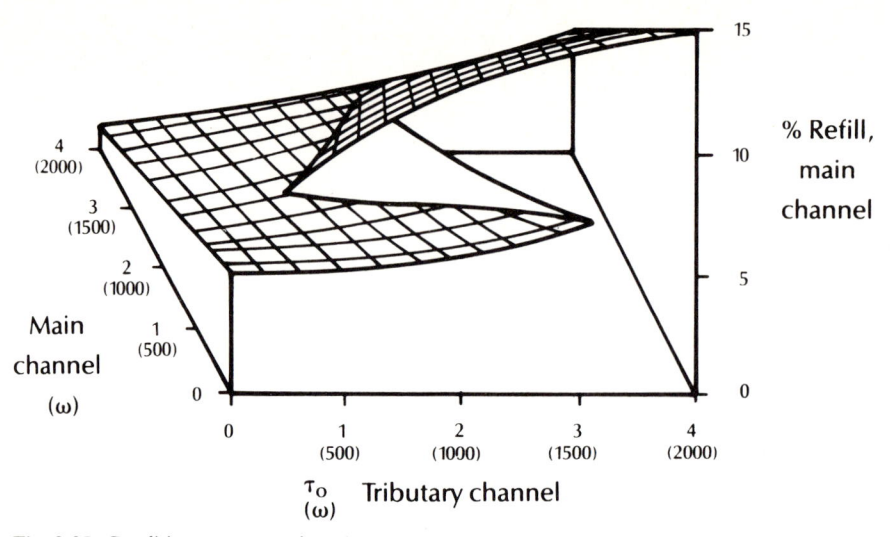

Fig. 9.25. Conditions at stream junctions in the northern Henry Mountains, Utah. *Horizontal axes* show the tractive force in the main tributary channels. *Vertical axis* shows the per cent filling of the arroyo in the main channel at each junction. Surface is smoothed for 28 points. (W. L. Graf 1982)

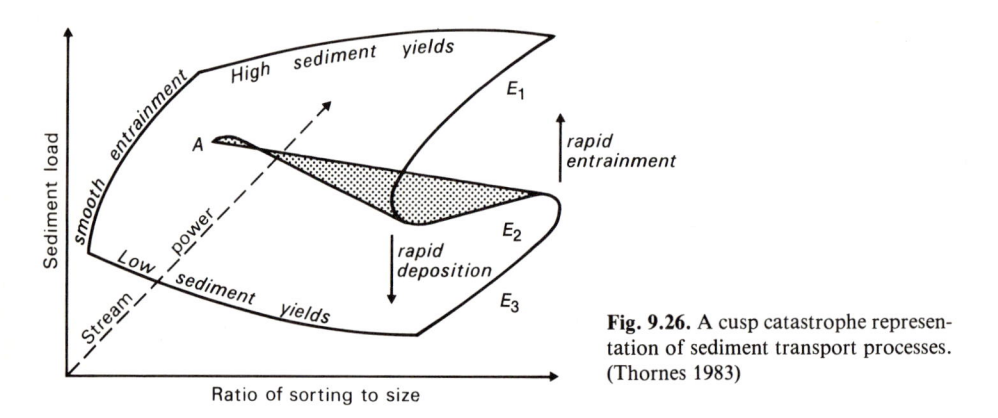

Fig. 9.26. A cusp catastrophe representation of sediment transport processes. (Thornes 1983)

intermediate force levels for both main and tributary channels can produce either large or small amounts of sediment in the main channel. W. L. Graf (1979) argues that the amount of sediment in the main channels under conditions of intermediate force in channels depends on whether the tributary valleys have or have not experienced large floods in the last 80 years. Junctions which have experienced large floods plot on the upper central surface of the cusp catastrophe: their sediment has been flushed into the the main channel. Junctions which have not experienced large floods in the last 80 years plot on the lower central surface of the cusp catastrophe because they do not, as yet, have major deposits in the main channel.

A cusp catastrophe representation of sediment transport in a river has been described by Thornes (1983). The model is portrayed in Fig. 9.26 in which the

Dissipative Structures

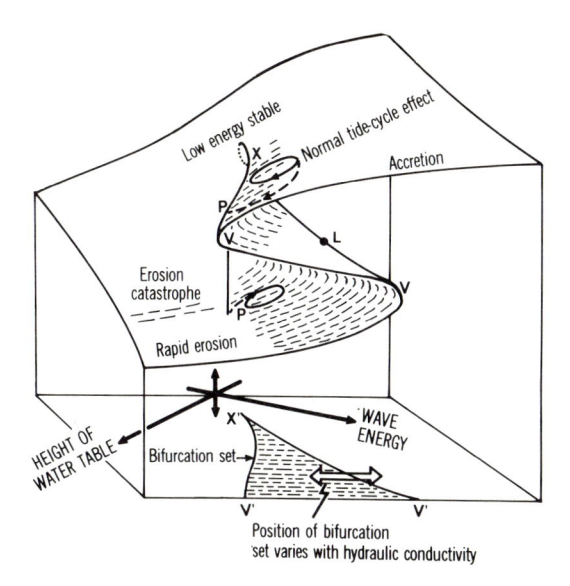

Fig. 9.27. A cusp catastrophe representation of the relationships between wave energy, water table height relative to the beach surface (*horizontal axes*), and erosion versus accretion (*vertical axis*). (Chappell 1978)

three-dimensional surface defines the equilibrium value of sediment load (the response variable) in relation to stream power and the ratio of sediment sorting to sediment size (the control variables). The ratio of sorting to size is, in the parlance of catastrophe theory, the splitting factor: for low values of the sorting-size ratio the equilibrium surface is a single, smooth curve; beyond the point where the surface splits (point A in the diagram) and becomes complex, three equilibria are possible (points E_1, E_2, and E_3 in the diagram) the middle of which is unstable. As indicated on Fig. 9.26, the system jumps from one of the equilibrium states to another; the transition is not smooth. The model shows that for mixed sediments of relatively small mean size the equilibrium sediment load is a smooth function of stream power. Larger and better sorted sediments are entrained in a discontinuous way, the sediment loads of the rising and falling limbs of a hydrograph being very different. A fuller account of a similar model applied to ephemeral channel behaviour is given by Thornes (1980).

Chappell (1978) develops a cusp catastrophe model to describe the sudden changes from accretion to degradation which take place on beaches. In the model, as Fig. 9.27 shows, accretion-degradation, dh/dt, is the response variable and the onshore flux of wave energy, E, and the height of the water table above low water datum, h_*, are the control variables. The equilibrium surface defined by $dh/dt = f(E, h_*)$ is discontinuous: catastrophic change from accretion to degradation may occur as a result of liquefaction within the intertidal beach being triggered by a combination of a high water table and the shock induced by waves breaking on the shore. Once rapid degradation has occurred, it tends to continue slowly because the vadose "buffer" zone above the water table is no longer there. The processes involved seem well described by the cusp catastrophe model. Chappell illustrates this by considering the changes occurring during a normal tide cycle on a mildly accreting beach. The sequence is represented by the small ellipse above the fold in Fig. 9.27. Moving clockwise round the ellipse, the cycle runs as follows: X marks the spot where the flood tide starts with the energy flux rising on the beach face as the depth over the nearshore zone increases. The rise of the water table lags behind the tide and rise of energy. During the ebb phase, the energy flux wanes, but the water table rises for a while before proceeding to fall slowly. The change in the accretion rate is represented by the arrow on the ellipse heading downslope during ebb and upslope during flood. If the normal cycle is interrupted by a moderate increase of wave energy with an accompanying rise of the water table, as, say, during a mild storm with heavy rain, the system follows the dashed line to P, at which point liquefaction and an erosion catastrophe occur, the system jumping to point P_1. The vadose buffering zone having been eliminated, the tidal cycle resumes its degradation regime as represented by the lower ellipse. An implication of the model is that a converse, "accretion catastrophe" should occur. Chappell believes that this may be represented by the final, rapid accretion to the beach face of shorewards-moving, transverse bars which may arise during fair weather and waning swell (see also Chappell and Eliot 1979, Chappell et al. 1979).

9.4.2 Thresholds

A threshold separates different states of a system. In this sense, it represents a discontinuity in a system's state space. Examples of discontinuities in everyday experience are common; water changing from liquid to gas in a kettle is an example. Chappell (1983) offers the instance of a block of ice inside a closed beaker which is itself placed in a box (Fig. 9.28a). The temperature inside the box creates the boundary condition for the beaker and its contents. If the temperature inside the box is held constant at $-1°C$, then the ice in the beaker will stay as ice, but if the temperature in the box is raised to $+1°C$ and held there, the ice will, given time, melt because the threshold temperature between the ice state

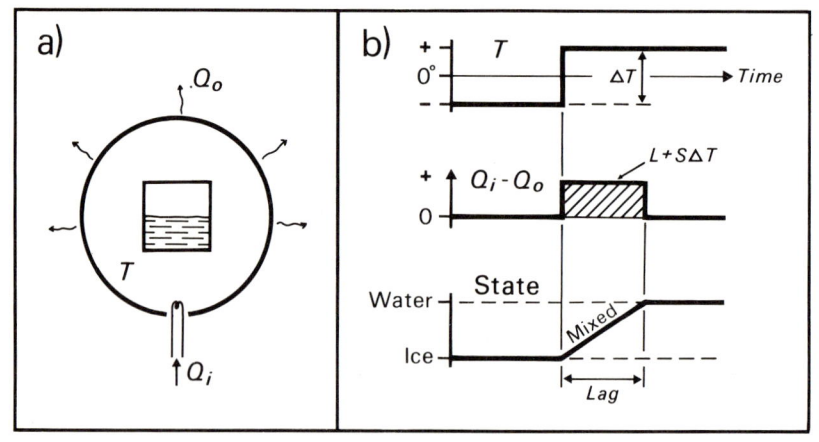

Fig. 9.28. a Closed vessel containing water inside lagged enclosure. Q_i is controlled heat input; Q_o is loss from enclosure. **b** Change of state from ice to water. During transition $Q_o < Q_i$ while latent and caloric heat is taken up by water. (Chappell 1983)

Dissipative Structures

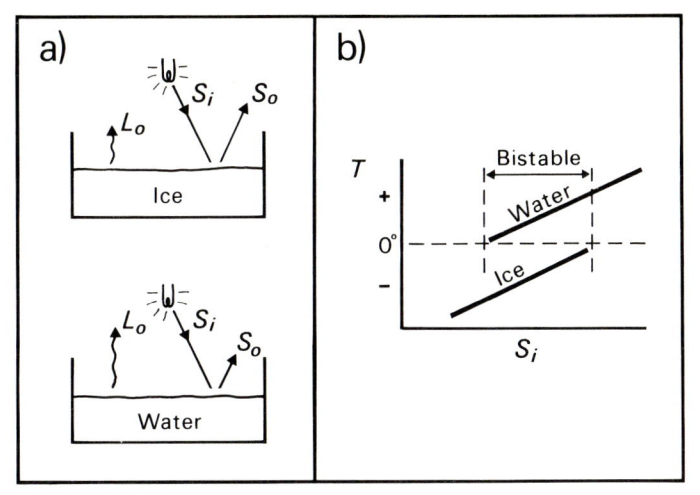

Fig. 9.29 a, b. Bistable behaviour of ice-water system. Ice albedo > water albedo, and either may exist within a certain range of input radiation, S_i. (Chappell 1983)

and the liquid water state has been crossed. The difference between the states is enormous though the change in boundary condition which caused the change is tiny. The time taken for the ice to melt following a disruption of the boundary condition is the lag of the system (Fig. 9.28b). Chappell uses a similar example to illustrate the important concept of a bistable system, that is, a system which can exist in two states under the same boundary conditions. He envisages a pan of water over which is a light source surrounded by nothing but space. Depending on the arrangement of the light source, the water will be frozen or unfrozen. Within a certain range of adjustments, however, the water has an equal likelihood of existing in a frozen or an unfrozen state. This is because when the water exists as ice, the higher albedo (compared with water) causes more light to be reflected allowing a lower temperature in the pan than if the water were liquid (Fig. 9.29). The boundary conditions in both cases are the same. The change from one state to another requires a temporary change of the energy source (a fluctuation in Prigogine's terminology).

The two examples just described serve to illustrate two basic types of threshold. Where two different states are separated by different boundary conditions, as with the "ice in a pan in a box" example, the threshold is transitive; where two different states exist with the same boundary conditions, as with the water heated by a light source, the threshold is intransitive (Fig. 9.30). The sudden retreat of a beach, involving the process of liquefaction, is an example of an intransitive change. The process of liquefaction, and the scarp retreat it produces is induced by the raising of the beach water table during stormy spells. A simple example of a transitive threshold is the change in sediments removed from a slope. Generally, fine sediments are produced by slope wash and creep, but occasionally coarse sediments are produced by slope failure. The result of this is the alternation of fine and coarse deposits in colluvial and alluvial sequences.

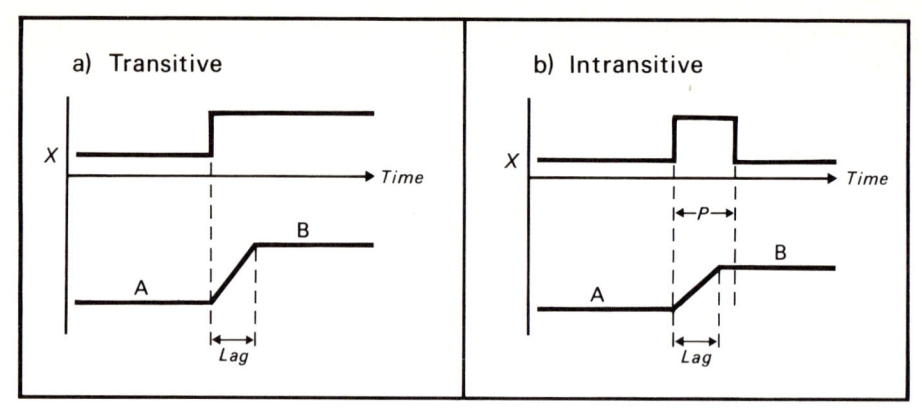

Fig. 9.30a, b. Transitive and intransitive changes of state. X represents boundary condition. (Chappell 1983)

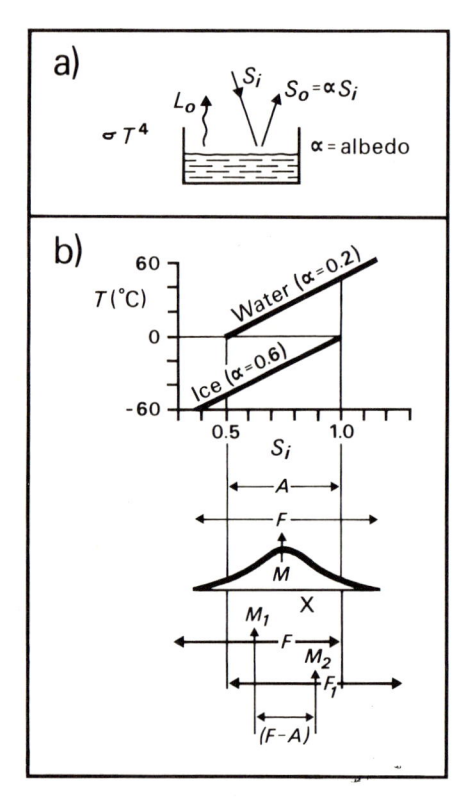

Fig. 9.31a, b. Definition model for parameters of transitive and intransitive thresholds. T is temperature of water or ice, σ is the Boltzman constant, S_i and S_0 are the incoming and outgoing shortwave radiation, L_0 is outgoing longwave radiation, and α is albedo. See text for details of other symbols. (Chappell 1983)

The distinction between transitive and intransitive threshold is, as Chappell claims, of great importance in understanding the behaviour of a number of Earth surface systems. If, for instance, in the case of landforms, boundary conditions are equated with "climate", then a change in landform from one distinct state to another may denote one of two things. Firstly, if the change involves the crossing of a transitive threshold, it indicates that climate has changed. Secondly, if the change involves the crossing of an intransitive threshold, it indicates the occurrence of a climatic perturbation (fluctuation) in an otherwise uniform boundary condition.

Chappell throws light on the occurrence of transitive and intransitive changes by considering in more depth the bistable, water-in-the-pan system. The energy balance, S_i, is depicted in Fig. 9.31a. The relationships between temperature and the energy balance for the cases of ice and of water are shown on Fig. 9.31b. Clearly, if the input radiation is such that ice temperature is less than $-43\,°C$ or water temperature is greater than $52\,°C$ then bistable behaviour cannot occur. Within the limits $0.5 < S_i < 1$, where S_i has been scaled to give $T_{ice} = 0$ at $S_i = 1$, either ice or water can exist. The bell-shaped curve on Fig. 9.31b represents the probability distribution of a boundary parameter, X, which influences S_i. X varies randomly according to the probability distribution. If a rare event occurs when $X < 0.5$, then the system can only be ice and if it occurs when $X > 1$ then the system can only be water. If the midpoint, M, of the frequency distribution is moved to M_1, then the system is always ice even though X varies randomly. Similarly, if the midpoint is shifted to M_2, then the system is always water. If $M_1 < M < M_2$, then the system changes irregularly from ice to water and vice versa through random events near the extremities of F. Thus the system changes randomly and intransitively, provided that an extreme event lasts long enough for a change to occur. If $M < M_1$ and then changes to $M > M_2$, then a transitive change from ice to water will occur. As can be seen in Fig. 9.31b, the minimum change in mean boundary condition to produce a transitive change is $|A - F|$ and change will be transitive only if $F > A$.

The bistable system, as Chappell shows, has two other interesting properties. Suppose that the system is ice with $M = M_1$, but then changes to M_2. The transitive state change to water does not start at once. It awaits the occurrence of a random event within F of sufficient magnitude to exceed the "always water" limit. The minimum change in S_i required for certainty of instantaneous state change is $|A - F/2|$; this is the boundary change necessary for a transitive change without a random waiting time. Two lags are therefore present in the system. Firstly, the system lag which is related to the time required for latent heat dissipation or absorption, and secondly, a statistical lag equal to the recurrence interval of the requisite extreme event which arises under a change of mean boundary conditions exceeding a minimum, $|A - F|$, but less than the necessary change, $|A - F/2|$. The duration of an extreme event which can cause change, either transitive or intransitive, is affected by feedback effects in the system itself during the change. In the example, the albedo changes towards its new value as the change proceeds. In general, positive or negative feedback can occur and, respectively, either shorten or lengthen the duration of the requisite extreme event. Another example, concerned with dune stabilization, is summarized in Fig. 9.32.

Thresholds, both transitive and intransitive, are believed to exist in the Earth-atmosphere system. Lorenz (1968) suggested that the Earth's climatic system does not have a unique, stable solution. In other words, there is not a macroscopic order which prevails for all time. Rather, Lorenz proposed, the Earth's climatic system behaves intransitively with one state persisting for some time and then suddenly, a change to a different state occurs. The system thus has

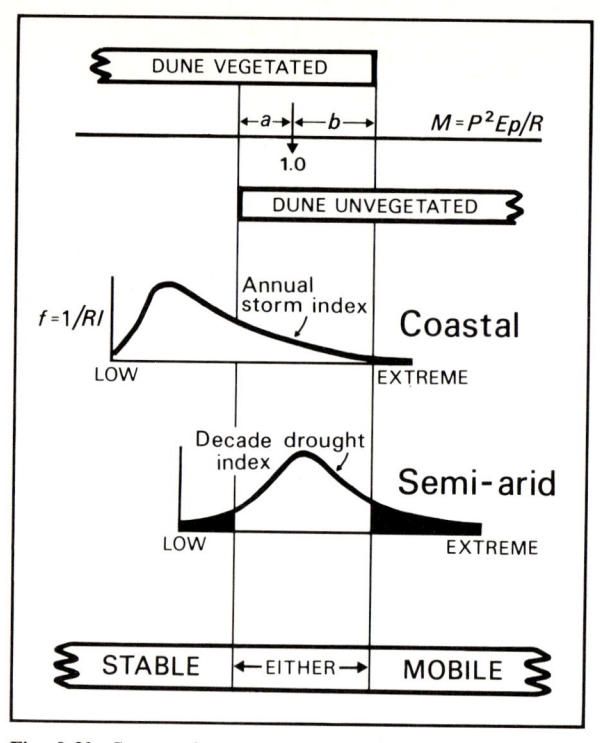

Fig. 9.32. Suggested relationship between dune stabilization and restabilization in terms of a modified Ash-Wasson (1983) index, $M = cP^2 Ep/R$ where: P is per cent occurrence of sand-shifting wind; Ep is potential evaportranspiration; R is rainfall; and c is a scaling constant. When $M < 1 - a$, then dune vegetation is stable; when $M > 1 + b$, then devegetation occurs. In the range $1 - a < M < 1 + b$, either vegetation condition can exist. In a coastal situation, where M normally lies below $1 - a$, the dunes are stable until an extreme storm which exceeds the vegetation stability threshold (middle curve: note frequency of occurrence on vertical axis). Semi-arid dunes become unstable when drought is extreme (right-hand region exceeding $M = 1 + b$ on bottom curve); restabilization requires drought-free condition at the other extreme. (Chappell 1983)

several solutions, each of which yields a distinctive global climatic pattern. All these patterns may exist within the same external boundary conditions. Many computer models of the global energy balance and the general circulation do indeed exhibit the switching behaviour envisaged by Lorenz. Ghil (1976), for instance, found that Seller's (1973, 1976) energy balance model gives three stable solutions with global mean temperatures of 288 K (present), 267 K (ice-age Earth), and 175 K (white Earth – completely ice-covered). Other models produce a "greenhouse state" as envisaged by Fischer (1981). Recently, Hunt (1984) has used a global climatic model to show that "icehouse" states are the norm, whether or not there is land at the poles, and that "greenhouse" states are the abnormal states which need to be explained. The transitions between the different global climatic states may be studied by catastrophe theory (Fraederick 1978).

Dissipative Structures

9.4.3 Dominance Domains

Related to the idea of thresholds and bifurcations is the idea that the dominance of different Earth surface processes will vary from place to place, that each Earth surface process is likely to have a dominance domain. This notion, discussed in a geomorphological context by Thornes (1979, 1983) and by Kirkby (1980), gives a new meaning to the old idea of geomorphological and soil regions. Thornes (1983) defines process domains by plotting the work done by several processes against an environmental factor such as rainfall intensity (Fig. 9.33a). In the example given on Fig. 9.33a, the domains are areas dominated by throughflow and by overland flow. Using two or more environmental factors, process domains can be defined in two or more dimensions (Fig. 9.33b). Dominance domains at a particular place may change with time. Two mechanisms seem to be involved in temporal changes (Fig. 9.34). Firstly, a shift in the total effective

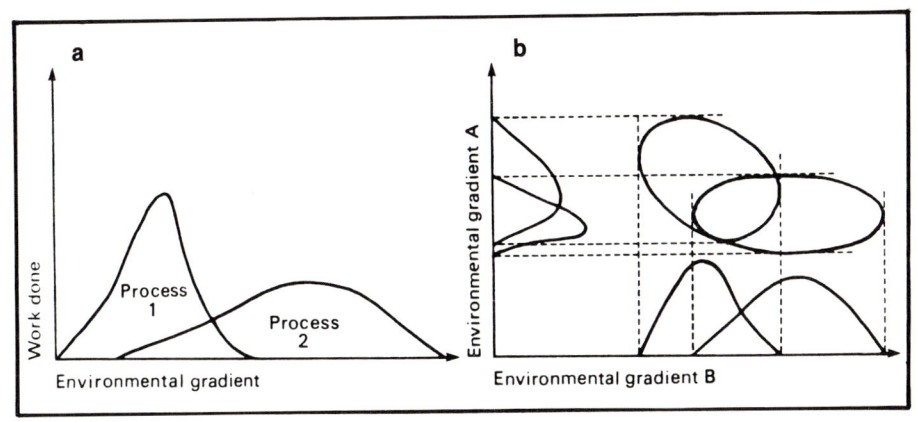

Fig. 9.33a, b. Dominance domains in **a** one-dimensional, and **b** two-dimensional state space. (Thornes 1983)

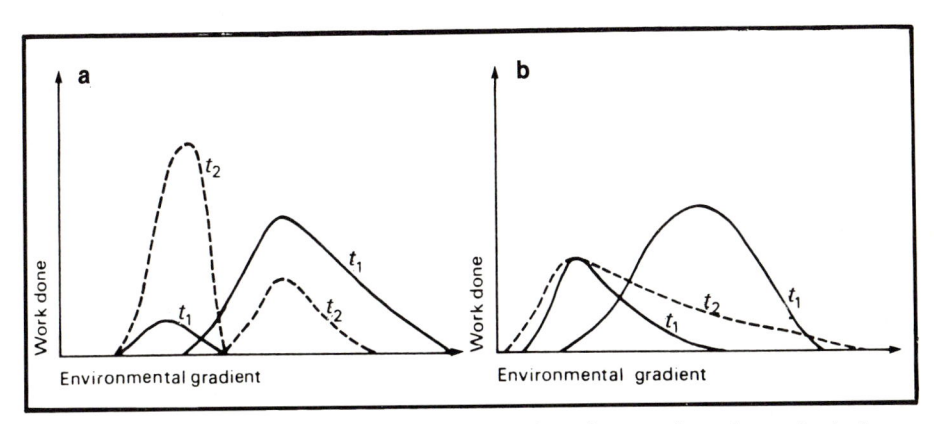

Fig. 9.34a, b. The shift in domain between two processes from time t_1 to time t_2 by **a** a simple change in process intensity and, **b** a change in the range of the process domain. (Thornes 1983)

energy curve may lead to a change in the relative importance of the processes (Fig. 9.34a). This mechanism would be induced by a swing of climatic belts. Secondly, the total effective energy curve of one process may change width and take over the domain of another process (Fig. 9.34b). The result may be either that both processes coexist or that one process replaces the other. An example of replacement is a regime of glacial processes surplanting a regime of subaerial processes.

CHAPTER 10
Conclusion and Prospect

It is perhaps only human nature that the promise of mathematical models in explaining phenomena at the Earth's surface is all too clear to the modellers but far from clear to those who reason on more "traditional" lines. Critics of mathematical models are quick to point out that, owing to oversimplification and geomorphological naivety, modelling can be of little use in understanding how real landscapes develop. Büdel (1982) is sceptical of quantitative models, both field and theoretical, which attempt to understand the complex processes of relief development on the grounds that, though they may clarify some details of the processes involved, on their own they rarely contribute much to an understanding of the entire relief. He writes:

"Certainly one can measure some of the basic components of the whole process, with all-to-delusive precision, but one can hardly use this method to formulate a balanced, comprehensive picture of the highly intricate relief-forming mechanisms developing out of the specific interaction of many already complex elements". (Büdel 1982, p. 348)

Pitty (1982, p. 54) condemns theoretical models in a string of somewhat unguarded and acerbic comments: they (theoretical models) often start with the untested assumption that land surface forms and processes may be defined in terms of the laws of physics; those based on the continuity equation tell us nothing about the surface form of the landscape; they depend on basic assumptions unacceptably simplified for workers with real world preferences; and they underestimate the role of chemistry, biology, and the complexities of underlying geology.

These charges, and others like them, against field and theoretical quantitative methods should not be dismissed lightly. It is no use theoreticians and quantitative field-persons feeling affronted at being misinterpreted or misunderstood but not responding constructively: they must try to convince the unbelievers that a numerate approach is salutary. To this end, the book will conclude by addressing the question: why model Earth surface systems mathematically? This is an important question because, as Kirkby (1976) holds, if mathematical modelling is to be more than just an intellectual game, its objectives must be clear. Three basic objectives will be considered: models as a complement to field studies; models as a testing ground for long-term change; and models as good predictors of complex situations.

10.1 Models as a Complement to Field Studies

Field and laboratory measurements of Earth surface processes provide a means of calibrating and testing theoretical models while, in turn, models provide a framework within which to study form and process in the field. Were it not for field measurements and testing, theoretical analysis would become increasingly abstract; were it not for theoretical models, it would be difficult to see any general application of particular observations (Huggett 1980, p. 20). Fashions in modelling change abruptly. Kennedy (1977) claims that these changes of paradigm, which seem to occasion so many alarums and excursions, are no more than exercises in reclassification which leave the basic manner of investigation virtually unchanged. To her, even the apparently traumatic change from erosion surfaces to systems is of little lasting account in terms of what geomorphologists study. Although some sympathy with this view may be felt, it can be demonstrated that each change does at least usher in a new rash of field studies which look at the same phenomena as before in a new light.

A major change was the advent of process studies, started in France by Tricart and Cailleux in 1945, but given much impetus and a theoretical base by Strahler's (1952a) statement on the dynamic basis of geomorphology. From these beginnings, process studies have followed several different lines. "Force-resistance" studies apply Newtonian principles to objects at the Earth's surface. The success of these studies, especially when applied to "simple" systems, cannot be denied. An exemplary case is Kirkby and Statham's (1975) process-response model of talus accumulation by rockfall based on the physics of individual particle movement and on the frictional properties of individual rock fragments. The recent work of Scheidegger (1979, 1983) takes force-resistance studies in a new and exciting direction. Scheidegger (1979), in a reinterpretation of the old idea that the Earth's surface features reflect the balance between endogenic and exogenic processes, promulgates a principle of antagonism. He argues that endogenic and exogenic effects are active at the same time and that the Earth's surface is the instantaneous result of their antagonistic action. Moreover, the endogenic processes are produced by a system of global tectonics and are distinctly non-random in their effects, whereas the exogenic processes, which lead to surface erosion, drainage basin development, and the large-scale decay of mountain ranges, are essentially random in their effects. Scheidegger and his coworkers have marshalled much evidence which supports this view (Chap. 4.3.3). An interest in tectonics and landforms is growing fast in other areas, too (Ollier 1981; L. C. King 1983). Perhaps a new paradigm is emerging which may give many of the older school of geomorphologists a sense of déjà vu.

Deductive stochastic process models, despite criticisms to the effect that they are models which build on ignorance, have a sound scientific base (Thornes and Ferguson 1981, p. 288). Random patterns and processes do appear to exist in the landscape. Indeed, Scheidegger (1983) considers that exogenic processes have an intrinsically stochastic nature and generally act in a random fashion. Wherever intrinsic or aggregate randomness is suspected in Earth surface systems, deductive stochastic models may be able to provide some insight as to the processes involved. W. E. H. Culling (1963), in his pioneering study of hillslope development, recognized the intrinsic randomness in the process of soil creep. By assuming that each soil particle moves at random under constraints imposed by the availability of pore spaces at different depths and by an overall drift downslope, he was able to deduce a diffusion equation which predicts, for given boundary conditions, the slope form produced by soil creep. Shreve (1966) recognized aggregate randomness in the pattern of drainage networks. By assuming that all topologically distinct ways of combining a given number of headwater streams into one main river are equally likely, he was able to give an explanation of Horton's laws of drainage composition. More generally, the notion of aggregate randomness in Earth surface systems is based on the assumption that there are too many deterministic influences in the development of a system for their effects to be studied individually; the only thing amenable to observation is the combined effect of the innumerable deterministic influences as represented by the average properties of the system in question. This line of argument leads to the analogy between landscape systems and thermodynamic systems and leads to statistical mechanics as the apposite vehicle of methodology (Chap. 4.6).

Inductive stochastic process models have been applied where series of temporal or spatial data seem to display a mixture of regular and random variation. Time series analysis in fluvial geomorphology and hydrology, the main areas of application, has been carried out on both process variables and form variables. The study of process variables includes hydrological series treated in much the same way as economic time series with prediction as the goal. Early examples of such work include Matalas (1963, 1967), Chow and Kareliotis (1970), Quimpo (1968), and Yevjevich (1971). Form variables are difficult to study directly using methods of time series analysis because most landforms change too slowly to be sampled over a time period that has any geomorphological significance (Anderson and Richards 1979, p. 206). However, Nordin and Algert (1966) studied the time series produced by the movement of sand dunes past a river cross-section and found that the series accords well with a series of bed elevations along a river at a given time. Just how useful such ergodic studies will prove in an understanding of other Earth surface systems still remains to be seen. Spatial series analysis has been used to analyse hillslopes (Thornes 1972; Nieuwenhuis and van den Berg 1971; Pike and Rozema 1975; Craig 1982b, 1984), meander patterns (Speight 1965; Thakur and Scheidegger 1970; Chang and Toebes 1970; Ferguson 1975, 1976, 1979; Trenhaile 1979), and channel bed forms including ripples and dunes (Crickmore 1970; Nordin and Algert 1966; Squarer 1970), and larger-scale features in gravel bed streams (Melton 1962; M. A. Church 1972; Richards 1976; Anderson and Richards 1979).

At present the most promising applications of inductive stochastic process models seem to be adaptive parameter modelling techniques (R. J. Bennett 1979) and the use of autoregressive moving-average models to infer process from landform (Craig 1982b, 1984). A good example of adaptive parameter modelling is R. J. Bennett's (1976) investigation of the control of sediment transport equations on river channel morphology. As a source of data, Bennett borrowed Yatsu's (1955) distance series of height loss along the courses of a number of Japanese rivers. Yatsu had found that sediment in the rivers could be divided into two regimes which influence the overall bedslope: an upper reach with mainly gravelly fractions and a lower reach with sand fractions. Estimates of the autocorrelation, partial autocorrelation, and spectrum of height loss along one of the rivers indicates that a first-order moving average process is a suitable description of the system:

$$\nabla H_s = d_1 e_{s-1} + e_s.$$

Estimates of the moving-average parameter, d_1, using a Kalman filter, confirmed Yatsu's findings that changes in sediment fractions from gravel to sand are linked with important changes in the river long profile and bed morphology (sinuosity). Craig (1982c) shows that many traverses of elevation in the eastern United States conform to an ARIMA (2,1,0) model and may be described by the equation

$$E'_i = \phi_1 E'_{i-1} + \phi_2 E'_{i-2} + e_i.$$

Craig demonstrates that the parameters ϕ_1 and ϕ_2 are equivalent to the a and b coefficients in deterministic models of slope development (Chap. 5.3.3). The a parameter determines the contribution of wash processes, the b parameter the contribution of creep and solifluxion processes, in slope development. Thus the values of ϕ_1 and ϕ_2 estimated from the elevation series indicate the relative importance of wash and creep in the development of the landscape. Specifically,

$$\phi_1 = b + a$$
$$\phi_2 = -a.$$

If $\phi_2 = 0$ there is no creep and the surface is shaped solely by wash. The equation in this case is

$$E'_1 = b E'_{i-1} + e_i.$$

A pure creep process will have the form

$$E'_i = a E'_{i-1} - a E'_{i-2} + e_i.$$

In the eight physiographic provinces he studied, Craig (1982c) found that process rates remain at a consistent level within a given province but may vary significantly from area to area within a province; and that the rate of slope wash greatly exceeds the rate of creep.

Statistical models are not so much a complement to field studies as a framework within which field work and laboratory work should be conducted if an hypothesis is to be tested using statistical arguments. A major problem with statistical models is that field conditions and limitations of time do not always enable all the stringent rules of sampling required by statistical models to be rigidly applied. This is particularly true in the case of parametric models of multivariate systems: the gigantic sampling designs required for multivariate models severely hamper their application to Earth surface systems (Terjung 1976). Other problems of multivariate models have been considered in Chaps. 6.2 and 6.3. It is probably true to say that statistical models are most successful when they tackle problems involving a few variables, the relationships between which can be given a fairly clear physical interpretation. The danger with multivariate models is that,

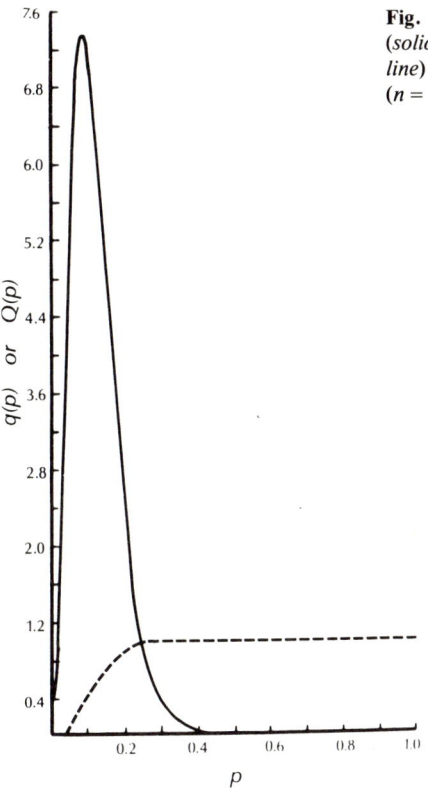

Fig. 10.1. Graph showing the probability density function (*solid line*) and cumulative distribution function (*broken line*) curves generated from proportional tree-ring data ($n = 34$). (Romesburg and DeGraff 1982)

through transformations and the extraction of "factors" or whatever, sight is lost of the original variables and interpretation is apt to become loose. An example of a successful statistical model is Romesburg and DeGraff's (1982) study of mass movement as evidenced in a sample of 20 trees on the lower slopes of Fishlake plateau near Mount Ferrell in central Utah. Careful study of cores from the trees enabled periods of eccentric growth caused by mass movements to be dated. A 34-year period, 1946 to 1979, was examined. Using proportional data expressing the number of trees among the 20 sampled showing mass movement disturbance in a given year, probability density and cumulative distribution functions using the normal generated distribution (Chiu 1974) were established (Fig. 10.1). These curves provide insight on mass movement activity during the last 34 years. Each tree can be assumed to represent 0.35 ha (one tree per 0.35 ha was sampled) or about 5 per cent of the study area. The fitted cumulative distribution function yields an estimated percentage of 42 that active mass movements affected 10 per cent or less of the area ($p \leqslant 0.1 = 0.42$); it yields an estimated percentage of 89 that active mass movement affected 20 per cent or less of the area ($p \leqslant 0.2 = 0.89$). Thus the cumulative distribution function enables and estimation of the degree of contribution active mass movement has made to landform development in the area over the last 34 year period.

10.2 Models as a Testing Ground for Long-Term Change

Earth surface processes can be studied by observation for but short lengths of time. Few records extend back more than a century and trustworthy measurements of many processes have been available for very few decades. Problems arise in trying to assess the effect of observed process rates on long term landscape development. Indeed, M. Church (1980) feels that process geomorphology provides little or no insight into the long-term development of landscapes. Büdel (1982) concurs. He argues that individual process studies, no matter how precise, can say nothing as to the role of the processes at the level at which they interact to produce relief-forming processes; and that:

"Even if it were possible for us to depict the balanced interaction of all the current mechanisms, simple projection of this picture millions of years into the future would not produce a relief corresponding to that around us, for most of the latter was largely produced by totally different relief-forming mechanism of the past". (Büdel 1982, p. 34)

Certainly, many processes operate too slowly for their effects on soil and landform development to be assessed except by recourse to simulation models. Empirical process studies can provide information on rates of sediment and solute transport as functions of topographic and soil variables. These relations can then be used in soil and landscape models to predict how soils and landscapes develop over thousands of years. These predictions can be tested to some extent by comparing the results with soil and landscape changes which can be inferred from the stratigraphic record.

A drawback with modelling long-term changes is the assumption that climate has remained constant. However, attempts are being made to remedy this limitation. Kirkby (1984), for instance, has allowed for a changing climate in his model of cliff retreat in South Wales, as originally studied by Savigear (1952) and often quoted as an exemplar of space – time substitution. The model was run for three phases. Firstly, for a period, starting 500,000 years ago and ending 50,000 years ago corresponding roughly to inland valley development with a fixed base level under mainly periglacial conditions (Fig. 10.2); secondly, a period of cliff retreat from 50,000 to 10,000 years ago (Fig. 10.3); and thirdly, a period of basal removal covering the last 10,000 years. The observed upper convexities of the slope profiles as surveyed by Savigear can, according to the model, only be formed during the periglacial phase and require at least 100,000 years to form. They are today relict features.

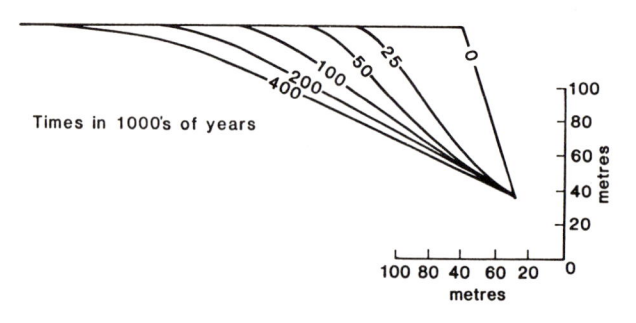

Fig. 10.2. Modelled slope decline with fixed basal point. Initial slope a 70° cliff cutting a summit plateau. Time in thousands of years. (Kirkby 1984)

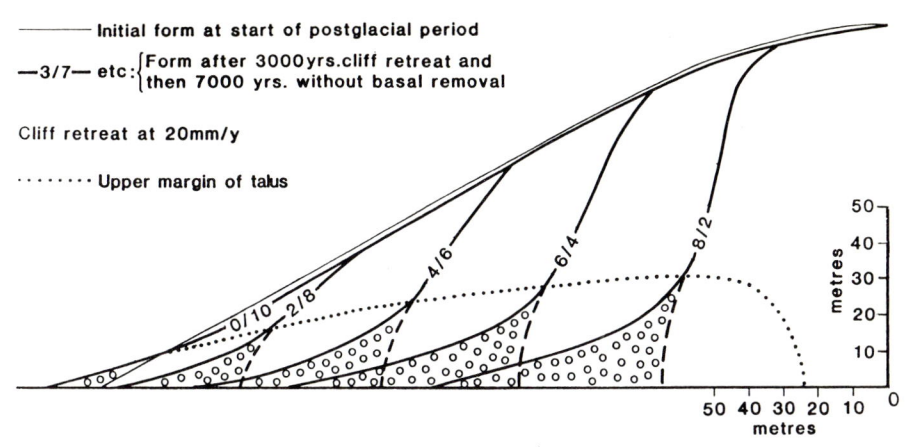

Fig. 10.3. Modelled Savigear set of profiles for 20 mm yr^{-1} basal retreat for periods of 0–8000 years and basal accumulation for the remainder of a 10,000 year period. (Kirkby 1984)

Another case in which landscape simulation has made a valuable contribution to studies of long-term change is Armstrong's (1980b) test of the ergodic (space–time substitution) hypothesis. This was discussed in Chap. 8.3.1. Armstrong found, it will be recalled, that sequences of a single slope profile through time are not interchangeable with down-valley sequences of profiles as might be observed at the present day. It is interesting that, independently of Armstrong, the ergodic hypothesis was brought into question by Church and Mark (1980), who compared static and dynamic allometry in landscapes.

10.3 Models as Good Predictors of Complex Situations

Relationships between climate, lithology, time, soils, and landforms are expressed in many traditional, conceptual models in a manner which, though inventive and imaginative, is not capable of giving good predictions and, more seriously, is often not capable of disproof. This last point is of paramount importance. Is Jenny's "clorpt" equation a truism or an hypothesis? The same question could (and has) been asked of Davis's model of landscape development. Mathematical models in contrast can sometimes give fairly exact predictions and the results can be refuted using independent tests. A case in point is Kirkby's (1985) attempt to model the relationship between rock type and landform, with rock type acting through the regolith and soil. The components and linkages in Kirkby's model (Fig. 10.4) are more precisely defined than in traditional models of landscape development. Rock type is thought to influence rates of denudation by solution, the geotechnical properties of soil, and the rates of percolation through the rock mass and its network of voids to groundwater. Climate is seen to act through its control of slope hydrology which in turn determines the partitioning of overland and subsurface flow. With suitable process equations fitted to the model, the development of slopes and soils can be simulated for a fixed

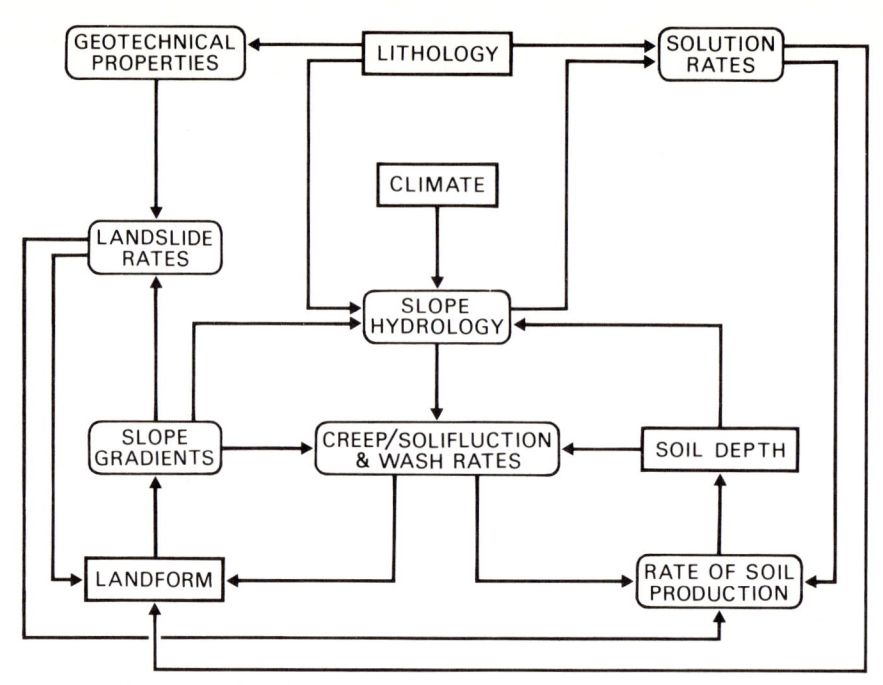

Fig. 10.4. Linkages considered in Kirkby's model. (Kirkby 1985)

basal point. Figure 10.5 is the outcome of a simulation which started with a gently sloping plateau ending in a steep bluff and a band of hard rock dipping at 10° into the slope.

The hard rock is less soluble and has a lower rate of landslide retreat, than the soft band but has the same threshold gradient for landsliding. Threshold gradients, or angles close to them, develop rapidly on the soft strata. The hard rock is undercut, forming a free face within a few hundred years. After some 20,000 years, a summit convexity begins to replace the threshold slope above the hard band, the process of replacement being complete by 200,000 years when the hard band has little or no topographic expression. The lower slope after 200,000 years stands at an almost constant gradient of 12.4°, just below the landslide threshold. Soil development involves initial thickening on the plateau and thinning by landslides on the scarp. Soil distribution is uneven owing to the localized nature of landslides. Once the slope stabilizes, thick soils form everywhere except over the hard band.

From this simulation and another in which solution is the sole process, Kirkby makes a number of deductions which appear to correspond to features in actual landscapes. Firstly, geotechnical properties of rock, in particular the rate of decline towards the threshold gradient of landslides, are more important than solution in determining slope form. Only on slopes of low gradient and after long times (200,000 years and more) do solutional properties play a dominant role in influencing slope form. Secondly, gradient steepening and soil thinning over "resistant" strata is strictly associated with the current location of an outcrop, though resistant beds, by maintaining locally steep gradients, tend to hold the less resistant beds close to the landslide threshold and so increase gradients everywhere. Thirdly, gradients close to landslide threshold gradients commonly outlive landslide activity by many thousands of years and, because of this, may play a dominant role in determining regional relief in a tectonically stable area. Fourthly, soils are generally thin under active landsliding and wash; thick soils tend to indicate the predominance of solution and creep or solifluxion processes. Catenas in humid climates can be expected to develop thicker soils in downslope positions but in semi-arid areas, where wash keeps soils thin except on the lowest gradients, catenas can be expected to have deeper soils upslope and thinner soils downslope.

Another situation in which models are good predictors of complex changes is in the case of thresholds and system stability. The idea of threshold behaviour is not new: Hjulstrom recognized it in his famous curves separating the erosion, transport, and deposition of uniform grains. What is new is the idea of intransitive changes of state, of a threshold inside the system being crossed without any change in a driving variable such as climate. The possibility of intransitive changes in the general circulation of the atmosphere had been commented on by Lorenz (1968), but it was Schumm (1973) who introduced the revolutionary concept on intrinsic thresholds in geomorphological systems. By way of example, Schumm referred to the effect of progressive channel aggradation on channel sinuosity which, in experiments and in the field, shows sudden changes at critical slope gradients. Seven years later, the concept of thresholds in Earth surface systems had been awarded the status of a paradigm (Coates and Vitek 1980), and rightly so, for its implications are far-reaching and fundamental, as the following passage suggests:

"The implications of intrinsic thresholds for inferential chronological studies are formidable. It can no longer be automatically assumed that changes in erosional and sedimentological history represent changes in climate, or sea level, or tectonic activity, unless they occur unequivocally over a relatively wide area". (Thornes 1983, p. 226).

Attention to thresholds and their associated parameters should, as Chappell (1983) claims, enable the question of whether change in landscape systems is due to climatic change or to extreme events to be resolved.

It was noted by A. D. Howard (1965) that the response of a system to change may involve a threshold which separates two rather different economies. A system can in fact possess many different economies, some of which may be stable, others unstable. The concept of multiple system-equilibria, the bifurcations between them, and the trajectories connecting them is, arguably, the most exciting and promising development in the study of Earth surface systems. Thornes (1983, p. 234) believes that it will produce major new insights into historical problems such as river terrace formation, drainage initiation, and drainage basin evolution. It has certainly given the steady-state (equilibrium) concept a good shaking. Theoretical work by Smith and Bretherton (1972) specifies the conditions under which a hollow or rill on a hillslope will fill with sediment, restoring the hillslope to a uniform condition, or will grow to form a channel and eventually a valley. The point at which channel initiation is imminent

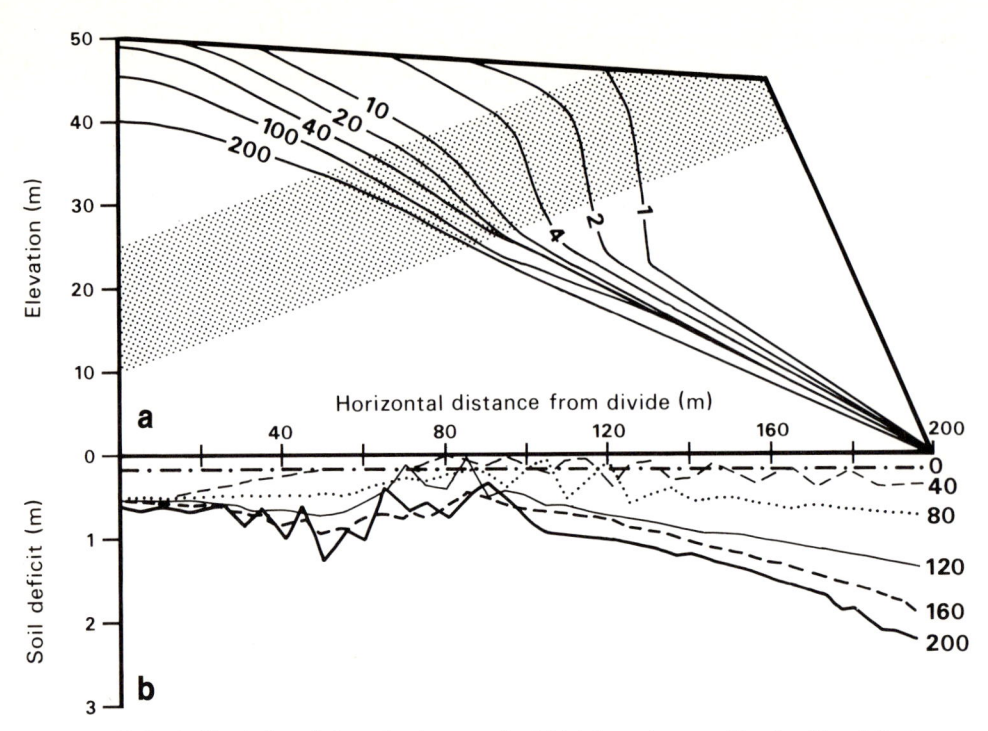

Fig. 10.5 a, b. Simulation of slope development for initial form shown, with a hard band dipping at 10° into the slope. **a** Slope profile development. **b** Soil deficit development. (Kirkby 1985)

is a threshold separating a negative feedback regime from a positive feedback regime. Kirkby (1980), pursuing this approach, looks at the initiation and development of drainage networks in the light of slope hydrology and sediment transport, and provides some concrete ideas about the relationship between drainage density and climate. Slingerland (1981) considers the fate of a random deviation in a dynamic landscape system in terms of stable and unstable states. Scheidegger (1983) takes this idea a step further and lists many examples of features in the landscape that are inherently unstable, the exogenic processes responsible for their formation acting most vigorously where a deviation has occurred owing to some earlier accidental fluctuation. The list includes stepped valleys, benches in mountain valleys, river terraces, pools and riffles, tuff terracettes, river meanders, cirques, and karst depressions.

To prognosticate about the direction a discipline might fruitfully take is a perilous occupation: new developments in science often come from unexpected quarters. None the less, the new generation of mathematical models does seem to hold out much promise for uniting at long last the "timeless" and "timebound" aspects of Earth surface systems. This is not to say that mathematical models are sufficient for a full understanding and explanation of Earth surface systems; but they are necessary for understanding and explanation. Unless the theoretical consequences of simple assumptions about form and process interactions are known, then it is not possible rationally to interpret complex field situations.

References

Ahnert F (1970) A comparison of theoretical slope models with slopes in the field. Z Geomorph Suppl 9:88–101
Ahnert F (1976) Brief description of a comprehensive three-dimensional process-response model of landform development. Z Geomorph Suppl 25:29–49
Ahnert F (1977) Some comments on the quantitative formulation of geomorphological processes in a theoretical model. Earth Surface Processes 2:191–201
Ai N, Scheidegger AE (1981) Valley trends in Tibet. Z Geomorph 25:203–212
Ai N, Scheidegger AE (1982) The tectonic stress field in north China. J Lanzhou Univ (Nat Sci) 18:157–176
Ai N, Liang Guozhao, Scheidegger AE (1982) The valley trends and neotectonic stress field of southeast China. Acta Geogr Sin 37:111–122 (in Chinese)
Allègre C (1964) Vers une logique mathématique des séries sédimentaires. Bull Soc Géol Fr 6:214–218
Anderson EW, Cox NJ (1978) A comparison of different instruments for measuring soil creep. Catena 5:81–93
Anderson KE, Furley PA (1975) An assessment of the relationship between surface properties of chalk soils and slope form using principal components analysis. J Soil Sci 26:130–143
Anderson MG, Richards KS (1979) Statistical modelling of channel form and process. In: Wrigley N (ed) Statistical applications in the spatial sciences. Pion, London, pp 205–228
Anderson MP (1979) Using models to simulate the movement of contaminants through groundwater flow systems. Crit Rev Environ Control 9:97–156
Andrews JT (1970) A geomorphological study of post-glacial uplift with particular reference to arctic Canada. Inst Brit Geogr Spec Publ No 2, London
Armstrong AC (1976) A three-dimensional simulation of slope forms. Z Geomorph Suppl 25:20–28
Armstrong AC (1980a) Soils and slopes in a humid temperate environment: a simulation study. Catena 7:327–338
Armstrong AC (1980b) Simulated slope development sequences in a three-dimensional context. Earth Surface Processes 5:265–270
Armstrong AC (1982) A comment on the continuity equation model of slope profile development and its boundary conditions. Earth Surface Processes Landforms 7:283–284
Arnett RR (1974) Environmental factors affecting the speed and volume of topsoil interflow. In: Gregory KJ, Walling DE (eds) Fluvial processes in instrumented watersheds. Inst Br Geogr Spec Publ No 6, London, pp 7–22
Ash JE, Wasson RJ (1983) Vegetation and sand stability in the Australian desert dunefield. Z Geomorph 45:7–25
Bagnold RA (1954) The physics of blown sand and desert dunes, 2nd edn. Methuen, London (1st edn 1941)
Bagnold RA (1956) The flow of cohesionless grains in fluids. Philos Trans R Soc Lond A249: 235–297
Bagnold RA (1960) Theoretical model of energy loss in curved channels. US Geol Surv Prof Pap 282-D, pp 122–134
Bagnold RA (1966) An approach to the sediment transport problem from general physics. US Geol Surv Prof Pap 282-E, pp 135–144
Bakker JP, Le Heux JWN (1946) Projective-geometric treatment of O. Lehmann's theory of the transformation of steep mountain slopes. K Ned Akad Wet Amsterdam 49:533–547

Bakker JP, Le Heux JWN (1947) Theory of central rectilinear recession of slopes, I and II. K Ned Akad Wet Amsterdam 50:959–966 and 1154–1162

Bakker JP, Le Heux JWN (1950) Theory of central rectilinear recession of slopes, III and IV. K Ned Akad Wet Amsterdam 53:1073–1084 and 1364–1374

Barron EJ (1983) A warm, equable Cretaceous: the nature of the problem. Earth-Sci Rev 19:305–338

Barron EJ, Washington WM (1982) Cretaceous climate: a comparison of atmospheric simulations with the geologic record. Palaeogeogr Palaeoclimatol Palaeocol 40:103–133

Barron EJ, Sloan JL, Harrison CGA (1980) Potential significance of land-sea distribution and surface albedo variations as a climatic forcing factor: 180 m.y. to the present. Paleogeogr Paleoclimatol Paleoecol 30:17–40

Barron EJ, Thompson SL, Schneider SH (1981) An ice-free Cretaceous? Results from climate model simulations. Science 212:501–508

Bassett K, Chorley RJ (1971) An experiment in terrain filtering. Area 3:78–91

Beasley DB, Huggins LF (1981) ANSWERS users manual. Agricultural Engineering Department, Purdue University, West Lafayette, Indiana. United States Environmental Protection Agency Region V

Beaumont JR (1982) Towards a conceptualization of evolution in environmental systems. Int J Man-Mach Stud 16:133–145

Beede JW (1911) The cycle of subterranean drainage as illustrated in the Bloomington, Indiana, quadrangle. Proc Indiana Acad Sci 20:81–111

Beer T (1983) Australian estuaries and estuarine modelling. Search 14:136–140

Bennett JP (1974) Concepts of mathematical modelling of sediment yield. Water Resour Res 10:485–493

Bennett RJ (1974) Process identification for time series modelling in urban and regional planning. Region Stud 8:157–174

Bennett RJ (1976) Adaptive adjustment of channel geometry. Earth Surface Processes 1:136–150

Bennett RJ (1979) Spatial time series: analysis, forecasting and control. Pion, London

Bennett RJ, Chorley RJ (1978) Environmental systems: philosophy, analysis, and control. Methuen, London

Benson MA (1965) Spurious correlation in hydraulics and hydrology. Proc American Society of Civil Engineers. J Hydraul Div 91:35–43

Berner RA, Lasaga AC, Garrels RM (1983) The carbonate-silicate geochemical cycle and its effects on atmospheric carbon dioxide over the past 100 million years. Am J Sci 283:641–683

Bertalanffy L von (1932) Theoretische Biologie. Borntraeger, Berlin

Bertalanffy L von (1950) The theory of open systems in physics and biology. Science 111:23–29

Beven K, Kirkby MJ (1979) A physically based variable contributing area model of basin hydrology. Hydrol Sci Bull 24:43–69

Beven K, Gilman K, Newson M (1979) Flow and flow routing in upland channel networks. Hydrol Sci Bull 24:303–325

Bidwell OW, Hole FD (1964) Numerical taxonomy and soil classification. Soil Sci 97:58–62

Blackie JR (1972) Hydrological effects of a change of land use from rain forest to tea plantation in Kenya. In: Proceedings of Wellington Symposium, Results of research in representative and experimental basins, vol 2. International Association of Scientific Hydrology, UNESCO, Paris, pp 312–329

Blackith RE, Reyment RA (1971) Multivariate morphometrics. Academic, London

Blalock HM (1964) Causal inferences in non-experimental research. University of North Carolina Press, Chapel Hill, North Carolina

Blume H-P (1968) Die pedogenetische Deutung einer Catena durch die Untersuchung der Bodendynamik. Trans Ninth Int Congr Soil Sci, Adelaide, 4:441–449

Blume H-P, Schlichting E (1965) The relationships between historical and experimental pedology. In: Hallsworth EG, Crawford DV (eds) Experimental pedology. Butterworths, London, pp 346–353

Boast CW (1973) Modeling the movement of chemicals by soil water. Soil Sci 115:224–230

Bohm D (1980) Wholeness and the implicate order. Routledge and Kegan Paul, London

Bolin B (ed) (1981) Carbon cycle modelling. Scope 16. Wiley, Chichester

Bonham-Carter G, Sutherland AJ (1968) Mathematical model and Fortran IV program for computer simulation of deltaic sedimentation. Kans State Geol Surv, Univ Kans, Comput Contrib No 24

References

Boots BN, Murdoch DJ (1983) The spatial arrangement of random Voronoi polygons. Comput Geosci 9:351–365

Box GEP, Jenkins G (1976) Time series analysis: forecasting and control, 2nd edn. Holden-Day, San Francisco (1st edn 1970)

Box GEP, Tiao GC (1975) Intervention analysis with applications to economic and environmental problems. J Am Stat Assoc 70:70–79

Briggs LI, Pollack HN (1967) Digital model of evaporite sedimentation. Science 155:453–456

Brunsden D (1973) The application of system theory to the study of mass movement. Geol Appl Idriogeol, Bari, 7:185–207

Bryan K (1923) Erosion and sedimentation in Papago Country. US Geol Surv Bull 730:19–90

Bryan K (1940) The retreat of slopes. Ann Assoc Am Geogr 30:254–267

Bryan RB (1974) A simulated rainfall test for the prediction of soil erodibility. Z Geomorph Suppl 21:138–150

Bryant E (1983) Coastal erosion and accretion, Stanwell Park Beach, N.S.W., 1890–1980. Aust Geogr 15:382–390

Büdel J (1982) Climatic geomorphology. Princeton University Press, Princeton. New Jersey. (Translated by Fisher L, Busche D)

Bull WB (1962) Relations of alluvial-fan size and slope to drainage-basin size and lithology in western Fresno County, California. US Geol Surv Prof Pap 450-B

Bull WB (1975) Landforms that do not tend toward a steady state. In: Melhorn, WN, Flemal RC (eds) Theories of landform development. George Allen and Unwin, London, pp 111–128

Bull WB (1977) The alluvial-fan environment. Prog Phys Geogr 1:222–270

Buol SW, Hole FD, McCracken RJ (1980) Soil genesis and classification, 2nd edn. Iowa State University Press, Ames, Iowa

Butler BE (1959) Periodic phenomena in landscapes as a basis for soil studies. CSIRO Australia, Soil Publ 14

Butzer KW (1976) Geomorphology from the Earth. Harper and Row, New York

Cambers G (1976) Temporal scales in coastal erosion systems. Trans Inst Brit Geogr, New Ser, 1:246–256

Camp CR, Gill WR (1969) The effect of drying on soil strength parameters. Proc Soil Sci Soc Am 33:641–644

Campbell WJ, Rasmussen LA (1969) Three-dimensional surges and recoveries in a numerical glacier model. Can J Earth Sci 6:979–986

Carey SW (1976) The expanding Earth. Elsevier, Amsterdam

Carr DD, Horowitz A, Hrabar SV, Ridge KF, Rooney R, Staw WT, Webb W, Potter PE (1966) Stratigraphic sections, bedding sequences, and random processes. Science 154:1162–1164

Carson MA (1981) Influence of porefluid salinity on instability of sensitive marine clays: a new approach to an old problem. Earth Surface Processes Landforms 6:499–515

Carson MA, Kirkby MJ (1972) Hillslope form and process. Cambridge University Press, Cambridge

Cattell RB (1965) Factor analysis: an introduction to essentials. Biometrics 21:190–215 and 405–435

Chang TP, Toebes GH (1970) A statistical comparison of meander planforms in the Wabash basin. Water Resour Res 6:557–578

Chapman GP (1977) Human and environmental systems: a geographer's appraisal. Academic, London

Chappell J (1974) The geomorphology and evolution of small valleys in dated coral reef terraces, New Guinea. J Geol 82:795–812

Chappell J (1978) On process-landform models from Papua New Guinea and elsewhere. In: Davies JL, Williams MAJ (eds) Landform evolution in Australasia. Australian National University Press, Canberra

Chappell J (1983) Thresholds and lags in geomorphologic changes. Aust Geogr 15:357–366

Chappell J, Eliot IG (1979) Surf beach dynamics in time and space – an Australian case study and a predictive model. Mar Geol 32:231–250

Chappell J, Eliot IG, Bradshaw MP, Lonsdale E (1979) Experimental control of beach face dynamics by water table pumping. Eng Geol 14:29–41

Chatfield C, Collins AJ (1980) Introduction to multivariate analysis. Chapman and Hall, London

Chiu WK (1974) A new prior distribution for attributes sampling. Technometrics 16:73–102

Chorley RJ (1962) Geomorphology and general systems theory. US Geol Surv Prof Pap 500-B
Chorley RJ (1964) Geography and analogue theory. Annals of the Assoc Am Geogr 54:127 – 137
Chorley RJ (1967a) Models in geomorphology. In: Chorley RJ, Haggett P (eds) Models in geography. Methuen, London, pp 59 – 96
Chorley RJ (1967b) Application of computer techniques in geography and geology. Abstr Proc Geol Soc, London 1642:183 – 186
Chorley RJ (1969a) The drainage basin as the fundamental geomorphic unit. In: Chorley RJ (ed) Water, Earth, and Man. Methuen, London, pp 77 – 99
Chorley RJ (1969b) The elevation of the Lower Greensand ridge, South-East England. Geol Mag 106:231 – 248
Chorley RJ, Beckinsale RP (1980) G. K. Gilbert's geomorphology. In: Yochelson EL (ed) The scientific ideas of G. K. Gilbert. Geol Soc Am Spec Pap 183, pp 129 – 142
Chorley RJ, Kennedy BA (1971) Physical geography: a systems approach. Prentice-Hall, London
Chorley RJ, Stoddart DR, Haggett P, Slaymaker HO (1966) Regional and local components in the areal distribution of surface sand facies in the Breckland, Eastern England. J Sediment Petrol 36:209 – 220
Chow VT (1959) Open-channel hydraulics. McGraw-Hill, Tokyo
Chow VT, Kareliotis SJ (1970) Analysis of stochastic hydrologic systems. Water Resour Res 6:1569 – 1582
Church M (1980) Records of recent geomorphological events. In: Cullingford RA, Davidson DA, Lewin J (eds) Timescales in geomorphology. Wiley, Chichester, pp 13 – 29
Church M, Mark DM (1980) On size and scale in geomorphology. Prog Phys Geogr 4:342 – 390
Church MA (1972) Baffin island sandar: a study of Arctic fluvial processes. Bull Geol Soc Can 216:208
Cleaves ET, Godfrey AE, Bricker OP (1970) Geochemical balance of a small watershed and its geomorphic implications. Bull Geol Soc Am 81:3015 – 3032
Coates DR, Vitek JD (1980) Perspectives on geomorphic thresholds. In: Coates DR, Vitek JD (eds) Thresholds in geomorphology. George Allen and Unwin, London, pp 3 – 23
Commoner B (1972) The closing circle: confronting the environmental crisis. Cape, London
Conacher AJ, Dalrymple JB (1977) The nine-unit landsurface model: an approach to pedogeomorphic research. Geoderma 18:1 – 154
Cook R (1977) Raymond Lindeman and the trophic-dynamic concept in ecology. Science 198:22 – 26
Cox DR (1962) Renewal theory. Methuen, London
Cox NJ (1977) Allometric change of landforms: discussion and reply. Bull Geol Soc Am 88:1199 – 1202
Cox NJ (1980) On the relationship between bedrock lowering and regolith thickness. Earth Surface Processes 5:271 – 274
Cox NJ (1983) On the estimation of spatial autocorrelation in geomorphology. Earth Surface Processes Landforms 8:89 – 93
Craig RG (1981) Natural systems. In: Craig RG, Labovitz ML (eds) Future trends in geomathematics. Pion, London, pp 265 – 274
Craig RG (1982a) Criteria for constructing optimal digital terrain models. In: Craig RG, Craft JC (eds) Applied geomorphology. George Allen and Unwin, London, pp 108 – 130
Craig RG (1982b) The ergodic principle in erosion models. In: Thorn CE (ed) Space and time in geomorphology. George Allen and Unwin, London, pp 81 – 115
Craig RG (1982c) Evaluation of terrain complexity by autocorrelation. Final Rep Nat Aeronaut Space Administr. Contract NAG 5-165. Dep Geol, Kent State Univ, Kent, Ohio
Craig RG, Labovitz ML (1981) Introduction. In: Craig, RG, Labovitz ML (eds) Future trends in geomathematics. Pion, London, pp 1 – 2
Crain IK (1972) Monte Carlo simulation of random Voronoi polygons: preliminary results. Search 3:220 – 221
Crain IK (1976) Statistic analysis of geotectonics. In: Merriam DF (ed) Random processes in geology. Springer, Berlin Heidelberg New York, pp 3 – 15
Crain IK (1978) The Monte-Carlo generation of random polygons. Comput Geosci 4:131 – 141
Crary AP, Robinson ES, Bennett HF, Boyd WW (1962) Glaciological studies of the Ross Ice Shelf, Antarctica, 1957 – 60. Am Geophys Soc NY, IGY Glaciology Report No 6
Crickmay CH (1959) A preliminary inquiry into the formulation and applicability of the geological principle of uniformity. Published by the author, Calgary, Alberta

Crickmay CH (1960) Lateral activity in a river of northwestern Canada. J Geol 68:377 – 381
Crickmay CH (1975) The hypothesis of unequal activity. In: Melhorn WN, Flemal RC (eds) Theories of landscape development. George Allen and Unwin, London, pp 103 – 109
Crickmore MJ (1970) Effect of flume width on bed-form characteristics. Proceedings of the American Society of Civil Engineers, J Hydraul Div 96:473 – 496
Culling E (1981) Stochastic processes. In: Wrigley N, Bennett RJ (eds) Quantitative geography: a British view. Routledge and Kegan Paul, London, pp 202 – 211
Culling WEH (1963) Soil creep and the development of hillside slopes. J Geol 71: 127 – 161
Culling WEH (1983a) Slow particulate flow in condensed media as an escape mechanism: mean translation distance. Catena Suppl 4:161 – 190
Culling WEH (1983b) Rate process theory of geomorphic soil creep. Catena Suppl 4:191 – 214
Curl RL (1959) Stochastic models of cavern development. Bull Geol Soc Am 70:1802 (Abstract)
Dacey MF (1971) Probability distribution of number of networks in topologically random network patterns. Water Res Res 7:1652 – 1657
Dacey MF (1972) Some properties of link magnitude for channel networks and network patterns. Water Resour Res 8:1106 – 1111
Dacey MF (1976) Summary of magnitude properties of topologically distinct channel networks and network patterns. In: Merriam DF (ed) Random processes in geology. Springer, Berlin Heidelberg New York, pp 16 – 38
Dacey MF, Krumbein WC (1976) Three growth models for stream channel networks. J Geol 84:153 – 163
Dacey MF, Krumbein WC (1979) Models of breakage and selection for particle size distributions. Math Geol 11:193 – 222
Dacey MF, Krumbein WC (1981) Similarities between models for particle-size distributions and stream-channel networks. In: Craig RG, Labovitz ML (eds) Future trends in geomathematics. Pion, London, pp 179 – 200
Dalrymple JB, Blong RJ, Conacher AJ (1968) A hypothetical nine-unit landsurface model. Z Geomorph 12:60 – 76
Dalrymple T (1960) Flood frequency analysis: Manual of hydrology, part 3, flood flow techniques. US Geol Surv Water Supply Pap 1543A:1 – 80
Dan J, Yaalon DH (1968) Pedomorphic forms and pedomorphic surfaces. Trans Ninth Int Congr Soil Sci, Adelaide, 4:577 – 584
Daniels RB, Gamle EE, Cady JG (1971) The relationship between geomorphology and soil morphology and genesis. Adv Agron 23:51 – 88
Davidson DA (1977) The subdivision of a slope profile on the basis of soil properties: a case study from mid-Wales. Earth Surface Processes 2:55 – 61
Davidson DA (1978) Science for physical geographers. Edward Arnold, London
Davidson-Arnott RGD (1981) Computer simulation of nearshore bar formation. Earth Surface Processes Landforms 6:23 – 34
Davis WM (1889) The rivers and valleys of Pennsylvania. Nat Geogr Mag 1:183 – 253 (also in Geographical essays)
Davis WM (1899) The geographical cycle. Geogr J 14:481 – 504 (also in Geographical essays)
Davis WM (1900) Glacial erosion in France, Switzerland, and Norway. Proc Boston Soc Nat Hist 29:273 – 322 (also in Geographical essays)
Davis WM (1903) The mountain ranges of the Great Basin. Bull Harvard Univ Mus Comp Zool 42:129 – 177 (also in Geographical essays)
Davis WM (1905) The geographical cycle in an arid climate. J Geol 13:381 – 407 (also in Geographical essays)
Davis WM (1906) The sculpture of mountains by glaciers. Scott Geogr Mag 22:76 – 89 (also in Geographical essays)
Davis WM (1909) Geographical essays. Ginn, Boston
Davis WM (1912) Die erklärende Beschreibung der Landformen. Teubner, Leipzig
Davis WM (1930) Rock floors in arid and humid climates. J Geol 38:1 – 27 and 136 – 158
Davy BW, Davies TRH (1979) Entropy concepts in fluvial geomorphology. Water Resour Res 15:103 – 106
Dawdy DR, O'Donnell T (1965) Mathematical models of catchment behaviour. Proc Am Soc Civil Eng, J Hydraul Div 91:123 – 137
Denbigh KG (1951) The thermodynamics of the steady state. Methuen, London

Denny CS (1965) Alluvial fans in the Death Valley region, California and Nevada. US Geol Surv Prof Pap 466
Denny CS (1967) Fans and pediments. Am J Sci 265:81–105
de Ploey J, Savat J (1968) Contribution à l'étude de l'érosion par le splash. Z Geomorph 12:174–193
Dillaha TA, Beasley DB, Huggins LG (1982) Using the ANSWERS model to estimate sediment yields on construction sites. J Soil Water Conserv 37:117–120
Di Toro DM, O'Connor DJ, Thomann RV, Mancini JL (1975) Phytoplankton-zooplankton-nutrient interaction model for western Lake Erie. In: Patten BC (ed) Systems analysis and simulation in ecology, vol 3. Academic, New York, pp 423–474
Dole SH (1970) Formation of planetary systems by aggregation: a computer simulation. Icarus 13:494–498
Drake JJ, Ford DC (1981) Karst solution: a global model for groundwater solute concentrations. Trans Jpn Geomorph Union 2:223–230
Draper NR, Smith H (1966) Applied regression analysis. Wiley, New York
Drewry DJ (1983) Antarctic ice sheet: aspects of current configuration and flow. In: Gardner R, Scoging H (eds) Mega-geomorphology. Clarendon, Oxford, pp 18–38
Edwards AMC (1973) Dissolved load and tentative solute budgets of some Norfolk catchments. J Hydrol 18:201–217
Elderton WP, Johnson NL (1969) Systems of frequency curves. Cambridge University Press, Cambridge
Elsasser WM (1971) Two-layer model of upper-mantle circulation. J Geoph Res 76:4744–4753
Engman ET, Rogowski AS (1974) A partial area model for storm flow synthesis. Water Resour Res 10:464–472
Erhart H (1967) La genèse des sols en tant que phénomène géologique, 2nd edn. Masson, Paris
Fairbridge RW (ed) (1981) Neotectonics. Z Geomorph Suppl 40
Ferguson RI (1973) Channel pattern and sediment type. Area 5:38–41
Ferguson RI (1975) Meander irregularity and wavelength estimation. J Hydrol 26:315–333
Ferguson RI (1976) Disturbed periodic model for river meanders. Earth Surface Processes 1:337–347
Ferguson RI (1979) River meanders: regular or random? In: Wrigley N (ed) Statistical applications in the spatial sciences. Pion, London, pp 229–241
Fetter CW (1980) Applied hydrogeology. Merrill, Columbus, Ohio
Fischer AG (1981) Climatic oscillations in the biosphere. In: Nitecki MH (ed) Biotic crises in ecological and evolutionary time. Academic, New York, pp 103–131
Fisher O (1866) On the disintegration of a chalk cliff. Geol Mag 3:354–356
Fleming G (1975) Computer simulation techniques in hydrology. Elsevier, New York
Ford DC (1980) Threshold and limit effects in karst geomorphology. In: Coates DR, Vitek JD (eds) Thresholds in geomorphology. George Allen and Unwin, London, pp 345–362
Ford DC, Drake JJ (1982) Spatial and temporal variations in karst solution rates: the structure of variability. In: Thorn CE (ed) Space and time in geomorphology. George Allen and Unwin, London, pp 147–170
Forsyth D, Uyeda S (1975) On the relative importance of the driving forces of plate motion. Geophys J R Astronom Soc 43:163–200
Fox WT, Davis RA (1973) Simulation model for storm cycles and beach erosion on Lake Michigan. Bull Geol Soc Am 84:1769–1790
Fraederick K (1978) Structural and stochastic analysis of a zero-dimensional climate system. Q J R Meteorol Soc 104:461–474
Freeze RA (1971) Three-dimensional, transient, saturated-unsaturated flow in a ground water basin. Water Resour Res 7:347–366
Freeze RA, Cherry JA (1979) Groundwater. Prentice-Hall, Englewood Cliffs, New Jersey
Friedkin JF (coordinator) (1945) A laboratory study of the meandering of alluvial rivers. War Department, United States Corps of Engineers, Mississippi River Commission, United States Waterways Experiment Station, Vicksburg, Mississippi, 40 pp
Funderlic RE, Heath MT (1971) Linear compartmental analysis of ecosystems. ORNL-IBP-71-4, Oak Ridge National Laboratory, Oak Ridge, Tennessee
Furley PA (1968) Soil formation and slope development. 2. The relationship between soil formation and gradient in the Oxford area. Z Geomorph 12:25–42
Furley PA (1971) Relationship between slope form and soil properties over chalk parent materials. In: Brunsden D (ed) Slopes: form and process. Inst Br Geogr Spec Publ No 3, pp 141–163

Gabriel KR, Neumann J (1962) A Markov chain model for daily rainfall occurrence at Tel Aviv. Q J R Meteorol Soc 88:90–95

Garrels RM, Mackenzie FT (1971) Evolution of sedimentary rocks. Norton, New York

Garrels RM, Lerman A, Mackenzie FT (1976) Controls of atmospheric O_2 and CO_2: past, present, and future. Am Sci 63:306–315

Gerrard AJ (1981) Soils and landforms: An integration of geomorphology and pedology. George Allen and Unwin, London

Getzen RT (1977) Analog-model analysis of regional three-dimensional flow in the ground-water reservoir of Long Island, New York. US Geol Surv Prof Pap 982

Ghil M (1976) Climate stability for a Sellers-type model. J Atmos Sci 33:3–20

Gilbert GK (1877) Report on the geology of the Henry Mountains. US Geol Surv Rocky Mountain Region (Powell), Washington, D.C.

Glazovskaya MA (1968) Geochemical landscapes and geochemical soil sequences. Trans Ninth Int Congr Soil Sci, Adelaide, 4:303–312

Gonzales RC, Winz P (1977) Digital image processing. Addison-Wesley, Reading, Massachusetts

Goodall DW (1954) Objective methods for the classification of vegetation: III, an essay in the use of factor analysis. Aust J Bot 2:304–324

Goodman A (1983) Compare: a FORTRAN IV program for the quantitative comparison of polynomial trend surfaces. Comput Geosci 9:417–454

Gossman H (1970) Theorien zur Hangentwicklung in verschiedenen Klimazonen. Würzburg Geogr Arb, Heft 31, 146 pp

Gossman H (1975) L'importance des processus se déroulant à la ligne de partage locale des eaux pour l'évolution des versant sous la dominance du ruissellement pluvial (à l'aide des formules mathématiques élémentaires). Actes du Symposium sur les Versant en Pays Meditérranées, Aix-en-Provence, 1975

Gossman H (1976) Slope modelling with changing boundary conditions – effects of climate and lithology. Z Geomorph Suppl 25:72–88

Gossman H (1981) Fragen und Einsichten zum Einsatz von Hangmodellen in der geomorphologischen Analyse. Geoökodynamik 2:205–218

Gould SJ (1977) Eternal metaphors of palaeontology. In: Hallam A (ed) Patterns of evolution as illustrated by the fossil record. Elsevier, Amsterdam, pp 1–26

Gower JC (1966) Some distance properties of latent root and vector methods used in multivariate analysis. Biometrika 53:325–338

Graf WH (1971) Hydraulics of sediment transport. McGraw-Hill, New York

Graf WL (1979) Catastrophe theory as a model for changes in fluvial systems. In: Rhodes DD, Williams GP (eds) Adjustments of the fluvial system. Kendall Hunt, Dubuque, Iowa, pp 13–32

Graf WL (1982) Spatial variation of fluvial processes in semi-arid lands. In: Thorn CE (ed) Space and time in geomorphology. George Allen and Unwin, London, pp 193–217

Gray JM (1978) Low-level shore platforms in the south-west Scottish Highlands: altitude, age and correlation. Trans Inst Br Geogr, New Ser, 3:151–164

Gray JR (1967) Probability. Oliver and Boyd, Edinburgh

Gretener PE (1967) Significance of the rare event in geology. Bull Am Assoc Petrol Geol 51:2197–2206

Gumbel EJ (1958) Statistics of extreme values. Columbia University Press, New York

Gutenberg B, Richter CF (1954) Seismicity of the Earth and associated phenomena, 2nd edn. Princeton University Press, Princeton, New Jersey

Haan CT (1977) Statistical methods in hydrology. Iowa University Press, Ames, Iowa

Haan CT, Allen DM, Street JO (1976) A Markov chain model of daily rainfall. Water Resour Res 12:443–449

Hack JT (1960) Interpretation of erosional topography in humid temperate regions. Am J Sci 258A:80–97

Hack JT (1965a) Geomorphology of the Shenandoah Valley, Virginia and West Virginia, and origin of residual ore deposits. US Geol Surv Prof Pap 484

Hack JT (1965b) Postglacial drainage evolution and stream geometry in the Ontonagon Area, Michigan. US Geol Surv Prof Pap 504-B

Haggett P (1965) Locational analysis in human geography, 1st edn. Edward Arnold, London

Haken H (ed) (1982) Evolution of order and chaos in physics, chemistry and biology. Springer, Berlin Heidelberg New York

Hales ZL, Shindala A, Denson KH (1970) River bed degradation prediction. Water Resour Res 8:1530 – 1540
Harbaugh JW, Bonham-Carter G (1970) Computer simulation in geology. Wiley, New York
Harradine, F, Jenny H (1958) Influence of parent material and climate on texture and nitrogen and carbon content of virgin Californian soils. Soil Sci 85:235 – 243
Harrison PW (1957) New techniques for three-dimensional fabric analysis of till and englacial debris containing particles from 3 to 40 mm in size. J Geol 65:98 – 105
Harrison W, Krumbein WC (1964) Interactions of the beach-ocean-atmosphere system at Virginia beach, Virginia. US Army Coast Eng Res Cent, Technical Memo No 7
Haskell EF (1940) Mathematical systemization of "Environment", "Organism" and "Habitat". Ecology 21:1 – 16
Hawley JW, Wilson WE (1965) Quaternary geology of the Winnemucca area, Nevada. Univ Nevada Desert Res Inst Techn Rep No 5
Hess SL (1959) Introduction to theoretical meteorology. Henry Holt, New York
Hett JM, O'Neill RV (1974) Systems analysis of the Aleut ecosystem. Arctic Anthropology 11:31 – 40
Hickin EJ, Nanson GC (1975) The character of channel migration on the Beatton river, north-east British Columbia, Canada. Bull Geol Soc Am 86:487 – 494
Higgins CG (1975) Theories of landscape development: a perspective. In: Melhorn WN, Flemal RC (eds) Theories of landform development. George Allen and Unwin, London, pp 1 – 28
Hills ES (1956) A contribution to the morphotectonics of Australia. J Geol Soc Aust 3:1 – 15
Hipel KW, McLeod AI (1981a) Box-Jenkins modelling in the geophysical sciences. In: Craig RG, Labovitz ML (eds) Future trends in geomathematics. Pion, London, pp 65 – 86
Hipel KW, McLeod AI (1981b) Time series modelling for water resources and environmental engineers. Elsevier, Amsterdam
Hipel KW, Lennox WC, Unny TE, McLeod AI (1975) Intervention analysis in water resources. Water Resour Res 11:855 – 861
Hipel KW, McLeod AI, Lennox WC (1977) Advances in Box-Jenkins modelling, part one, model construction. Water Resour Res 13:567 – 575
Hirano M (1968) A mathematical model of slope development – an approach to the analytical theory of erosional topography. J Geosci, Osaka City University 2:13 – 52
Hirano M (1975) Simulation of development process of interfluvial slopes with reference to graded form. J Geol 83:113 – 123
Hirano M (1976) Mathematical model and the concept of equilibrium in connection with slope shear ratio. Z Geomorph Suppl 25:50 – 71
Hirsch MW, Smale S (1974) Differential equations, dynamical systems and linear algebra. Academic, New York
Holliman J (1974) Consumer's guide to the protection of the environment. Pan-Ballantine, London
Hooke R Le B (1968) Steady-state relationships on arid-region alluvial fans in closed basins. Am J Sci 266:609 – 629
Hooke R Le B, Rohrer WL (1979) Geometry of alluvial fans: effect of discharge and sediment type. Earth Surface Processes 4:147 – 166
Horton RE (1945) Erosional development of streams and their drainage basins: hydrophysical approach to quantitative morphology. Bull Geol Soc Am 56:275 – 370
Howard AD (1965) Geomorphological systems – equilibrium and dynamics. Am J Sci 263:302 – 312
Howard AD (1971a) Simulation of stream networks by headward growth. Geogr Anal 3:29 – 50
Howard AD (1971b) Simulation model of stream capture. Bull Geol Soc Am 82:1355 – 1376
Howard RA (1971) Dynamic probabilistic systems. Volume 1: Markov models. Wiley, New York
Hoyle F, Wickramasinghe NC (1979) Diseases form space. Dent, London
Huggett RJ (1973) Soil landscape systems: theory and field evidence. Ph. D thesis, London University, London
Huggett RJ (1975) Soil landscape systems: a model of soil genesis. Geoderma 13:1 – 22
Huggett RJ (1976) Lateral translocation of soil plasma through a small valley basin in the Northaw Great Wood, Hertfordshire. Earth Surface Processes 1:99 – 109
Huggett RJ (1980) Systems analysis in geography. Clarendon, Oxford
Huggett RJ (1982) Models and spatial patterns of soils. In: Bridges EM, Davidson DA (eds) Principles and applications of soil geography. Longman, London, pp 132 – 170
Hughes TJ (1979) Reconstruction and disintegration of ice sheets for the CLIMAP 18 000 and 125 000 years B. P. experiments: theory. J Glaciol 24:493 – 495

References

Hunt BG (1984) Polar glaciation and the genesis of ice ages. Nature 308:48 – 51
Imbrie J (1963) Factor and vector programs for analyzing geologic data. Off Naval Res Tech Rep 6, ONR Task No 389-135. Dep Geol, Columbia Univ NY
Imbrie J, Purdie EG (1962) Classification of modern Bahamian carbonate sediments. Am Assoc Petrol Geol Mem 1:253 – 272
Ingram HAP (1982) Size and shape in raised mire ecosystems: a geophysical model. Nature 297:300 – 303
Isermann R (1975) Modelling and identification of dynamic processes – an extract. In: Vansteenkiste GC (ed) Modeling and simulation of water resources systems. North-Holland, Amsterdam, pp 7 – 37
Ishaq AM, Huff DD (1979) Hydrologic source areas. B. Runoff simulations. In: Morel-Seytoux HJ, Salas JD, Sanders TG, Smith RE (eds) Modeling hydrologic processes. Water Resour Publ, Fort Collins, Colorado, pp 511 – 523
Jacobs CE (1943) Correlation of ground-water levels and precipitation on Long Island, New York. Trans Am Geophys Union 564 – 573
Jahn A (1967) Some features of mass movement of Spitsbergen slopes. Geogr Ann 49A:213 – 225
Jeffreys H (1918) Problems of denudation. Philos Mag 36:179 – 190
Jenny H (1941) Factors of soil formation. A system of quantitative pedology. McGraw-Hill, New York
Jenny H (1958) The role of the plant factor in pedogenic functions. Ecology 39:5 – 16
Jenny H (1961) Derivation of state factor equations of soils and ecosystems. Proc Soil Sci Soc Am 25:385 – 388
Jenny H (1980) The soil resourece: origin and behaviour. Springer, Berlin Heidelberg New York (Ecological Studies 37)
Jenny H, Bingham F, Padilla-Saravina B (1948) Nitrogen and organic carbon contents of equatorial soils of Colombia, South America. Soil Sci 66:173 – 186
Jenssen D (1977) A three-dimensional polar ice-sheet model. J Glaciol 18:373 – 389
Johnson DW (1919) Shore processes and shoreline development. Wiley, New York
Jordan CF, Kline JR, Sasscer DS (1973) A simple model of strontium and manganese dynamics in a tropical rain forest. Health Phys 24:477 – 489
Karcz I (1980) Thermodynamic approach to geomorphic thresholds. In: Coates DR, Vitek JD (eds) Thresholds in geomorphology. George Allen and Unwin, London, pp 209 – 226
Kemmerly PR (1976) Definitive doline characteristics of the Clarksville quadrangle, Tennessee. Bull Geol Soc Am 87:42 – 46
Kendall MG (1939) The geographical distribution of crop productivity in England. J R Stat Soc 102:21 – 48
Kendall MG (1957) A course in multivariate analysis. Griffin, London
Kendall MG (1975) Multivariate analysis. Griffin, London
Kennedy BA (1977) A question of scale? Prog Phys Geogr 1:154 – 157
Kennedy BA (1983) On outrageous hypotheses in geography. Geography 68:326 – 330
Kennedy BA, Melton MA (1972) Valley asymmetry and slope form of a permafrost area in the Northwest Territories, Canada. In: Price RJ, Sugden DE (eds) Polar geomorphology. Inst Br Geogr Spec Publ No 4, pp 107 – 121
Keulegan GH (1944) Spatially variable discharge over a sloping plane. Trans Am Geophys Union 956 – 959
Keulegan GH (1948) Gradual damping of solitary waves. J Res, National Bureau of Standards 40:487 – 498
King CAM, McCullagh MJ (1971) A simulation model of a complex recurved spit. J Geol 79:22 – 37
King LC (1953) Canons of landscape evolution. Bull Geol Soc Am 64:721 – 752
King LC (1963) South African scenery. Oliver and Boyd, Edinburgh
King LC (1967) The morphology of the Earth, 2nd edn. Oliver and Boyd, Edinburgh (1st edn 1962)
King LC (1983) Wandering continents and spreading sea floors on an expanding Earth. Wiley, Chichester
Kirkby, MJ (1967) Measurement and theory of soil creep. J Geol 75:359 – 378
Kirkby MJ (1971) Hillslope process-response models based on the continuity equation. In: Brunsden D (ed) Slopes: form and process. Inst Br Geogr Spec Publ No 3, pp 15 – 30
Kirkby MJ (1976) Deterministic continuous slope models. Z Geomorph Suppl 25:2 – 19

Kirkby MJ (1980) The stream head as a significant geomorphic threshold. In: Coates DR, Vitek JD (eds) Thresholds in geomorphology. George Allen and Unwin, London, pp 53 – 73

Kirkby MJ (1984) Modelling cliff development in South Wales: Savigear re-viewed. Z Geomorph 28:405 – 426

Kirkby MJ (1985) A model for the evolution of regolith-mantled slopes. In: Woldenberg MD (ed) Models in geomorphology. George Allen and Unwin, London (Forthcoming)

Kirkby MJ, Statham I (1975) Surface stone movement and scree formation. J Geol 83:349 – 362

Kleiss HJ (1970) Hillslope sedimentation and soil formation in northeastern Iowa. Proc Soil Sci Soc Am 34:287 – 290

Klovan JE (1966) The use of factor analysis in determining depositional environments from grain-size distributions. J Sediment Petrol 36:115 – 125

Klute A, Scott EJ, Whistler FD (1965) Steady state water flow in a saturated inclined soil slab. Water Resour Res 1:287 – 294

Knapp BJ (1974) Hillslope throughflow observation and the problem of modelling. In: Gregory KJ, Walling DE (eds) Fluvial processes in instrumented watersheds. Inst Br Geogr Spec Publ No 6, pp 23 – 31

Knighton AD (1975) Channel gradient in relation to discharge and bed material characteristics. Catena 2:263 – 274

Knisel WG (ed) (1980) CREAMS: A field-scale model for chemicals, runoff, and erosion from agricultural management systems. US Dep Agric, Conserv Serv Rep No 26, 640 pp

Kolmogorov AN (1933) Grundbegriffe der Wahrscheinlichkeitsrechnung. Springer, Berlin (Translated by Morrison N as Foundations of the theory of probabilitiy, Chelsea, New York)

Komar PD (1973) Computer models of delta growth due to sediment input from rivers and longshore transport. Bull Geol Soc Am 84:2217 – 2226

Koreleski K (1975) Types of soil degradation on loess near Kraków. J Soil Sci 26:44 – 52

Krumbein WC (1967) FORTRAN IV computer programs for Markov chain experiments in geology. Kans Geol Surv Comput Contrib No 13, Kansas

Krumbein WC (1976) Probabilistic modeling in geology. In: Merriam DF (ed) Random processes in geology. Springer, Berlin Heidelberg New York, pp 39 – 54

Krumbein WC, Graybill FA (1965) An introduction to statistical models in geology, McGraw-Hill, New York

Krzysztofowicz R (1983a) Why should a forecaster and a decision maker use Bayes' theorem? Water Resour Res 19:327 – 336

Krzysztofowicz R (1983b) A Bayesian Markov model of the flood forecast process. Water Resour Res 19:1455 – 1465

Lachenbruch AH (1962) Mechanics of thermal contraction cracks and ice-wedge polygons in permafrost. Geol Soc Am Spec Pap 70

Lai PW (1979) Transfer function modelling. Concepts and Techniques in Modern Geography No 22, Geo Abstracts, Norwich

Langbein WB, Leopold LB (1966) River meanders: theory of minimum variance. US Geol Surv Prof Pap 422-H

Laplace PS de (1951) A philosophical essay on probabilities. English edn, Dover Books, New York

LaValle PD (1967) Geographical processes and the analysis of karst depressions within limestone regions. Ann Assoc Am Geogr 57:794 (Abstract)

Lehmann O (1933) Morphologische Theorie der Verwitterung an Steinschlagwänden. Vierteljahresschr Naturforsch Ges Zür 78:83 – 126

Lehre AK (1982) Sediment budget of a small coastal range drainage basin in north-central California. In: Swanson FJ, Janda RJ, Dunne T, Swanston DN (eds) Sediment budgets and routing in forested drainage basins. US For Serv Pac Northwest For Range Exp St Gen Techn Rep PNW-141, pp 67 – 77

Leighly JB (1936) Meandering arroyos of the dry Southwest. Geogr Rev 26:270 – 282

Leopold LB (1962) Rivers. Am Sci 50:511 – 537

Leopold LB, Langbein WB (1962) The concept of entropy in landscape evolution. US Geol Surv Prof Pap 500-A

Leopold LB, Wolman MG, Miller JP (1964) Fluvial processes in geomorphology. Freeman, San Francisco

Lerman A (1979) Geochemical processes: water and sediment environments. Wiley, New York

Lerman A, Mackenzie FT, Garrels RM (1975) Modeling of geochemical cycles: phosphorus as an example. In: Whitten EHT (ed) Quantitative studies in the geological sciences. A memoir in honor of William C. Krumbein. The Geological Society of America Memoir 142, Boulder, Colorado, pp 205 – 218

Lewis WV, Miller MM (1955) Kaolin model glaciers. J Glaciol 2:533 – 538

Li Y-H (1981) Geochemical cycles of elements and human perturbation. Geochim Cosmochim Acta 45:2073 – 2084

Likens GE, Bormann FH, Pierce RS, Eaton JS, Johnson NM (1977) Biogeochemistry of a forested ecosystem. Springer, Berlin Heidelberg New York

Lindeman RL (1942) The trophic-dynamic aspect of ecology. Ecology 23:399 – 418

Lindgren BW (1975) Basic ideas of statistics. Macmillan, New York

Lindgren BW (1976) Statistical theory, 3rd edn. Macmillan, New York

Lloyd EH (1967) Stochastic reservoir theory. Adv Hydrosci 4:281 – 335

Lorenz EN (1968) Climatic determinism. Meteorological Monographs 8:1 – 3

Lotka AG (1925) Elements of physical biology. Williams and Wilkins, Baltimore

Lovelock JE (1979) Gaia: a new look at life on Earth. Oxford University Press, Oxford

Luk S-H (1982) Variability of rainwash erosion within small sample areas. In: Thorn CE (ed) Space and time in geomorphology. George Allen and Unwin, London, pp 243 – 268

Lustig LK (1965) Clastic sedimentation in Deep Springs Valley, California. US Geol Surv Prof Pap 352-F

MacKay JR (1965) Glacier flow and analogue simulation. Geogr Bull 7:1 – 6

Mackenzie FT, Garrels RM (1966) Chemical mass balance between rivers and oceans. Am J Sci 264:507 – 525

Mahaffy MW (1976) A three-dimensional numerical model of ice sheets: tests on the Barnes Ice Cap, Northwest Territories. J Geophys Res 81:1059 – 1066

Manabe S, Stouffer RJ (1980) Sensitivity of a global climate model to an increase of CO_2 concentration in the atmosphere. J Geophys Res 85:5529 – 5554

Mansell RS, Selim HM, Fiskell JGA (1977) Simulated transformations and transport of phosphorus in soil. Soil Sci 124:102 – 109

March L, Steadman P (1974) The geometry of environment. Massachusetts Institute of Technology Press, Cambridge, Massachusetts

Matalas NC (1963) Autocorrelation of rainfall and streamflow minimums. US Geol Surv Prof Pap 434-B

Matalas NC (1967) Time series analysis. Water Resour Res 3:817 – 829

Mather PM (1981) Factor analysis. In: Wrigley N, Bennett RJ (eds) Quantitative geography: a British view. Routledge and Kegan Paul, London, pp 144 – 150

McClellan PH (1984) Earthquake seasonality before the 1906 San Francisco earthquake. Nature 307:153 – 156

McConnell H, Horn JM (1972) Probabilities of surface karst. In: Chorley RJ (ed) Spatial analysis in geomorphology. Methuen, London, pp 111 – 134

McCullagh MJ, Hardy NE, Lockman WO (1972) Formation and migration of sand dunes: a simulation of their effect in the sedimentary environment. In: Merriam DF (ed) Mathematical models of sedimentary processes. Plenum, New York, pp 175 – 190

McIntosh RP (1980) The background and some current problems of theoretical ecology. In: Saarinen E (ed) Conceptual issues in ecology. Reidel, Dordrecht, pp 1 – 61

McLeod AI (1977) Improved Box-Jenkins estimation. Biometrika 64:531 – 534

McLeod AI, Hipel KW (1978) Developments in monthly autoregressive modelling. Technical Report No 45-XM-011178, Department of Systems Design Engineering, University of Waterloo, Waterloo, Ontario, Canada

McLeod AI, Hipel KW, Lennox WC (1977) Advances in Box-Jenkins modelling, part two, applications. Water Resour Res 13:577 – 586

McQuitty LL (1957) Elementary linkage analysis for isolating orthogonal and oblique types and typal relevancies. Educ Psychol Measur 17:207 – 229

Meijering JL (1953) Interface area, edge length, and number of vertices in crystal aggregates with random nucleation. Phillips Res Rep 8:270 – 290

Melhorn WN, Edgar DE (1975) The case for episodic, continental-scale erosion surfaces: a tentative geodynamic model. In: Melhorn WN, Flemal RC (eds) Theories of landform development. George Allen and Unwin, London, pp 243 – 276

Melton MA (1958a) Geometric properties of mature drainage systems and their representation in E_4 phase space. J Geol 66:35 – 54
Melton MA (1958b) Correlation structure of morphometric properties of drainage systems and their controlling agents. J Geol 66:442 – 460
Melton MA (1960) Intravalley variation in slope angles related to microclimate and erosional environment. Bull Geol Soc Am 71:133 – 144
Melton MA (1962) Methods for measuring the effect of environmental factors on channel properties. J Geophys Res 67:1485 – 1490
Mercer JW, Faust CR (1981) Ground-water modeling. National Water Well Association, Worthington, Ohio
Meybeck M (1982) Carbon, nitrogen, and phosphorus transport by world rivers. Am J Sci 282:401 – 450
Miall AD (1973) Markov chain analysis applied to an ancient alluvial plain succession. Sedimentology 20:347 – 364
Miall AD (1977) A review of the braided-river depositional environment. Earth Sci Rev 13:1 – 62
Miles RE (1964a) Random polygons determined by random lines in a plane. I. Proc Nat Acad Sci (USA) 52:901 – 907
Miles RE (1964b) Random polygons determined by random lines in a plane. II. Proc. Nat Acad Sci (USA) 52:1157 – 1160
Miles RE (1970) On the homogeneous planar Poisson point process. Math Biosci 6:85 – 127
Miles RE (1971) Random points, sets, and tessellations on the surface of a sphere. Sankhya, A33:145 – 174
Mills HH, Starnes DD (1983) Sinkhole morphometry in a fluviokarst region: eastern Highland Rim, Tennessee, USA. Z Geomorph 27:39 – 54
Milne G (1935a) Composite units for the mapping of complex soil associations. Trans Third Int Congr Soil Sci 1:345 – 347
Milne G (1935b) Some suggested units of classification and mapping particularly for East African soils. Soil Res, Berlin, 4:183 – 198
Mitchell CW, Willimott SG (1974) Dayas of the Moroccan Sahara and other arid regions. Geogr J 140:441 – 453
Miyashiro A, Aki K, Celâl Sengör AM (1982) Orogeny. Wiley, Chichester
Morel-Seytoux HJ, Salas JD, Sanders TG, Smith RE (eds) (1979) Modeling hydrologic processes. Proceedings of the Fort Collins Third International Hydrology Symposium on Theoretical and Applied Hydrology, Water Resources Publications, Fort Collins, Colorado
Morisawa M (1975) Tectonics and geomorphic models. In: Melhorn WN, Flemal RC (eds) Theories of landform development. George Allen and Unwin, London, pp 199 – 216
Morison CGT, Hoyle AC, Hope-Simpson JF (1948) Tropical soil-vegetation catenas and mosaics. A study in the south-western part of the Anglo-Egyptian Sudan. J Ecol 36:1 – 84
Mörner N-A (1983) Sea levels. In: Gardner R, Scoging H (eds) Mega-geomorphology. Clarendon, Oxford, pp 73 – 91
Mosley MP (1981) Semi-determinate hydraulic geometry of river channels, South Island, New Zealand. Earth Surface Processes Landforms 6:127 – 137
Mosley MP, Zimpfer GL (1978) Hardware models in geomorphology. Progr Phys Geogr 2:438 – 461
Moultrie W (1970) Systems, computer simulation, and drainage basins. Bull Ill Geogr Soc 12:29 – 35
Mulcahy MJ, Churchward HM, Dimmock GM (1972) Landforms and soils on an uplifted peneplain in the Darling Range, Western Australia. Austr J Soil Res 10:1 – 14
Nash DB (1980a) Forms of bluffs degraded for different lengths of time in Emmet County, Michigan, U.S.A. Earth Surface Processes 5:331 – 345
Nash DB (1980b) Morphologic dating of degraded normal fault scarps. J Geol 88:353 – 360
Nash DB (1981) Fault: a FORTRAN program for modeling the degradation of active normal fault scarps. Comput Geosci 7:249 – 266
Nieuwenhuis JD, van den Berg JA (1971) Slope investigations in the Morvan (Haut Folin area). Rev Géomorph Dyn 20:161 – 176
Nikiforoff CC (1959) Reappraisal of the soil. Science 129:186 – 196
Nordin CF, Algert JH (1966) Spectral analysis of sand waves. Proceedings of the American Society of Cicil Engineers, J Hydaul Div 92:95 – 114
Nye JF (1951) The flow of glaciers and ice sheets as a problem in plasticity. Proc R Soc London A207:554 – 572

Nye JF (1957) The distribution of stress and velocity in glaciers and ice sheets. Proc R Soc London A239:113 – 133
Nye JF (1958) A theory of wave formation on glaciers. International Assoc Sci Hydrol Publ 47:139 – 154
Nye JF (1959) The motion of ice sheets and glaciers. J Glaciol 3:493
Nye JF (1960) The response of glaciers and ice sheets to seasonal and climatic changes. Proc R Soc London A256:559 – 584
Nye JF (1961) The influence of climatic variations on glaciers. Int Assoc Sci Hydrol Publ 54:397 – 404
Nye JF (1963a) On the theory of the advance and retreat of glaciers. Geophys J R Astronom Soc 7:431 – 456
Nye JF (1963b) The response of a glacier to changes in the rate of nourishment and wastage. Proc R Soc London A275:87 – 112
Nye JF (1965a) The frequency response of glaciers. J Glaciol 5:567 – 587
Nye JF (1965b) A numerical method of inferring the budget history of a glacier from its advance and retreat. J Glaciol 5:589 – 607
Nye JF (1969) The effect of longitudinal stress on the shear stress at the base of an ice sheet. J Glaciol 8:207 – 213
Odum HT (1960) Ecological potential and analogue circuits for the ecosystem. Am Sci 48:1 – 8
Odum HT (1971) Environment, power, and society. Wiley, London
Odum HT (1983) Systems ecology: an introduction. Wiley, New York
Odum HT, Pinkerton RC (1955) Time's speed regulator: the optimum efficiency for maximum power output in physical and biological systems. Am Sci 43:331 – 343
Ollier CD (1981) Tectonics and landforms. Longman, London
Olson JS (1963) Energy storage and the balance of producers and decomposers in ecological systems. Ecology 44:322 – 331
Öpik EJ (1958) On the catastrophic effects of collisions with celestial bodies. Ir Astronom J 5:34 – 36
Öpik EJ (1973) Our cosmic destiny. Ir Astronom J 11:113 – 124
Owen HG (1976) Continental displacement and expansion of the Earth during the Mesozoic and Cenozoic. Philos Trans R Soc, London A281:223 – 291
Owen HG (1981) Constant dimensions or an expanding Earth? In: Cocks LRM (ed) The evolving Earth. Cambridge University Press, Cambridge, pp 179 – 192
Palmer J (1956) Tor formation at the Bridestones in north-east Yorkshire and its significance in relation to problems of valley-side development and regional glaciation. Trans Inst Br Geogr 22:55 – 71
Palmquist RC (1979) Geologic controls on doline characteristics in mantled karst. Z Geomorph Suppl 32:90 – 106
Parkhurst DL, Thorstenson DC, Plummer LN (1980) PHREEQE – A computer program for geochemical calculations. US Geol Surv Water-Resour Invest 80 – 96
Parkhurst DL, Plummer LN, Thorstenson DC (1982) BALANCE – A computer program for calculating mass transfer for geochemical reactions in ground water. US Geol Surv Water-Resour Invest 82 – 14
Paterson WSB (1972) Laurentide ice sheet: estimated volumes during late Wisconsin. Rev Geophys Space Phys 10:885 – 917
Paterson WSB (1981) The physics of glaciers, 2nd edn. Pergamon, Oxford
Patten BC (1971) A primer for ecological modelling and simulation with analog and digital computers. In: Patten BC (ed) Systems analysis and simulation in ecology, vol 1. Academic, New York pp 3 – 121
Pearson K (1901) On the lines and planes of closest fit to systems of points in space. Philos Mag 6:320 – 331
Peltier LC (1950) The geographical cycle in periglacial regions as it is related to climatic geomorphology. Ann Assoc Am Geogr 40:214 – 236
Peltier LC (1975) The concept of climatic geomorphology. In: Melhorn WN, Flemal RC (eds) Theories of landform development. George Allen and Unwin, London, pp 129 – 143
Penck W (1924) Die morphologische Analyse, ein Kapitel der physikalischen Geologie. Engelhorn, Stuttgart (Translated and edited in 1953 as Czech H, Boswell KC, Morphological analysis of landforms. Macmillan, London)
Pickup G (1975) Downstream variations in morphology, flow conditions and sediment transport in an eroding channel. Z Geomorph 19:443 – 459

Pickup G (1977) Simulation modelling of river channel erosion. In: Gregory KJ (ed) River channel changes. Wiley, Chichester, pp 47–60

Pickup G, Chewings VH (1983) The hydrology of the Purari and its environmental implications. In: Petr T (ed) The Purari – Tropical environment of a high rainfall river basin. Junk, The Hague, pp 123–139

Pickup G, Rieger WA (1979) A conceptual model of the relationship between channel characteristics and discharge. Earth Surface Processes 4:37–42

Pickup G, Higgins RJ, Grant I (1983) Modelling sediment transport as a moving wave – the transfer and deposition of mining waste. J Hydrol 60:281–301

Pike RJ, Rozema WJ (1975) Spectral analysis of landforms. Ann Assoc Am Geogr 65:499–516

Pitman WC (1978) Relationship between eustacy and stratigraphic sequences of passive margins. Bull Geol Soc Am 89:1289–1403

Pitty AF (1979) Conclusions. In: Pitty AF (ed) Geographical approaches to fluvial processes. Geo Books, Norwich, pp 261–280

Pitty AF (1982) The nature of geomorphology. Methuen, London

Playfair J (1802) Illustrations of the Huttonian theory of the Earth. W. Creech, Edinburgh

Plummer LN, Parkhurst DL, Thorstenson DC (1983) Development of reaction models for groundwater systems. Geochim Cosmochim Acta 47:665–686

Pollack HN (1969) A numerical model of the Grand Canyon. In: Baars DC (ed) Geology and natural history of the Grand Canyon region. Four Corners Geological Society Guidebook to Fifth Field Conference, pp 61–62

Potter PE, Blakely RF (1968) Random processes and lithologic transitions. J Geol 76:154–170

Powell RL (1975) Theories of the development of karst topography. In: Melhorn WN, Flemal RC (eds) Theories of landform development. George Allen and Unwin, London, pp 217–242

Price RK (1973) Flood routing methods for British rivers. Proc Inst Civ Eng 55:913–930

Price WE (1976) A random-walk simulation model of alluvial-fan deposition. In: Merriam DF (ed) Random processes in geology. Springer, Berlin Heidelberg New York, pp 55–62

Prigogine I (1947) Étude thermodynamique des phénomènes irréversibles. Dunod, Paris

Prigogine I (1980) From being to becoming: time and complexity in the physical sciences. Freeman, San Francisco

Prior DB, Stephens N, Douglas GR (1971) Some examples of mudflow and rockfall activity in north-east Ireland. In: Brunsden D (ed) Slopes: form and process. Inst Br Geogr Spec Publ No 3, pp 129–140

Quimpo RG (1968) Autocorrelation and spectral analyses in hydrology. Proceedings of the American Society of Civil Engineers, J Hydraul Div 94:363–373

Rao AR (1980) Stochastic analysis of thresholds in hydrologic time series. In: Coates DR, Vitek JD (eds) Thresholds in geomorphology. George Allen and Unwin, London, pp 179–208

Rasmussen LA, Campbell WJ (1973) Comparison of three contemporary flow laws in a three-dimensional, time-dependent glacier model. J Glaciol 12:361–373

Raudkivi AJ (1982) Hydrology: An advanced introduction to hydrological processes and modelling. Pergamon, Oxford

Raup DM (1981) Introduction: what is a crisis? In: Nitecki MH (ed) Biotic crises in ecological and evolutionary time. Academic, New York, pp 1–12

Rayner JH (1966) Classification of soils by numerical methods. J Soil Sci 17:79–92

Reeh N (1982) A plasticity theory approach to the steady-state shape of a three-dimensional glacier. J Glaciol 28:431–455

Reichmann WJ (1961) Use and abuse of statistics. Methuen, London

Reid JM, Macleod DA, Cresser MS (1981) The assessment of chemical weathering rates within an upland catchment in north-east Scotland. Earth Surface Processes Landforms 6:447–457

Reyment RA (1969) Statistical analysis of some volcanological data regarded as a series of point events. Pure Appl Geophys 74:57–77

Reyment RA (1976) Analysis of volcanic earthquakes of Asamayama (Japan). In: Merriam DF (ed) Random processes in geology. Springer, Berlin Heidelberg New York, pp 87–95

Rich PH (1984) Trophic-detrital interactions: vestiges of ecosystem evolution. Am Nat 123:20–29

Richards KS (1976) The morphology of riffle-pool sequences. Earth Surface Processes 1:71–88

Richards KS (1982) Rivers: form and process in alluvial channels. Methuen, London

Robinson G, Petersen JA, Anderson PM (1971) Trend surface analysis of corrie altitudes in Scotland. Scott Geogr Mag 87:142–146

References

Rode AA (1961) The soil-forming process and soil evolution. Israel Programme for Scientific Translation, Jerusalem

Romesburg HC, DeGraff JV (1982) Using the normal generated distribution to analyze spatial and temporal variability in geomorphic processes. In: Thorn CE (ed) Space and time in geomorphology. George Allen and Unwin, London pp 315–325

Rose AW, Hawkes HE, Webb JS (1979) Geochemistry in mineral exploration, 2nd edn. Academic, London

Roy AG, Jarvis RS, Arnett RR (1980) Soil-slope relationships within a drainage basin. Ann Assoc Am Geogr 40:214–236

Ruhe RV (1960) Elements of the soil landscape. Trans Seventh Int Congr Soil Sci, Madison, 4:165–170

Ruhe RV, Walker PH (1968) Hillslope models and soil formation. I. Open systems. Trans Ninth Int Congr Soil Sci, Adelaide, 4:551–560

Runge ECA (1973) Soil development sequences and energy models. Soil Sci 115:183–193

Rushton KR, Redshaw SC (1979) Seepage and groundwater flow. Wiley, New York

Salter PJ, Williams JB (1967) The influence of texture on the moisture characteristics of a sandy loam soil. J Soil Sci 18:174–181

Sasscer DS, Jordan CF, Kline JR (1971) Mathematical model of tritiated and stable water movement in an old-field ecosystem. In: Nelson DJ (ed) Radionuclides in ecosystems. Proceedings of the Third National Symposium on Radioecology, CONF-710501-P1, United States Atomic Energy Commission, pp 915–923

Savigear RAG (1952) Some observations on slope development in South Wales. Trans Inst Br Geogr 18:31–52

Savigear RAG (1965) A technique of morphological mapping. Ann Assoc Am Geogr 55:514–538

Scheidegger AE (1954) Statistical hydrodynamics in porous media. Appl Phys J 25:994–1001

Scheidegger AE (1960) Analytical theory of slope development by undercutting. J Alberta Soc Petrol Geol 8:202–206

Scheidegger AE (1961a) On the statistical properties of some transport equations. Can J Phys 39:1573–1582

Scheidegger AE (1961b) Mathematical models of slope development. Bull Geol Soc Am 72:37–49

Scheidegger AE (1961c) General theory of dispersion in porous media. J Geophys Res 66:3273–3278

Scheidegger AE (1964a) Some implications of statistical mechanics in geomorphology. Bull Int Assoc Sci Hydrol 9:12–16

Scheidegger AE (1964b) Lithological variations in slope development theory. US Geol Surv Circ No 485

Scheidegger AE (1967a) A stochastic model for drainage patterns into an intramontane trench. Bull Int Assoc Sci Hydrol 12:15–20

Scheidegger AE (1967b) A thermodynamic analogy for meander systems. Water Resour Res 3:1041–1046

Scheidegger AE (1970) Theoretical geomorphology, 2nd edn. Springer, Berlin Heidelberg New York

Scheidegger AE (1979) The principle of antagonism in the Earth's evolution. Tectonophysics 55:T7–T10

Scheidegger AE (1981) The stress-field in the Alpine-Mediterranean region. Geophys Surv 4:233–253

Scheidegger AE (1983) Instability principle in geomorphic equilibrium. Z Geomorph 27:1–19

Scheidegger AE, Langbein WB (1966) Probability concepts in geomorphology. US Geol Surv Prof Pap 500-C

Scheidegger AE, Padale JG (1982) A geodynamic study of peninsular India. Rock Mech 15:209–241

Schenck H (1963) Simulation of the evolution of drainage basin networks with a digital computer. J Geophys Res 68:5739–5745

Schilling W (1947) A frequency distribution represented as the sum of two Poisson distributions. J Am Stat Assoc 42:407–424

Schlesinger WH (1977) Carbon balance and terrestrial detritus. Annu Rev Ecol Syst 8:51–81

Schrödinger E (1944) What is life? The physical aspect of the living cell. Cambridge University Press, London

Schumm SA (1956) The evolution of drainage basin systems and slopes in badands at Perth Amboy, New Jersey. Bull Geol Soc Am 67:597–646

Schumm SA (1973) Geomorphic thresholds and complex response of drainage systems. In: Morisawa M (ed) Fluvial geomorphology. State University of New York, Binghampton, Publications in Geomorphology, pp 299–310

Schumm SA (1977) The fluvial system. Wiley, New York

Schumm SA, Lichty RW (1965) Time, space, and causality in geomorphology. Am J Sci 263:110–119

Schwarzacher W (1976) Stratigraphic implications of random sedimentation. In: Merriam DF (ed) Random processes in geology. Springer, Berlin Heidelberg New York, pp 96–111

Selby MJ (1982) Hillslope materials and processes. Oxford University Press, Oxford

Sellers WD (1973) A new global climatic model. J Atmos Sci 12:241–254

Sellers WD (1976) A two-dimensional global climate model. Mon Weather Rev 104:233–248

Shakal AF, Toksöz MN (1977) Earthquake hazard in New England. Science 195:171–173

Shane, RM, Lynn WR (1964) Mathematical model for flood risk evaluation. Proceedings of the American Society of Civil Engineers, J Hydraul Div 90:1–20

Sharpe CFS (1938) Landslides and related phenomena. Columbia University Press, New York

Shaw CF (1930) Potent factors in soil formation. Ecology 11:239–245

Shepherd RG, Schumm SA (1974) Experimental study of river incision. Bull Geol Soc Am 85:257–268

Shreve RL (1966) Statistical law of stream numbers. J Geol 74:17–37

Shreve RL (1967) Infinite topologically random channel networks. J Geol 75:178–186

Shreve RL (1969) Stream lengths and basin areas in topologically random channel networks. J Geol 77:397–414

Shumskiy PA (1975) Mechanisms and causes of glacier variations. Int Assoc Hydrol Sci Publ 104:318–332

Sibson R (1985) Multidimensional scaling. Wiley, Chichester

Sillén LG (1961) The physical chemistry of sea water. In: Sears M (ed) Oceanography. Am Assoc Adv Sci, Washington, D.C., pp 549–581

Sillén LG (1963) How the sea got its present composition. Svensk Kemisk Tidskrift 75:161–177

Sillén LG (1967) The ocean as a chemical system. Science 156:1189–1197

Silver LT, Schultz PH (eds) (1982) Geological implications of impacts of large asteroids and comets on the Earth. Geol Soc Am Spec Pap 190

Simonson RW (1959) Outline of a generalized theory of soil genesis. Proc Soil Sci Soc Am 23:152–156

Simonson RW (1978) A multiple-process model of soil genesis. In: Mahaney WC (ed) Quaternary soils. Geo Abstracts, Norwich, pp 1–25

Simpson GG (1952) Probabilities of dispersal in geologic time. Bull Am Mus Nat Hist 99:163–176

Singh VP (1979) A linear dynamic model for prediction of surface runoff. In: Morel-Seytoux HJ, Salas JD, Sanders TG, Smith RE (eds) Modeling hydrologic processes. Water Resources Publications, Fort Collins, Colorado, pp 369–384

Skempton AW (1964) The long-term stability of clay slopes. Géotechnique 14:77–104

Slingerland R (1981) Qualitative stability analysis of geologic systems with an example of river hydraulic geometry. Geology 9:491–493

Small RJ, Clark MJ (1982) Slopes and weathering. Cambridge University Press, Cambridge

Smalley IJ (1966) Contraction crack networks in basalt flows. Geol Mag 103:110–114

Smart JS, Surkan AJ, Considine JP (1967) Digital simulation of channel networks. International Association of Scientific Hydrology, General Assembly of Berne, September-October 1967, Symposium on River Morphology, pp 87–98

Smith DI, Atkinson TC (1976) Process, landforms and climate in limestone regions. In: Derbyshire E (ed) Geomorphology and climate. Wiley, Chichester, pp 367–409

Smith TR, Bretherton FP (1972) Stability and the conservation of mass in drainage basin evolution. Water Resour Res 8:1506–1529

Smith WD (1978) Earthquake risk in New Zealand: statistical estimates. N Z J Geol Geophys 21:313–327

So CL (1974) Some coast changes around Aberystwyth and Tanybwlch. Trans Inst Br Geogr 62:115–128

Solohub JT, Klovan JE (1970) Evaluation of grain-size parameters in lacustrine environments. J Sediment Petrol 40:81–101

Solomon SC, Sleep NH, Richardson RM (1975) On the forces driving plate tectonics: inferences from absolute plate velocities and intraplate stress. Geophys J R Astronom Soc 42:769–801

Solomon SC, Sleep NH, Jurdy DM (1977) Mechanical models for absolute plate motions in the Early Tertiary. J Geophys Res 82:203–212

Soni JP, Garde RJ, Ranga Raju KG (1980) Aggradation in streams due to overloading. Proceedings of the American Society of Civil Engineers, J Hydraul Div 106:117-132
Sorenson HW (1970) Least squares estimation from Gauss to Kalman. IEEE Spectrum 7:63-68
Southam RJ, Hay WW (1977) Time scales and dynamic models of deep-sea sedimentation. J Geophys Res 82:3825-3842
Speight JG (1965) Meander spectra of the Angabunga river, Papua. J Hydrol 3:1-15
Speiss FN et al. (1980) East Pacific Rise: hot springs and geophysical experiments. Science 207:1421-1433
Sprunt B (1972) Digital simulation of drainage basin development. In: Chorley RJ (ed) Spatial analysis in geomorphology. Methuen, London, pp 371-389
Squarer D (1970) Friction factors and bedforms in fluvial channels. Proceedings of the American Society of Civil Engineers, J Hydraul Div 96:995-1017
Starkel L (1976) The role of extreme (catastrophic) meteorological events in contemporary evolution of slopes. In: Derbyshire E (ed) Geomorphology and climate. Wiley, Chichester, pp 203-246
Starkel L (1979) The role of extreme meteorological events in the shaping of mountain relief. Geogr Pol 41:13-20
Statham I (1975) Slope instabilities and recent slope development in Glencullen, County Wicklow. Ir Geogr 8:42-54
Statham I (1977) Earth surface sediment transport. Clarendon, Oxford
Stephens FR (1969) Source of cation exchange capacity and water retention in southeast Alaskan Spodosols. Soil Sci 108:429-431
Stoddart DR (1969) Climatic geomorphology. In: Chorley RJ (ed) Water, Earth, and Man. Methuen, London, pp 473-485
Strahler AN (1950) Equilibrium theory of erosional slopes, approached by frequency distribution analysis. Am J Sci 248:673-696 and 800-814
Strahler AN (1952a) Dynamic basis of geomorphology. Bull Geol Soc Am 63:923-938
Strahler AN (1952b) Hypsogeometric (area-altitude) analysis of erosional topography. Bull Geol Soc Am 63:1117-1142
Strahler AN (1956) Quantitative slope analysis. Bull Geol Soc Am 67:571-596
Strahler AN (1958) Dimensional analysis applied to fluvially eroded landforms. Bull Geol Soc Am 69:279-300
Strahler AN (1980) Systems theory in physical geography. Phys Geogr 1:1-27
Strahler AN, Strahler AH (1973) Environmental geoscience. Hamilton, Santa Barbara, California
Strahler AN, Strahler AH (1974) Introduction to environmental science. Hamilton, Santa Barbara, California
Sudgen DE, Hamilton P (1971) Scale, systems and regional geography. Area 3:139-144
Sumner GN (1978) Mathematics for physical geographers. Edward Arnold, London
Sunamura T (1975) A laboratory study of wave-cut platform formation. J Geol 83:389-397
Surkan AJ, Kan J van (1969) Constrained random walk meander generation. Water Resour Res 5:1343-1352
Sweeting MM (ed) (1981) Karst geomorphology. Benchmark Papers in Geology, Vol 59, Hutchinson Ross, Stroudsburg, Pennsylvania
Takeshita K (1963) Theoretical analysis of slope evolution based on laboratory experiments and relative consideration. Bull Fukuoka-ken For Exp St 16:115-136
Tamburi AJ (1974) Creep of single rocks on bedrock. Bull Geol Soc Am 85:351-356
Tansley AG (1935) The use and abuse of vegetational concepts and terms. Ecology 16:284-307
Terjung WH (1976) Climatology for geographers. Ann Assoc Am Geogr 66:199-222
Thakur TR, Scheidegger AE (1968) A test for the statistical theory of meander formation. Water Resour Res 4:317-329
Thakur TR, Scheidegger AE (1970) Chain model of river meanders. J Hydrol 12:25-47
Thom R (1975) Structural stability and morphogenesis. Benjamin, New York
Thomas RW (1977) An introduction to quadrat analysis. Concepts and Techniques in Modern Geography No 12, Geo Abstracts, Norwich
Thomas RW (1981) Point pattern analysis. In: Wrigley N, Bennett RJ (eds) Quantitative geography: a British view. Routledge and Kegan Paul, London, pp 164-176
Thomas RW, Huggett RJ (1980) Modelling in geography: a mathematical approach. Harper and Row, London

Thornes JB (1972) Debris slopes as series. Arct Alp Res 4:337 – 342
Thornes JB (1973) Markov chains and slope series: the scale problem. Geogr Ann 5:322 – 328
Thornes JB (1979) Processes and interrelationships, rates and changes. In: Embleton C, Thornes JB (eds) Processes in geomorphology. Edward Arnold, London, pp 378 – 387
Thornes JB (1980) Structural instability and ephemeral channel behaviour. Z Geomorph 36:233 – 244
Thornes JB (1982) Problems in the identification of stability and structure from temporal data series. In: Thorn CE (ed) Space and time in geomorphology. George Allen and Unwin, London, pp 327 – 353
Thornes JB (1983) Evolutionary geomorphology. Geography 68:225 – 235
Thornes JB, Ferguson RI (1981) Geomorphology. In: Wrigley N, Bennett RJ (eds) Quantitative geography: a British view. Routledge and Kegan Paul, London, pp 284 – 293
Thornes JB, Jones DKC (1969) Regional and local components in the physiography of the Sussex Weald. Area 1:13 – 21
Torgerson WS (1952) Multidimensional scaling. I. Theory and method. Psychometrika 17:401 – 419
Torgerson WS (1958) Theory and methods of scaling. Wiley, New York
Toth J (1962) A theory of groundwater motion in small drainage basins in central Alberta, Canada. J Geophys Res 67:4375 – 4387
Trenhaile AS (1979) The morphometry of valley steps in the Canadian Cordillera. Z Geomorph 23:27 – 44
Tricart J, Cailleux A (1972) Introduction to climatic geomorphology. Longman, London (Translated from the French original of 1965 by C. Kiewiet de Jonge)
Trimble SW (1981) Changes in sediment storage in the Coon Creek Basin, Driftless Area, Wisconsin, 1853 – 1975. Science 214:181 – 183
Trimble SW (1983) A sediment budget for Coon Creek basin in the driftless area, Wisconsin. Am J Sci 283:454 – 474
Unwin DJ (1981) Introductory spatial analysis. Methuen, London
Van Bemmelen RW (1967) The importance of geonomic dimensions for geodynamic concepts. Earth-Sci Rev 3:79 – 110
Van Dijk W, Le Heux JWN (1952) Theory of parallel rectilinear slope recession. K Ned Akad Wet Amsterdam B55:115 – 129
Van Dyne GM (1980) Reflections and projections. In: Breymeyer AI, Van Dyne GM (eds) Grasslands, systems analysis and man. Cambridge University Press, Cambridge, pp 881 – 921 (Int Biol Programme 19)
Velikovsky I (1950) Worlds in collision. Victor Gollancz, London
Vincent PJ, Clarke V (1980) Terracette morphology and soil properties: a note on a canonical correlation study. Earth Surface Processes 5:291 – 295
Vincent PJ, Howarth JM, Griffiths JG, Collins R (1976) The detection of randomness in plant patterns. J Biogeogr 3:373 – 380
Vistelius AB (1949) On the question of the mechanism of formation of strata. Dokl Akad Nauk SSSR 65:191 – 194
Vreeken WJ (1973) Soil variability in small loess watersheds: clay and organic matter content. Catena 1:181 – 196
Vreeken WJ (1975) Variability of depth to carbonates in fingertip loess watersheds in Iowa. Catena 2:321 – 336
Waddington CH (1977) Tools for throught. Paladin, St. Albans
Walker PH (1966) Postglacial environments in relation to landscape and soils on the Cary drift, Iowa. Iowa State Univ Exp St Res Bull 549:838 – 875
Walker PH, Ruhe RV (1968) Hillslope models and soil formation. 2. Closed systems. Trans Ninth Int Congr Soil Sci, Adelaide, 4:561 – 568
Walker PH, Hall GF, Protz R (1968a) Soil trends and variability across selected landscapes in Iowa. Proc Soil Sci Soc Am 32:97 – 101
Walker PH, Hall GF, Protz R (1968b) Relation between landform parameters and soil properties. Proc Soil Sci Soc Am 32:101 – 104
Walling DE, Webb BW (1975) Spatial variation of river water quality: a survey of the River Exe. Trans Inst Br Geogr 65:155 – 171
Wang HF, Anderson MP (1982) Introduction to groundwater modeling: finite difference and finite element methods. Freeman, San Francisco

References

Watson KK, Curtis AA (1975) Numerical analysis of vertical water movement in a bounded profile. Aust J Soil Res 13:1–11

Weaver W (1958) A quarter century in the natural sciences. Annual Report of the Rockefeller Foundation, New York, pp 7–122

Webster R (1977) Canonical correlation in pedology: how useful? J Soil Sci 28:196–221

Webster R (1979) Exploratory and descriptive uses of multivariate analysis in soil survey. In: Wrigley N (ed) Statistical applications in the spatial sciences. Pion, London, pp 286–306

Webster R, Butler BE (1976) Soil classification and survey studies at Ginnindera. Aust J Soil Res 14:1–24

Weertman J (1958) Travelling waves on glaciers. Int Assoc Sci Hydrol Publ 47:162–168

Weertman J (1961) Equilibrium profile of ice caps. J Glaciol 3:953–964

Weisskopf VF (1975) Of atoms, mountains, and stars: a study of qualitative physics. Science 187:605–612

Werner C (1971) Expected number and magnitudes of stream networks in random drainage patterns. Proc Assoc Am Geogr 3:181–185

Whipkey RZ (1965) Subsurface stormflow from forested slopes. Bull Int Assoc Sci Hydrol 10:74–85

Whistler FD, Klute A (1967) Rainfall infiltration into a vertical soil column. Trans Am Soc Agric Eng 10:391–395

Whitfield WAD, Furley PA (1971) The relationship between soil patterns and slope form in the Ettrick Association, south-east Scotland. In: Brunsden D (ed) Slopes: from and process. Inst Br Geogr Spec Publ No 3, pp 165–167

Wickman FE (1976) Markov models of response-period patterns of volcanoes. In: Merriam DF (ed) Random processes in geology. Springer, Berlin Heidelberg New York, pp 135–161

Wilson AG (1981) Geography and the environment: systems analytical methods. Wiley, Chichester

Wolman MG, Gerson R (1978) Relative scales of time and effectiveness of climate in watershed geomorphology. Earth Surface Processes 3:189–208

Wolman MG, Miller JP (1960) Magnitude and frequency of forces in geomorphic processes. J Geol 68:54–74

Wood A (1942) The development of hillside slopes. Proc Geol Assoc, London, 53:128–139

Wurm A (1935) Morphologische Analyse und Experiment Hangentwicklung, Einebnung Piedmonttreppen. Z Geomorph (Old ser) 9:57–89

Yaalon DH (1960) Some implications of fundamental concepts of pedology in soil classification. Trans Seventh Int Congr Soil Sci, Madison, 4:119–123

Yaalon DH (1965) Downward movement and distribution of anions in soil profiles with limited wetting. In: Hallsworth EG, Crawford DV (eds) Experimental pedology. Butterworths, London, pp 157–164

Yaalon DH (1971) Soil-forming processes in time and space. In: Yaalon DH (ed) Paleopedology. International Society of Soil Science and Israel Universities Press, Jerusalem, pp 29–39

Yaalon DH (1975) Conceptual models in pedogenesis: can soil-forming functions be solved? Geoderma 14:189–205

Yaalon DH, Brenner I, Koyumdjisky H (1974) Weathering and mobility sequence of minor elements on a basaltic pedomorphic surface. Geoderma 12:233–244

Yalin MS (1977) Mechanics of sediment transport. Pergamon, Oxford

Yang CT (1972) Unit stream power and sediment transport. Proceedings of the American Society of Civil Engineers, J Hydraul Div 98:1805–1826

Yatsu E (1955) On the longitudinal profile of the graded river. Trans Am Geophys Union 36:655–663

Yevjevich V (1971) Properties of river flows of significance to river mechanics. In: Shen HW (ed) River mechanics. Colorado State University, Fort Collins, pp 1.1–1.28

Yevjevich V, Jeng RI (1969) Properties of non-homogeneous hydrologic series. Col State Univ Hydrol Pap 32, Fort Collins, Colorado

Young A (1963a) Deductive models of slope evolution. International Geographical Union, Slopes Commission Report No 3, Nachrichten Akad Wiss Göttingen, Math-Physik Klasse, Jahrg 1963, 5:45–66

Young A (1963b) Some field observations of slope form and regolith, and their relation to slope development. Trans Inst Br Geogr 32:1–29

Young A (1972) Slopes. Oliver and Boyd, Edinburgh

Yule GU (1927) On a method of investigating periodicities in disturbed series with special reference to Wolfer's sunspot numbers. Philos Trans R Soc A226:267–298

Yuretich RF, Cerling TE (1983) Hydrogeochemistry of Lake Turkana, Kenya: mass balance and minerals reactions in an alkaline lake. Geochim Cosmochim Acta 47:1099–1109

Zaghloul NA (1979) Stormwater routing models. In: Morel-Seytoux HJ, Salas JD, Sanders TG, Smith RE (eds) Modeling hydrologic processes. Water Resour Publ, Fort Collins, Colorado, pp 670–681

Zaslavsky D, Rogowski AS (1969) Hydrologic and morphologic implications of anisotropy and infiltration in soil profile development. Proc Soil Sci Soc Am 33:594–599

Subject Index

accounting model (see model)
addition axiom, for mutually exclusive events (see axiom)
Afon Elan, Wales (see river)
aggregate measures 5, 22
aggregate randomness (see randomness)
Aire river, England (see river)
Alberta, Canada 145
Aleutian Islands 200, 201
Aleut ecosystem (see ecosystem)
alluvial fan (see fan, alluvial)
analogue model (see model)
analysis of variance 102
Angabunga river, Papua New Guinea (see river)
ANSWERS model (see model)
antagonism, principle of 234
area patterns 31, 60
Argonne, Illinois 216
Ashland, Kentucky 51
Aswan Dam, Egypt 78
autocorrelation (see correlation)
autoregressive integrated moving-average (ARIMA) models (see model)
autoregressive moving-average (ARMA) models (see model)
axiom 49–51
  addition, for mutually exclusive events 49
  multiplication, for dependent or general events 50, 51
  multiplication, for independent events 50

balance equations (see equation)
Barnes Ice Cap, Baffin Island, Northwest Territories 140–142
basin hydrograph (see hydrograph)
Bayes's theorem 51
beach 5, 7, 39, 113, 117, 192, 225, 226
benches in mountain valleys 242
Bernoulli process 52
beta distribution (see distribution)
bifurcation 192, 222–226
bifurcation point 222

binomial coefficient 52
binomial distribution (see distribution)
binomial formula 52, 55
binomial processes 51
binomial theorem 51
biogeochemical cycle 4, 8, 23, 28, 206–221
biostasie 44
bistable system (see system)
Bollin river, Cheshire (see river)
boundary conditions (see system)
box-and-arrow diagram 9
Box and Jenkins's models (see model)
Brandywine Creek, Pennsylvania (see river)

canonical correlation (see correlation)
canonical structures 9, 32
canonical vector 129–133
capture model (see model)
carbon cycle 36, 200, 201
carbon cycle cascade (see cascade)
carbon dioxide 108, 155, 156, 206–214
cascade
  carbon cycle 35
  debris 33
  endogenic 35
  exogenic 35
  landslide 37
  land-surface 33
  sediment 35
  soil material 37
  solute 35
  stream channel 33
  valley glacier 35
  water 35
cascading system (see system)
catastrophe surface 223
catastrophe theory 204, 230
catena (see soil catena)
causal analysis 123
Caydale, Yorkshire 120
channel form variable (see variable)
Chézy equation (see equation)
chi-square test 102

circular distribution (see distribution)
cirques 242
cliff 3, 7, 37, 238
climatic geomorphology 10, 40, 41
"clorpt" equation (see equation)
closed system (see system)
coefficient of correlation
  (see correlation coefficient)
coefficient of determination 104, 109
coefficient of diffusion
  (see diffusion coefficient)
coefficient of recession
  (see recessional coefficient)
coefficient of retreat 93, 167
coefficients of regression (see regression)
collapse sink 56
compartment 198
compartment model (see model)
complementary event (see event)
complex disorder, system of (see system)
complex order, system of (see system)
compound event (see event)
conceptual model (see model)
conservation, laws of (see law)
constant of debris diffusion 169
continuity condition 24, 161
continuity equation (see equation)
continuity of sediment transport
  (see sediment transport)
control variable (see variable)
convective cell 26, 222
Coon Creek, Wisconsin (see river)
corrasion 40, 169, 170
corrasion model (see model)
correlation
  autocorrelation 85, 88, 92–95, 103, 236
  canonical 102, 112, 129–134
  cross-correlation 83, 96–100
  multiple 117, 121
  partial autocorrelation 85, 88, 92–95, 100, 236
  problems of 106, 107
  simple 30, 103–112
  spurious 106
correlation coefficient 104–106, 109, 122
correlation matrix 105, 122, 126, 129
correlogram 90–92, 100
CREAMS model (see model)
creep (see soil creep and rock creep)
Crickmay's model of landscape development 40
cross-correlation (see correlation)
cusp catastrophe 223, 224

Darcy's law (see law)
Davis's model of landscape development 9, 10, 39, 43, 239

debris cascade (see cascade)
debris-flow deposit 68
Dee catchment, North Wales 129
delta 7, 10, 23, 192
denudation 161–164
de Saint Venant equation (see equation)
deterministic model (see model)
  steps in building 136
Detroit river (see river)
difference equation (see equation)
differencing 84, 87
differential equation (see equation)
diffusion 20, 25, 152, 217, 220, 235
diffusion coefficient 25, 152, 169
diffusion-decay model (see model)
diffusivity equation (see equation)
digital terrain model (see model)
dimensional analysis 39
discrete component model of hillslopes
  (see model)
discharge 48, 76, 78, 81, 85, 88, 109, 111
dispersion 34, 35
dissipative structures 27, 198, 204, 222–232
distance series 12, 78, 88–95, 97–100
distribution
  beta 55
  binomial 52, 56
  circular 55
  double Poisson 56, 57
  exponential 55, 67
  extreme value 54
  gamma 55, 57
  geometric 52, 53
  hypergeometric 53
  lognormal 55
  negative binomial 53, 57
  negative exponential 65
  normal 55, 126
  normal generated 237
  Poisson 53, 54, 56
disturbed periodic model, of river meanders
  (see model)
doline 39, 56, 111
dominance domain 231, 232
double log function (see function)
double Poisson distribution (see distribution)
double Poisson model (see model)
drainage basin 31, 33–35, 37, 39, 43, 78, 81, 82, 114, 158
drainage basin simulation 185–189
drought 50, 54
drumlin 31
Dun Moss, Perthshire 147
Dunedin, New Zealand 78
Dupuit-Forchheimer assumptions 146
Dye river, northeast Scotland (see river)
dynamic equilibrium (see equilibrium)

Subject Index

dynamic geomorphology 39
dynamic variable (see variable)
dynamical system model (see model)

earthflow 18
earthquake 54, 55, 57, 65
ecosystem 5, 11, 23, 24, 28, 44
   Aleut 200, 201, 204, 205
   Galapagos rift, weird ecosystem in 8
   Lake Erie 218–221
   old-field, Argonne, Illinois 216
   raised mire 146, 147
   tropical rain forest, Puerto Rico 214–216
ecosystem development 42
electrical analogue model (see model)
elementary event (see event)
elevation series 92, 236
Emporium Quadrangle, Pennsylvania 114
endogenic cascade (see cascade)
endogenic processes 234
energy 4, 5, 21–23, 26
energy conservation, law of (see law)
energy flow (see flow)
entropy 20, 21, 25
entropy balance equation (see equation)
entropy function (see function)
entropy maximization 74, 75, 89
entropy, maximum 21, 75
entropy minimization 75, 76
entropy, minimum 26
entropy model (see model)
entropy transport 25, 26
environment (see system)
equation
   balance 22
   Chézy 144
   "clorpt" 42, 239
   continuity 23, 136–139, 142, 144, 145, 148, 149, 151, 157, 173, 174, 217, 233
   de Saint Venant 142
   difference 170, 179
   differential 17, 39, 90, 95, 143, 151, 198, 215, 223
   diffusivity 152
   entropy balance 22, 25
   Laplace 23, 145, 146, 153
   Manning 144
   momentum 142–144
   phenomenological 22, 24, 136
   physical-chemical state 22, 24
   Poisson 146, 153
   process 75, 76, 136, 173, 239
   reaction-diffusion 223
   sediment transport 235
   state 192
   transfer 201, 202
   transport 136, 173, 181

equilibrium
   dynamic 27, 76, 146, 183, 184
   metastable 203, 204
   multiple 241
   neutrally stable 203
   stable 203
   thermodynamic 20, 21, 26
   unstable (see also steady state) 203, 204
ergodic (space-time substitution)
   hypothesis 239
erosion (see also soil erosion) 40, 41, 43, 44, 56, 67, 69, 70, 122, 124, 128, 132, 169, 170, 175, 181, 183, 189, 192, 197, 206, 209, 241
estuarine models (see model)
eustacy 8
eustatic curve 8
event
   complementary 49
   compound 47, 48
   elementary 47, 48
   independent in space 51, 55
   independent in time 51
   rare (extreme) 55, 229
"evolution" system (see system)
exogenic cascade (see cascade)
exogenic processes 234
exponential distribution (see distribution)
exponential function (see function)
extreme (rare) event (see event)
extreme value distribution (see distribution)
"eyeballing" 85, 101

factor analysis 112, 125–128
fan, alluvial 9, 10, 31, 65–70, 109, 204
fault 31, 57, 167–169
feedback 30, 206, 229
   loop 39, 122, 123
   negative 122, 241
   positive 122, 241
Fick's law (see law)
Fishlake plateau, central Utah 237
flood 50–53, 55
flood routing 144, 151
flow 32
   energy 21, 28
   in porous media 22, 25, 145–151
   saturated 145–147
   unsaturated 148–151
flow system (see system)
force-resistance studies 18, 234
form system (see system)
Fourier function (see function)
Fourier's law (see law)
free energy 21
function
   double or log-log 109–111
   entropy 20

Subject Index

expanded entropy 25
exponential 109
Fourier 114
linear 107, 108
polynomial 11, 112, 114
power 109–111
quadratic 111, 112, 140
semi-log 109

Galapagos Rift, weird ecosystem in (see ecosystem)
gambler's ruin problem (see model)
gamma distribution (see distribution)
geochemical cycle 8, 35, 198
geochemical reaction model (see model)
geometric distribution (see distribution)
geometry variable (see variable)
Ginninderra, Australian Capital Territory 125, 129
glacier 18, 137 (see also Valley glacier)
Grand Canyon, Arizona 169
gravitational potential 149
Great Plains, USA 10
groundwater system (see system)

Hambleton Hills, Yorkshire 120
Henry mountains, Utah 27, 223
heuristic models, of hillslopes (see model)
hierarchy, of systems (see system)
Highland Rim, Tennessee 111, 126
hillslope 3–5, 7, 31, 37, 47, 78, 135, 234, 235, 241
hillslope model (see model)
holistic 28
Huon Peninsular, Papua New Guinea 169
Hurst Castle, Hampshire 192
hydraulic conductivity 25, 145, 150
hydraulic head 145
hydrodynamic model (see model)
hydrodynamic system (see system)
hydrograph 81, 82, 144
hypergeometric distribution (see distribution)

ice 137–142
ice sheet 11, 18, 138–142
independent events (see event)
independent events model (see model)
infiltration 3, 122, 149, 150
input variable (see variable)
inselberg 177
intramontane trench 64
intransitive threshold (see threshold)
intrinsic randomness (see randomness)
intrinsic threshold (see threshold)
irreversible processes 20, 24–26
isolated system (see system)

jump (step), in time series 78

karst depression 31, 56, 57, 111, 242
karst landscape 39, 146
Kawerong river, Papua New Guinea (see river)
K-cycle concept 44
King's system of geomorphology 40
Kraków, Poland 111
Krishna river, India (see river)

lag, in inductive stochastic models 81, 85, 87
lag, in system thresholds 227–229
Lagan drainage basin, Kenya 78
Lagan rainfall and runoff series 78, 87
Lake Duluth, Michigan 63
Lake Erie (see ecosystem)
lake system (see system)
landscape development, models of 39–41
landscape simulation 181–185, 237
landscape variable (see variable)
landslide 3, 18, 37, 240, 241
landslide cascade (see cascade)
land-surface cascade (see cascade)
Laplace equation (see equation)
Laplacian operator 145
La Porte, Indiana 85
law
    conservation 22
    Darcy's 25, 145, 146, 148
    energy conservation 22
    Fick's 25, 152
    Fourier's 24
    mass conservation 23
    mechanical 5
    Newton's of motion 22, 24
    phenomenological 24
    slope transport 174
    Stokes's 13
    thermodynamic 20–22, 24, 25, 28
    transport 201
lead, in inductive stochastic models 81
least-squares regression (see regression)
linear relation (see relation)
linear versus nonlinear relation (see relation)
line patterns 31, 57
linkage analysis 123
log-log or double log function (see function)
lognormal distribution (see distribution)
Long Island, New York 146
Lucore hollow, Pennsylvania 115

macroscopic variable (see variable)
Manabe and Stouffer's CO_2-climate model (see model)
manganese 214–216

Subject Index

Manitoba, Canada   30, 31
Manning equation (see equation)
Markov chain model (see model)
mass balance   23, 137, 138, 157, 210, 218, 220
mass conservation   22–24, 32, 136, 177, 179
mass conservation, law of (see law)
mass extinction   54
mass-flow variable (see variable)
mass movement   169, 170, 237
mass movement model (see model)
mass transfer   161
material-property variable (see variable)
mathematical model (see model)
matric potential   149
Maumee river (see river)
maximum entropy (see entropy, maximum)
meander (see river meander)
mechanical laws (see law)
mechanical system (see system)
metastable equilibrium (see equilibrium)
Michigan basin, USA   152
minimum entropy (see entropy, minimum)
Mitchell plain, Indiana   56
model   7
  accounting   22
  analogue   11
  ANSWERS   189
  autoregressive integrated moving-average (ARIMA)   84, 85, 92–95, 236
  autoregressive moving-average (ARMA)   83–95, 235
  Box and Jenkins's   12, 78–101
  capture   64, 65, 187
  compartment   216, 219
  conceptual   8–13, 17, 39, 41, 239
  corrasion   169
  CREAMS   189
  deterministic   12, 13, 39, 47
  diffusion-decay   184
  digital terrain   32
  discrete component, of hillslopes   161, 162
  disturbed periodic, of meanders   90–92
  double Poisson   56
  dynamical system   13, 198–232
  electrical analogue   11
  entropy   51
  estuarine   192
  gambler's ruin   60, 61
  geochemical reaction   154, 157
  heuristic, of hillslopes   162–173
  hillslope   44, 47
  hydrodynamic   139
  independent events   12
  Manabe and Stouffer CO_2 climate   212
  Markov chain   12, 51, 70–73
  mass movement   169

  mathematical   10–12, 17, 47–232, 233
  morphometric   31, 32
  multivariate   112, 134
  negative binomial   56
  nine-unit landsurface   44
  Poisson process   54
  probabilistic   51
  process-response   234
  random-walk   12, 51, 60–70, 89, 90, 92, 165
  scale   10, 11
  scarp rounding   167
  simulation   180–197, 240
  solution   169, 170
  solution-corrasion-mass movement   170
  statistical   12, 30, 39, 102–134, 236, 237
  statistical mechanics   12
  stochastic   12, 47, 78
  theoretical   10
  transfer function (TF)   83, 95–101
momentum balance   24
momentum conservation   32, 177
momentum equation (see equation)
Monte Carlo simulation   64, 73
morphoclimatic regions   41
morphological (form) system (see system)
morphological variable (see variable)
morphology   30
morphometric model (see model)
mountain, height of   19
mudflow   107
multidimensional scaling   102
multiple correlation (see correlation)
multiple equilibria (see equilibrium)
multiple regression (see regression)
multiplication axiom, for dependent (general) events (see axiom)
multiplication axiom, for independent events (see axiom)
multivariate model (see model)

nearshore bar   192
negative binomial distribution (see distribution)
negative binomial model (see model)
negative entropy (negentropy)   26
negative exponential distribution (see distribution)
negative feedback (see feedback)
neotectonics   57
neutrally stable equilibrium (see equilibrium)
Newton, laws of motion (see law)
Newtonian system (see system)
nickpoint   179
Niger river, Mali (see river)
Nile river, Egypt (see river)
nine-unit landsurface model (see model)

nitrate nitrogen   221
nonequilibrium state (see state)
nonequilibrium system (see system)
Norfolk, England   111
normal distribution (see distribution)
normal generated distribution
   (see distribution)

ocean, as a system (see system)
old-field ecosystem (see ecosystem)
Ontonagon plain, Michigan   63
open system (see system)
output variable (see variable)
overland flow   142–144, 169, 173, 189, 231, 239
oxygen   206–214

paradigm, changes of   234
partial autocorrelation (see correlation)
pediplain   40
pedomorphic surface   44
Peltier's theory of quantitative
   geomorphology   41
Penck's model of landscape development
   10, 39, 166
peneplain   40, 177
Perth Amboy badlands, New Jersey   10
phenomenological equation (see equation)
phenomenological laws (see law)
phosphorus   159, 160, 206
physical-chemical state equation
   (see equation)
plate tectonics   40, 77
point patterns   31, 55
Poisson distribution (see distribution)
Poisson equation (see equation)
Poisson process model (see model)
Poisson processes   51, 53, 54, 60, 65
polygon   60
polynomial function (see function)
pool   198
pools and riffles   76, 99, 204, 242
positive feedback (see feedback)
power function (see function)
pressure potential   149
principal component analysis   102, 112, 123–125
principal coordinate analysis   125
probabilistic model (see model)
probability
   a priori definition   48
   concept of   47
   classical view   48
   Laplacian   48
   relative frequency view   48, 102
   space   47
   theory   47

process equation (see equation)
process system (see system)
process-form system (see system)
process-response model (see model)
process-response system (see system)
Purari river, Papua New Guinea (see river)

quadratic function (see function)
quantum mechanics   18

rainfall   51, 72, 78, 87, 88, 97, 121, 144, 150
raised mire (see ecosystem)
ramp, in time series   78
randomness
   aggregate   22, 234, 235
   intrinsic   22, 234, 235
random component   12, 13
random polygon   60
random process   12
random walk (see model)
rare (extreme) event (see event)
rate constant   160, 199, 200, 214
reaction-diffusion equation (see equation)
recessional coefficient   93, 165
reduced major axis method (see regression)
regression   102
   coefficients   103
   least-squares   103
   line   103, 104, 108, 114
   multiple   39, 102, 112–122
   reduced major axis method   104
   simple   30, 103–112
   "simple" multiple   113, 114
   stepwise   117–121
relation (see also equation)
   linear   107, 108
   linear vesus nonlinear   108–112
relativistic dynamics   18
reservoirs   198, 205, 206
residence time   199
reversible process   20, 21
rhexistasie   44
Rhône valley   64
Riesel (Waco), Texas   144
river   5, 18, 24, 31, 34, 173, 235
   Aire, England   91
   Angabunga, Papua New Guinea   89
   Bollin, Cheshire   109
   Brandywine Creek, Pennsylvania   109
   Coon Creek, Wisconsin   33
   Detroit   218
   Dye, Northeast Scotland   109
   Elan, Wales   97–100
   Kawerong, Papua New Guinea   180
   Krishna, India   78
   Maumee   218
   Niger, Mali   78

Nile, Egypt   78, 85
Purari, Papua New Guinea   85, 88
Trent, England   91
river channel variable (see variable)
river discharge (see discharge)
river downcutting   69, 165, 166, 184
river long profile   75, 109, 236
river meander   10, 61, 77, 89–92, 99, 204, 235, 242
river network   31
rock creep   35, 107
rock cycle   8

sample space   47, 48
sand dune   10, 18, 193–197, 229, 235
saturated flow (see flow)
scale model (see model)
scale problem   7
scaling problem in scale models   11
scarp rounding model (see model)
scree   32, 37, 161, 162
"scree" test   123
sea-floor spreading rate   213
sediment cascade (see cascade)
sediment transport   18, 75, 223, 224
  continuity of   23, 179
sediment transport equation (see equation)
sediment yield   189
sedimentary cycle   4, 23
semi-log function (see function)
sensitivity analysis   204, 205
simple correlation (see correlation)
simple regression (see regression)
"simple" multiple regression (see regression)
simple system (see system)
simulation model (see model)
sine-generated curve   89, 90
sinkhole   126–128
slope transport laws (see law)
slope system (see system)
soil   5, 10, 27, 41–44, 102, 104, 107–109, 111, 114, 117, 120, 122, 124, 125, 129–132, 134, 145, 148–151, 156–161, 198, 199, 201, 239–241
soil catena   37, 43, 124, 158, 241
soil creep   3, 18, 22, 35, 95, 124, 166, 173, 175, 181, 235, 236, 241
soil depth   204, 241
soil development   41–43, 237
soil erosion   18, 102
soil-forming factors   41, 42
soil-landscape development, concepts of   43, 44
soil-landscape system (see system)
soil material cascade (see cascade)
soil processes   159

soil profile   152, 159, 217
soil system (see system)
soilscape   43
solifluxion   3, 35, 175, 236, 241
solute   24, 25, 151–160
  flow rate   25
  in groundwater   154–157
  in lakes and seas   151–154
  in rivers   109, 111, 151
  in soils   157–160
solute cascade (see cascade)
solution   3, 93, 169, 240
solution-corrasion-mass movement model (see model)
solution model (see model)
source area concept   144
South Island, New Zealand   109, 117
space-time substitution   185, 238, 239
  (see also ergodic hypothesis)
spatial domain   135, 136
spatial system (see system)
spectra   89, 91, 92
spit   192
splash (rainsplash)   18, 173
spurious correlation (see correlation)
stable equilibrium (see equilibrium)
Stanwell Park Beach, New South Wales   117
state
  nonequilibrium   21
  steady   26, 75, 145, 150, 152, 154, 199, 222, 241
state equation (see equation)
state plane (space)   202–204
state transition function   198
state variable (see variable)
statistical mechanics   12
statistical mechanics model (see model)
statistical model (see model)
steady state (see state)
steady-state system (see system)
stepped valleys   242
steps, in time series   78
stepwise regression (see regression)
stochastic event   47
stochastic occurrence   47
stochastic model (see model)
Stokes's law (see law)
storm   65, 67, 69
stream channel cascade (see cascade)
stream channel network   59, 61
streamflow   35, 48 (see also discharge)
strontium   214–216
Student's t-test   102
subduing coefficient   165
surface wash (see wash)
system
  bistable   227, 229

boundary conditions   136, 166, 174, 175,
    177, 184, 221, 227, 229, 230
cascading   4, 9, 17, 32
closed   5, 21, 22, 25, 26
correlation   32, 122–134
environment of   5, 201, 202
"evolution"   155, 156
flow (= cascading, see entry above)
form (= morphological, see entry below)
groundwater   145, 146, 152
hierarchy of   7
hydrodynamic   23
isolated   5, 20–23
lake   28
morphological   4, 31, 39
mechanical   17
Newtonian (= mechanical, see entry
    above)
nonequilibrium   25–27
ocean   28
of complex disorder   5, 20, 22
of complex order   5, 25
open   5, 13, 25–27, 32, 156
processes   26
process-form   4, 5, 9, 37, 39, 134
process-response (= process-form,
    see entry above)
simple   5, 17, 18, 20, 234
slope   39
soil   27, 28, 41, 43
soil-landscape   28, 35, 43
spatial   23
steady-state   26, 28, 206
thermodynamic   235
transfer function of   81, 82, 95, 96
typology   4, 8, 17
valley-side slope   30
system constitution   30
system definition   3, 81
system stability   202–205, 241
system variable (see variable)
systems analysis   82

talus   35, 37, 161, 234
tectonics   40, 57
terracettes   133, 134, 242
theoretical model (see model)
thermodynamic analogy of landscapes   24,
    74, 77
thermodynamic branch   222
thermodynamic equilibrium
    (see equilibrium)
thermodynamic system (see system)
thermodynamic variable (see variable)
thermodynamics   7, 20–22, 24–26, 28,
    74–77
    laws of (see law)
Thiessen polygon   60
Thoms watershed, Iowa   114

threshold   222, 226–230, 241
    intransitive   227
    intrinsic   241
    transitive   227
throughflow   43, 150, 158, 169, 231
time constant   199
time series   78, 83–88, 95–97, 235
topologically distinct channel networks   53,
    59
topologically distinct network patterns   59
transfer equation (see equation)
transfer function (TF) models (see model)
transfer function of a system (see system)
transition probability   70
transitive threshold (see threshold)
transport equation (see equation)
trend, in a time series   78
trend surface analysis   114–117
Trent river, England (see river)
tritium   217
tropical rain forest ecosystem, Puerto Rico
    (see ecosystem)
turnover time   199

unsaturated flow (see flow)
unstable equilibrium (see equilibrium)
uplift   39, 40, 41, 47, 65, 165

valley glacier   10, 35, 139
valley glacier cascade (see cascade)
valley-side slope system (see system)
variable
    channel form   78
    control   223, 225
    dynamic   30, 39
    geometry   30
    input   81
    macroscopic   24
    mass-flow   30
    material-property   30
    morphological   30, 39
    output   81
    response   223, 225
    state   24, 136, 198–201
    system   3, 10
Verdugo Hills, California   102
volcano   31, 73
Voronoi polygon   60

wash   3, 35, 93, 95, 166, 173, 175, 236, 241
water cascade (see cascade)
water cycle   4, 8, 35
water-flow deposits   68, 70
water-flow rate   25
water potential   25, 145, 148
water table   56, 146, 149
weathered layer   67
weathering   18, 181, 210
Wungong Brook, Western
    Australia   130–132

/551H891E>C1/

DATE DUE